Exploring Nature's Dynamics

E. Atlee Jackson

A Wiley-Interscience Publication
JOHN WILEY & SONS, INC.
New York • Chichester • Weinheim • Brisbane • Singapore • Toronto

This text is printed on acid-free paper. ⊗

Copyright © 2001 by John Wiley & Sons, Inc.

All rights reserved. Published simultaneously in Canada.

No part of this publication may be reproduced, stored in a retrieval system or transmitted in any form or by any means, electronic, mechanical, photocopying, recording, scanning or otherwise, except as permitted under Section 107 or 108 of the 1976 United States Copyright Act, without either the prior written permission of the Publisher, or authorization through payment of the appropriate per-copy fee to the Copyright Clearance Center, 222 Rosewood Drive, Danvers, MA 01923, (978) 750-8400, fax (978) 750-4744. Requests to the Publisher for permission should be addressed to the Permissions Department, John Wiley & Sons, Inc., 605 Third Avenue, New York, NY 10158-0012, (212) 850-6011, fax (212) 850-6008, E-Mail: PERMREQ @ WILEY.COM.

For ordering and customer service, call 1-800-CALL-WILEY.

Library of Congress Cataloging is available.

Jackson, E. Atlee
   Exploring nature's dynamics / E. Atlee Jackson.
     p. cm.
   "A Wiley-Interscience Publication."
   ISBN 0-471-19146-9.

Printed in the United States of America.

10 9 8 7 6 5 4 3 2 1

*To my lieveling Cindi,
ever the wind beneath my wings;
and to Eric and Mark,
our lights to the future*

Contents

Internet Access of Programs ... xvii

Preface ... xix

Introduction ... xxi

1

"Understanding" Nature's Dynamics   1

1.1 A Look at Nature's Dynamic Diversity   1
Nature's amazing diversity of dynamic phenomena; Considerations of time spans and understanding; The challenge: How much can we scientifically understand?
References

*1.2 The Early Scientific Studies of Dynamics (300 BCE–1800)   6
The birth of empirical Western Science; the scientific dynamic studies by Pythagoras, Archimedes, Tycho, Kepler, Galileo, Harvey, and D. Bernoulli; Their unique modes of understanding dynamics; Newton's mathematical recursive description of dynamics, the universal law of celestial and terrestrial gravitational dynamics; applications to fluids.
References

1.3 An Example of Linear and Nonlinear and Uncertain Dynamics: Money in Two Very Different Banks!   20

The recursive dynamics of deposits in the Safe Bank and the Swift Bank (not Swiss); The implicit assumptions when modeling aspects of reality; The eponential growth from linear dynamics; The reality of limited resources produces nonlinear dynamics; Nonlinear recursive dynamics can only be explored with the help of computers. **References**

1.4 A First Look at Exploring with Basic Programs   28

An elementary introduction to the use of simple (Basic!) programs, which will empower you to explore the dynamic content of a wide variety of nonlinear dynamics found in nature; the generation of both numerical and graphical representations of dynamics. **References**

1.5 Implicit Assumptions in Models: Real "Swift Bank" Dynamics (and Human Populations?)   35

Three possible cases: The Swifts are: (1) philanthropic, (2) are crooks! (3) make a mildly dishonest promotion; Dynamic models are only as good as the implicit assumptions; Association with the exponential estimates, and real dynamic uncertainties of human population dynamics, and its fragile ecological basis. **References**

1.6 Some Basic Dynamic Insights   43

A review of the dynamic features and limitations encountered in attempting to model some natural phenomena involving congnitive processes.

2

Dynamic Aspects of Modern Science   45

*2.1 The Newton-to-1950 Stages of Dynamic Developments   45

Newton's introduction of ordinary differential equations (ODE) to describe dynamics; Partial differential equation (PDE) used to describe fluid dynamics, and Maxwell's description of the dynamics of light; Nonlinear ODE description of chemistry and population ecology; Poincaré's discovery that Newton's deterministic ODE do not imply deterministic empirical observations; Boltzmann's gas dynamics; Darwin's varied evolutionary concepts which continued to evolve; Einstein's theories of Brownian motion, relativity, and photoelectric effect (quanta); Schrödinger's probabalistic description of quantum mechanics; Its dynamic-exclusionary interpretation; Cosmological

dynamics: life cycle of stars, nova, expanding universe
References.

2.2 The Changing Science of Nature's Dynamics (After 1950)   56
The early dynamic discoveries due to the digital computer (before 1980); The new bases for scientific understanding; Three distinct sources of scientific information: (1) natural phenomena, (2) mathematical models, (3) computer explorations; Feynman's characterization of science; the evolving limitations of scientific understanding. **References**

2.3 The Modern Uses of Models in the Science of Dynamics   63
Models and understanding; Their evolving character; No laws; Physical models: limited representation of nature; Extraction of specific phenomenon, for limited time duration and accuracy; Exploration models: dynamic models to explore the creation of new types of dynamics, possibly related to complicated forms of natural dynamics.
References

2.4 A "Random Plus Rule" Dynamics that Exhibits Holistic Features   67
Explores the dynamics of a recursive relation that contains one deterministic step and one random step (a random-plus-rule dynamics), yielding a holistic regularity. How general? In nature? **References**

3

Different Types of Population Dynamics   71

3.1 Discrete-Time Models of Population Dynamics   72
Models of dynamics of real and abstract living populations; The effects of generation overlaps, epidemics, and finite resources. **References**

3.2 The Logistics Map and its Fixed Points   76
A mathematically simple, and physically simplistic dynamic model, illustrating some amazing general forms of dynamics. **References**

3.3 Graphical Representation of Map Dynamics: Fixed-Point Stability   78
A simple graphical device makes the map dynamics quite understandable.

3.4 The Bifurcations of Fixed-Point Attractors of the Logistic Map — 81
The concept of a dynamic bifurcation; First example, when a stable fixed becomes unstable, generating a stable period-two dynamics.

3.5 The Logistic-Dynamics Jungle!! — 85
The period 2^n bifurcation sequence; Semiperiodicity, intermittency, windows, and chaos; A survey of all the dynamic bifurcations.

3.6 The Sounds as Well as Sights of Logistic Dynamics—Using Two Senses to "Understand" What is Going On — 93
Using two of our senses to better understand what is going on; Hearing: periodic, semiperiodic, intermittency, and chaos; The separation and reconvergence of nearby chaotic states.

3.7 More Empirically Realistic Models of Population Dynamics: Different "Universal Classes" — 100
(1) Insects whose populations are controlled by epidemics; (2) 28 seasonally breeding insects, constrained by environmental limited resources. **References**

4
Dynamic Models Based on Differential Equations — 105

4.1 From Difference Equations to Differential Equations, and Back Again! — 105
Newton's geometric approach; The Newtonian limiting method, $dt \to 0$, and the Eulerian discrete method; Derivatives, dx/dt, and differential equations, $dx/dt = G(x)$; Ecological example; Stability relationship between the Verhulst equation and the logistic map. **References**

4.2 Harmonic Oscillators: Their Properties, Phase Space Representation, and Computational Algorithms — 115
Systems of second-order equations; Harmonic oscillator dynamics; Analytic solutions; Amplitude independent frequency; Fixed point, and attractor characterization due to damping; Phase space representations; Uniqueness and nonintersection of orbits; Limitations of the Euler approximation; Fourth-order Runge–Kutta method for computer simulations of ordinary differential equations.

4.3 Properties of Nonlinear Oscillators — 125
The Duffing oscillator; Amplitude dependent frequencies; Comparison of theory with computational dynamics; Phase space features; Energy constant of the motion; Hard and soft oscillators; Only nonlinear systems can have more than one fixed point; Separatrix orbits; Saddle points; Topologically equivalent orbits; Bifurcation points.

4.4 Dynamics in a "Double-Well" Potential — 133
Dynamics in a region with two adjacent "valleys"; Saddle points, elliptic points, and nodes; Damping yields two intertwining basins of attraction.

4.5 A Pendulum that Can *Really* Swing (Sometimes) — 138
A pendulum that can rotate around its pivot point; "Wrap-around" phase space representation; Constant of the motion; Damping and intertwining basins of attraction; Adding a coil at the pivot produces only a finite number of attractors in the infinite (θ, Ω) phase space.

5

Interacting Systems Producing (Near) Periodicity — 145

5.1 Predator–Prey Systems: The Lotka–Volterra Model — 145
Population limitations due to predator–prey interactions; The Lotka–Volterra differential equation model; The reality of the Canadian lynx–snowshoe hare complicated dynamics; Positive-variables phase space; Scaling concepts; Neutrally stable models. **References**

5.2 Host–Parasite Interactions: Uncovering Model Limitations — 153
The Nicholson–Bailey model; Their general mathematical deductions, and arithmetical analysis (1935); Their correct recognition of one physical limitation of the model, and the required generalization; Limitation expressed as nondeterministic computer N-B dynamics; "Extinction" modeling?; Inclusion of limited environmental resources for host; Fix point or limit cycle attractors; Andronov–Hopf bifurcation. **References**

5.3 Environmental Fluxes: Autocatalytic Oscillators — 161
Self-regulation of (energy/matter) input from, and output to, the system's environment to sustain dynamics; Examples, all

living systems, and many inanimate systems (e.g., bowed violins, clocks, pulsating stars, electrical circuits, engines, chemical-flux mixtures (Belousov oscillations); The aesthetic Japanese shishiodoshi; van der Pol oscillator; Liénhard phase space.

5.4 Excitable Neurons: They Pass on Information, Connect Us to Our Environment, Let Us Think, Keep Our Hearts Beating, Etc.! 168
Introduction to neuron structure, and connections; The Fitzhugh excitable Bonhoeffer–van der Pol (BvP) model of neuronal dynamics; Quiescent neural response to excitation pulses; The autocatalytic pacemaker dynamics. **References**

5.5 Reliable Pacemaker, Excitable Heart, and Arrhythmias 176
A BvP model of various pacemaker cells, and their reaction to environmental noise; the excitable BvP model of ventricular pulses of a heart when excited by the pacemaker sinus node in the heart; The response for different nodal periods; Arrhythmia. **References**

6

Temporal Chaos and Fractal Structures 183

6.1 The Computational Lyapunov Exponents of Maps 184
Lyapunov exponents as a measure of sensitivity of dynamics to initial conditions, or noise; The Lyapunov exponent, $L(x(0),c)$, for maps; Logistic map examples; Dependence on initial condition, $x(0)$, and on the parameter c?; Numerical precision?; Difficulties with empirical applications. **References**

6.2 Strange Sets Called Fractals, With Strange Dimensions 188
The topological and capacity dimensions; Strange geometrical sets: Cantor's middle-third; Cantor's comb (?); the Koch curve; The dynamic set generated by random-plus-rule dynamics; Its empirical capacity dimension; Physical fractals; observed fractals in life sciences, architecture, music, nature computations. **References**

6.3 "Systems" and "Environments": Roles of "Chaos," Fractals, and Functional and Adaptive Interactions 197
The chaotic development of systems from their environments

over billions of years; The earth system and its environment; Living systems adaptive functional interactions with environments; allometric scaling, physiological fractals, and chaotic contributions. **References**

7

Periodic Systems Coupled to Periodic Environments   207

7.1 The Concept of Systems Resonating With Their Environment   207
A weak periodic force acting on a nonlinear oscillator; The harmonic approximation; The concept of a resonant response to its environment; Comparison of theory with computed dynamics; System and environmental couplings: ice ages, pedagogical resonant effects. **References**

7.2 Moderate Periodic Forces Can Produce Bistable Oscillations   211
A moderate periodic force acting on a damped nonlinear oscillator; Bistability, and hysteresis effects due to the amplitude dependent natural frequency of a nonlinear oscillator; Dynamic catastrophes and the cusp catastrophe set; Multistability in many physiological contexts. **References**

7.3 Symmetry-Breaking and Chaos from Strong Forces   220
A strongly forced damped Duffing oscillator; Symmetry-breaking bifurcations yielding bistable attractors; Unique representation of dynamics in the extended (toroidal) phase space (EPS); Stronger forces produce chaotic attractors; A Poincaré-map in the EPS shows the fractal character of chaotic attractors in the EPS.

7.4 Pacemakers, Zeitbegers, Circadian Rhythms, and Entrainment   226
Biological clocks, or pacemakers, have periodic dynamics, and control a variety of periodic processes in living systems; Circadian rhythms are those which have periods near 24 hours; To ensure the synchronization of some pacemakers with the daily environment, they need to be "entrained" by a dynamic daily signal, a zeitgeber ("time giver"); Discoveries of many pacemakers, some genetic. **References**

8
A Variety of Dynamics in Space and/or Time 235

8.1 Spatial Pattern Dynamics of Excitation–Diffusion Systems 235
Cells with excited, refractory, and quiescent dynamic states, which interact with their nearest neighbors, yielding target and spiral spatial patterns of excitation; Similar spatial patterns in chemical neurons, biological, ecological, slime mold, localized neurological, and cardiac systems.
References

8.2 The Discovery of Chaotic Attractors in Meteorology 243
Edward Lorenz's computational discovery of sensitivity to initial conditions in meterological models; Introduced the simplified Lorenz equations, having the desired attraction to observed aperiodic dynamics; Lorenz's strange (chaotic) attractor, and many bifurcations, can best be understood with computer graphics in three-dimensions, with arbitrary perspectives; The use of a Poincaré map to uncover the structure of a standard and nonstandard fractal attractor.
References

8.3 A Symmetry-Breaking Bifurcation: Linked Limit Cycles 252
A unique dynamic bifurcation phenomenon of the Lorenz equations; An extension into parameter ranges that presently lack physical justification; The concept of symmetry-breaking autonomous dynamics, and its graphical detection; Three-dimensional survey of two stable linked limit cycles (like chain links).

8.4 Self-Organizing Ant Colony Dynamics: Social Insects 257
A model of an observed excitable ant colony, which has chaotic or self-organizing group excitations, depending on their density; References to differing ant societies, and the remarkable honey bees, and possible technological applications. **References**

Computer Appendixes 265

C0: Initial Actions 265
C1: A Place to Develop Your Idea 269
C2: Basic Commands, Interactions, and Saving Programs 271
C3: Some Graphical Information 276
C4: The Fourth-Order Runge–Kutta Interation Method 278
C5: Graphics Printout and Editing 281
C6: Let There Be Dynamic Tones! 282

Mathematical Appendixes — 285

M1: Properties of Derivatives and the Derivative of $a \cdot t^n$ — 285
M2: The Stability of Fixed Points of a One-Dimensional Map — 287
M3: The Derivatives of Exponential and Trigonometric Functions — 290
M4: The Nonlinear Frequency–Amplitude Relationship — 293
M5: The Stability of Fixed Points of Two-Dimensional Ordinary Differential Equations — 295

Index — 299

Internet Access of Programs

HOW TO TRANSFER THE INTERNET FILE OF THE DYNAMIC PROGRAMS TO YOUR COMPUTER'S DISK OR C DRIVE

If you prefer using the Internet, instead of the disk supplied with this book, the dynamic Basic programs used in this text can be downloaded from Wiley's ftp site. **These programs can either be put on a disk, or on the C drive of your computer.** Further modifications/additions of these programs may be put into this Internet file, and can be referenced at later dates.

In detail:

1. If you want the programs on a disk, put a formatted disk into drive A.
2. On the Internet enter

   ftp://ftp.wiley.com/public/sci_tech_med/dynamics

   noting the spaces. Here you will find a general message and a directory, with **dynamics.zip** highlighted. Click on this directory.
3. A green page will come up, offering a selection between "Save it to disk" or "Open it." Regardless of where you want to put these programs, select the command **"Open it."**

4. When you click on the button **OK**, a rather long process will begin to open these files. When it has finished, a screen will appear with all of the files displayed. Go to the top of this screen and **select Extract.** This decompresses the programs in this zip file to their original qbasic format.

5. You will see Extract to **C:\windows\temp,** and now can make the **two choices:**

   (a) If you want to save these **programs on the disk, change this command to A:**

   (b) If you want the programs **on the C drive, change this to C:** These will then be saved in the directory **C:\dynamics**

6. Now go to Start, Programs, and MS-DOS Prompt, where you will find **C:\WINDOWS>.**

   (a) If you made this selection, then <A:> to go to the disk, **A:\>**. By <dir>, you will find the DYNAMICS directory. To get all of these files out in plain sight, **<cd dynamics>,** yielding **A:\dynamics>,** and then **<move *.*..>**. This will move all of the programs outside of the DYNAMICS directory, where we can see them. When **A:\dynamics>** reappears, **<cd..>** (to get out of that directory). To remove this empty directory, **<rmdir dynamics>**. You know have the qbasic programs on the disk.

   (b) If you made this selection, type **cd\dynamics.** You will now run the **QBASIC program** by typing **qb** at the command line, and the programs can be selected from the **File** menu.

Preface

This book is intended to **introduce** a variety of examples of **natural dynamics** to anyone who would like to learn how these dynamics can arise from **natural interactions,** and to **see these dynamics** generated by the **graphics of computer programs.** The **disk** that comes with this book has **65 simple programs,** all related to natural activities (these programs can **also be obtained from the Internet,** as explained on page xvii). Using these programs, one can not only **appreciate what can cause these dynamic phenomena,** but also enjoy the opportunity to **explore possible modifications** of these dynamics through **use** of these programs (all of which **assumes no programing knowledge**). This opportunity to become **actively involved** in the use of **your imagination** (aided by questions), is one of the **unique features** of this book.

To assist in this process, each section contains two types of exercises. The **Think** exercise involves the exploration of **new concepts,** and how they might be **generalized.** You might even produce your own **Eureka!** The second type of exercise is called **Explore,** which makes use of **simple computer programs** to help you understand how the dynamics occurs, and how you can imagine possible modifications of these dynamics. To make this process **accessible** to those who know nothing about computer programs, there are several appendixes to introduce you to the few needed features of the **simple qbasic language**. It will be shown that not only is this language very simple and transparent (that's why I use it!), but that it also has great keyboard interactive capabilities. This makes it possible for **you to be actively involved** with the dynamics you are observing. Thus, these programs are not sophisticated; indeed some are rather crude, which gives you the opportunity to **"correct and improve"** them in the process of **exploring your own ideas**.

There are several sections of this book (**denoted by** *) that give **historical insights** into the way **science developed different forms of dynamic understanding,** which can only be appreciated by recounting historical examples. One area involves an early form of characterizing dynamics that was more **holistic in character,** and has recently been extended into mathematics (topology); it will very likely be of great value in the future. This history also illustrates the dangers of the cultural influences of the day, no less now than in the past. This also has had a retarding effect on the generalization of science into new areas of importance, particularly in the cognitive and social arenas. We will have only a little to say about this in this book, but it is important to be sensitive to these issues for your future development.

This book has only been made possible by the imagination and support of the editorial boards of John Wiley & Sons. This began about five years ago with the visionary assistance of Gregory Franklin. His appreciation of this new approach to understanding nature's dynamics was the genesis of this publication. After some delay, I was fortunate to become associated with another very competent and accommodating editorial staff, led by the Executive Editor, George Telecki. Lisa Van Horn, the Associated Managing Editor, edited the manuscript with impressive care and sensitivity, and made many helpful suggestions. I am also indebted to two friends, Donald Cooney and Dominic Skaperdas, who spent considerable time and effort trying to understand earlier versions of this book, which yielded many useful technical suggestions. Underlying all of this lies years of student interactions, which always inspired me to keep looking anew at our remarkable dynamic world. As a saying goes, "To teach is to learn twice."

E. ATLEE JACKSON

Urbana, Illinois
January 2001

Introduction

Science is in the early stages of trying to broaden its understanding of the amazing variety of dynamics that exits in the universe. It is dynamics that accounts for the formation and movement of all structures, and thereby all physical processes. The fact that we now have a meaningful opportunity to explore such complicated dynamic processes is due to the dramatic increase in the amount of dynamic and structural information that came about in the second half of the 20th century. First, there was the remarkable increase in **empirical information**. An explosion of this information occurred in biology, astronomy, geology, physics, chemistry, meteorology, economics, psychology, and in many areas of sociology, medicine, and technology. Added to this was the infusion of **mathematical insights** developed in the last century, which gave fresh perspectives to the concept of "understanding." And finally, there was the entirely **new world of computational information**, which arrived with digital computers. This wealth of information affords both opportunities and challenges to develop new forms and concepts of **"scientific understanding,"** which are needed to modernize the historically simpler basis of the **"scientific method."** Such issues as the temporal and spatial scales of dynamic processes, the categorical character of all observational and computational information, and the variety of physical and informational interactions between systems, conscious or not, present new challenges to the understanding of dynamic phenomena, due in part to the newly recognized general "sensitivity" of dynamic processes (depending on spatiotemporal scales).

Science has barely started this process of new construction, for it takes many years for it to be recognized that the comfortable, established methods and ideas do indeed need to be basically modified. To begin this appreciation, one must become familiar a variety of new dynamic concepts, which can be illustrated and examined without much difficulty, thanks to the opportunities that are offered by modern computers.

This book is an introduction into this **scientific exploration process**. It includes various ways of **characterizing dynamics**, some of the **common attributes** of the dynamics found **in Nature**, and how we can use the **computer** to assist in trying out **new ideas** related to **physical observations**, or to answer questions about **existing proposals**. The computer has become one of the basic logical sources of dynamic information, and now compliments the traditional mathematical methods. Thus, this book will emphasize the process of **exploring ideas** with the help of **simple computer programs**, that **you** can easily **modify** and then individually **generate**. This means that, in addition to learning some of the new dynamic phenomena that have been discovered during the past few decades, there is the challenge and satisfaction of exploring extensions of these concepts, and best of all, the fun of coming up with your own new dynamic ideas about how nature may work (your personal **Eureka! searches).** But this requires some discipline of thought, for if requires some thinking about the issues referred to by Richard Feynman, one of the most imaginative, and perceptive physicists of this century:

> Science is a way to teach how something gets to be known, what is not known, to what extent things are known (for nothing is known absolutely), how to handle doubt and uncertainty, what the rules of evidence are, how to think about things so that judgments can be made, how to distinguish truth from fraud, and from show.

This is a large and important set of issues, which we will consider in the beginning sections of this book. They are important for any responsible inquiry into how we in fact end up "knowing" something.

Among the new concepts that one needs to become familiar with is that of **nonlinear dynamics,** for most of Nature's dynamics is of this character. This fact was nicely captured in Stanislaw Ulam's observation:

> The study of nonlinear systems is like the study of nonelephant mammals.

You may need to think about this for a few moments, but it is worth the time.

We will learn about nonlinear dynamics early on, and see how it is central to one of the great truths about Nature, namely that **all resources in life are limited**. This is true for dynamics as simple as bank accounts and as complex as the Earth's ecosystem. Indeed it is one of the **great commonalities of all of Nature's dynamics.**

So let us begin this process of exploration!

1
"Understanding" Nature's Dynamics

1.1 A LOOK AT NATURE'S DYNAMIC DIVERSITY

In everyday life, it is certainly not common to stop and think about "dynamics." That is, we generally do not think about the different ways that things change around us in time. They just do, and we take it all for granted.

On the other hand, if **life** has any dominant characteristic it is certainly that it is a **dynamic affair,** with all sorts of activities going on all the time. Since life is clearly important, this great variety of "dynamics" can't be unimportant, to say nothing of **amazing.** However, the **general topic** of **dynamics as it occurs in nature,** referred to here as **nature's dynamics,** is not something that is usually discussed in any school or university at present. As interests are rapidly developing to extend our understanding of phenomena in biological, neurological, ecological, social, cognitive, and financial areas (to name but a few), it is clearly time to begin to learn some of the known features of nature's dynamics, and more importantly, to begin **explorations of your own.**

So let us take a few minutes to raise our sensitivity to this concept of dynamics by considering some of this variety of nature's dynamics that goes on around us. We, of course, do not know much

"Understanding" Nature's Dynamics

about the details of what causes many of these things to happen, but one of the objectives of science (in a generalized sense) in the next century will be to better "understand" a broader range of these phenomena.

Some of the dynamics in the world is suggested in the whimsical **Fig. 1.1** [1]. Hopefully, this may stimulate some thoughts, and better yet, your imagination to generalize. Some of the technical aspects of what may be going on in a small part of this activity will be explored in the following chapters, with the help of simple dynamic models that are used in computer programs.

A different perspective of this diversity is given in the following examples, which consider different activities, relationships and contrasts:

- A pendulum swings, a ball flies in an arch, but a leaf flutters to the ground.
- The moon circles the earth once a day, producing two high tides each day, causing the moon to slowly retreat from the earth [2].
- Our heart beats and our blood "regularly" pulsates, unless the heart fibrillates.

Fig. 1.1.

- A volcano erupts, magma pours from fissures in the oceans' floors, and earthquakes shake the ground as the tectonic plates move, all important influences on the Earth's biosystem [2].
- Birds flock in formation, a school of fish turns in unison, fireflies coherently light up trees, and ants can jointly oscillate in activity; all examples of social "self-organizations."
- Cars cluster along a highway, and ants along their chemical trails.
- Coherent hurricanes and tornadoes emerge out of apparently formless atmospheric conditions.
- Stock prices rise and fall, and social movements vacillate on collective predilections.
- On **progressively longer time scales,** species evolve [3], the continents move, and planetary systems, and their moons, chaotically form around new stars [2]. Stars evolve in life cycles and may violently collapse, producing a **supernova,** which give birth to **all the heavy elements** in the universe, including **essential ingredients on earth and within us [2, 4].** On larger spatial scales, stars collect by the billions into various galactic structures, which in turn form galactic groups, clusters, and superclusters, with walls and voids, which are the products of some cosmological dynamics [4, 5].
- But perhaps the most amazing example of dynamics is our ability to learn and remember, to reason and generate new ideas, to be introspective, and to interact and adapt to other people and environmental changes.

Some of these phenomena seem rather simple, whereas others appear to be beyond our ability to ever understand. The best that science can do, of course, is to try to "understand," at some level, as much of this activity as possible. But why do scientists now begin to believe that they may be able to understand some of these complicated forms of dynamics? This is due to a number of exciting dynamic discoveries that have been made during the 20th century which have resulted from a combination of developing several new ways of "understanding" dynamics, both empirically and conceptually, and the computational power of modern digital computers. In this book, a number of examples of these new dynamic concepts will be illustrated and **issues will be suggested for you to explore,** with the help of over **sixty programs** on the attached floppy disk.

So, in a real sense, **this can be the beginning of your adventure.** You have the **opportunity** to enjoy and develop some of the new ideas of the 21st century! This is not a fanciful statement. The field of dynamics is such a virgin area that, with a little background and training, anyone has the possibility of discovering new and significant insights into some aspect of nature's dynamics. However, to do this, you need to **accept this challenge: Think and explore, as well as read and learn.**

To assist in this process, this book contains both **Think** and **Explore exercises.** The **Think exercises** involve developing, extending, or clarifying concepts, whereas the **Explore exercises** involve the use of simple computer programs to see what happens when you probe for possible answers to some dynamic questions. The best situation is, of course, if you **generate your own questions,** think hard about the issues, and then **explore these ideas with some computer program** (either modified from one given in the attached floppy disk or written by yourself).

To start this process, begin by reviewing the above examples of dynamics and then consider the following.

Think 1.1

One very general way **to sensitize one's appreciation of the diversity of nature's dynamic phenomena** is to consider the concept of *time spans.* For example, consider **spans of time** of a few milliseconds, seconds, minutes, days, months, years, centuries, thousands of years, millions of years, and billions of years (say up to ten billion years). Now see how many dynamic processes in nature you can think of that "take place" in each such time span (that is, they are identifiable within those time spans, such as a hurricane, a bird song, or an ice age). Note that this influences the time and space scales over which data can be usefully taken. You might then think of **what type of information** you would need to start trying to understand some of the processes you list. Indeed, it is not too early to begin thinking about **what you personally mean by "understanding."**

Such issues are basic to the development of science, as will be outlined in **Sections *2.1 and 2.2.** Note that one of these sections contains the symbol *, as does the following section, *1.2. This designation indicates that the topics that are discussed in that section give a broader and deeper insight into the subject of the science of dynamics than is immediately required to proceed. However, references will be made to this material later in the book.

For some **stimulating reading** about a wide variety of scientific ideas, consisting of the best short essays and ideas of 100 prominent scientists and writers since the Renaissance, see reference **[6]**. To read about the wonder of dynamics of living cells and their development into people, read reference **[7]**. For a sense of the joys of a life of "finding things out" by the unique genius Richard P. Feynman, look into the collection of his short works **[8]**. For a widely varied group of opinions on the "impacts of foreseeable science," including dynamic **time scale issues,** as in **Think 1.1,** see reference **[9]**.

References and Notes

1. In a flight of fancy, I drew this figure for my book, *Perspectives of Nonlinear Dynamics*, Vol. 1. Cambridge University Press, Cambridge, 1989.
2. P. D. Ward and D. Brownlee. *Rare Earth: Why Complex Life is Uncommon in the Universe*. Copernicus; Springer-Verlag, New York, 2000.
3. R. Fortey. *Life: A Natural History of the First Four Billion Years of Life on Earth*. Vintage Books; Random House, New York, 1997.
4. *Scientific American*. Revolutions in Science. 1999. An anthology of *Scientific American* articles from 1948–1997. Scientific American, New York.
5. S. D. Landy. Mapping the Universe. *Scientific American*, June, 1999, pp. 38–45.
6. J. Carey (Ed.). *Science Eyewitness: Scientists and Writers Illuminate Natural Phenomena from Fossils to Fractals*. Harvard University Press, Cambridge, MA, 1997. A wonderfully diverse collection of around 100 essays that draws from the best scientific writings since the times of the Renaissance.
7. For an outstanding introduction to the variety of dynamics found within living cells, their divisions, embryonic development, genetic processes, etc., read B. Rensberger, *Life Itself: Exploring the Realm of the Living Cell*. Oxford University Press, New York, 1996.
8. J. Robbins (Ed.). *The Pleasure of Finding Things Out: The Best Short Works of Richard P. Feynman*. Perseus Books, Cambridge, MA, 1999. You should also **not overlook** the Foreword by Freeman Dyson, movingly illustrating the impact that Feynman had on prominent scientists.
9. Impacts of Foreseeable Science: Views from the *Nature* group of journals. Supplement to *Nature*, December 2, 1999.

*1.2 THE EARLY SCIENTIFIC STUDIES OF DYNAMICS (300 BCE–1800)

"The farther backward you can look, the farther forward you are likely to see"—Winston Churchill

In this section we will take a look at the very early (**embryonic**) stage of the **scientific** studies of **dynamic systems,** up to the time of birth of the modern science of dynamics, due to **Newton.** Several subsequent stages of the science of dynamics will be reviewed in **Section 2.1.** The purpose of reviewing this embryonic stage is to look for insights involving both the **pitfalls** in past scientific studies and **important viewpoints about dynamics,** which have largely been ignored in present day science, but which may help us see a little "farther forward" in the future of understanding some of nature's more complicated forms of dynamics.

All ancient cultures viewed natural events in the light of their mythical and spiritual traditions. It appears that it was the Greek philosopher **Thales of Miletus (640?–546? BCE)** who first proclaimed that the proper way to understand natural processes was to make systematic observations of nature itself, rather than to view events as a reflection of the actions of the Gods with some mythical purposes. This "simple" proposal had the profound impact of planting the seeds of **Western science,** distinguishing this form of studying nature from that in all other cultures [1].

Perhaps the earliest application of this philosophy was in several of the studies of **Pythagoras of Samos (580?–500 BCE)** and his secret society of disciples, in the sixth century BCE. Since no written records were kept by this society, the details vary according to the later sources of reference. Much of the focus of this society was on the mythical association of rational numbers and physical events. Therefore, it was of great interest when it was observed that the pitch of a note produced by a vibrating string (dynamics!) depends on the length of the string. The discovery was then made that **two harmonious tones** ("pleasing," "agreeable"?) are related to each other by the simple ratio of the lengths of the corresponding strings: 2:1 for the octave, 3:1 for a fifth, 4:3 for a fourth, and so on. This was apparently the first general **relationship** that was found between **two physical properties**; in this case the **qualitative properties** of two tones and the associated **quantitative values** of the two lengths of strings. As indicated by the above **question mark,** this actually involves subtle psychological factors as to what is "harmonious." Both the physical and psychological aspects of the musical scales and their production have fortunately been discussed with care [2]. Thus, this initial study of a dynamic phenomena began on quite a sophisticated scale, so to speak.

Think 1.2A

If you have interests in **psychological concepts,** you might mull over the difficulties of making such concepts "scientific." What about the concept of "harmonious" tones in Indian or Chinese music? Who are the "scientists" that would form a consensus in this matter? Must scientific understandings be universal?

Also, in his book *Sand-Reckoner* (216 BCE), Archimedes attributed to Pythagoras' disciple, **Aristarchus of Samos (c. 310–230 BCE)** the correct conclusion that all planets rotate on their axes as they revolve about the sun, creating day and night and changing seasons. **[6, p. 121]**. This correct **heliocentric concept of planetary motion** found no general acceptance, as did **Aristotle's (384–322 BCE) geocentric teachings** of planetary motions about a fixed earth. This misconception required nearly **two thousand years** to be fully disproved by **Kepler,** on the basis of **Tycho's** accurate planetary observations (see below).

Around three hundred years after Pythagoras, **Archimedes** ("the Divine" [3]) **(287?–212 BCE)** made numerous theoretical and practical contributions in the areas of mathematics, hydrology, and mechanics. Among these were the discovery of a variety of **static relationships** in both mechanical and hydrostatic systems (note that static (equilibrium) states are simply special cases of dynamics). Thus, he established that to balance two weights, W_1 and W_2, on a lever and fulcrum, **Fig. 1.2 (a),** their distances from the fulcrum, D_1 and D_2, must satisfy the **relationship**

$$D_1 \cdot W_1 = D_2 \cdot W_2 \qquad (1.2.1)$$

In other words, you can hold up a heavy weight with a small downward force on the other end of the lever, provided that the small force is applied at a much larger distance from the fulcrum than the heavy weight, as illustrated in **Fig. 1.2(a)**. Theoretically, a lever (if it doesn't break!) can be used to lift any weight.

Think 1.2B

Archimedes has been credited with the statement **"Give me a fulcrum on which to rest and I will move the World"** [4]. How

"Understanding" Nature's Dynamics

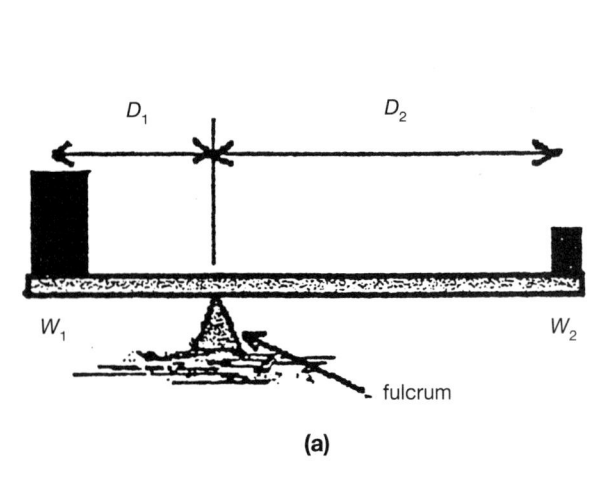

Fig. 1.2.
(a) The balanced lever and fulcrum. (b) A dynamic military application; the only engine of war to originate during the middle ages (the "trebuchet") [4].

does this idea differ fundamentally from the application in **Fig. 1.2(a)**?

The Egyptians had already realized this advantage of a lever for over two thousand years, but the **scientific relationship (1.2.1)** is due to Archimedes. The **dynamic application** of this "machine" apparently did not occur until the middle ages, **Fig. 2.1(b),** where it was used to hurl missiles at fortifications. Of course, there may well have been even earlier dynamic applications, as suggested in **Fig. 1.3 [4]**. Whatever the case, both the Egyptian usage and these dynamic applications represented **technological advances,** at which the ancient Chinese civilizations were the most highly skilled, but these technologies were not based on any general understanding of physical relationships, and therefore are **not a scientific form of understanding.**

Of greater importance were many discoveries that Archimedes made in the area of hydrology. An actual historic event is associated with one of the famous scientific discoveries made by Archimedes. Purportedly, when Archimedes sat in a full bath tub, he noted that water ran over the top of the tub as he sat down in it. From this observation, he got the idea that any body, when submerged into water, would displace a volume of water equal to the submerged volume. The story goes that he then joyously ran through the streets of Syracuse, au naturel, shouting ευρηκα!, ευρηκα! (**Eureka!, Eureka!**). This excitement may have been because it allowed Archimedes to establish for King Hieron of Syracuse the fact that his goldsmith had replaced some gold in his new crown with a cheaper metal. This is because he could now establish the volume of the irregular crown, and knowing the weight of pure gold per unit volume, he simply had to show that the crown did not have the same density (weight per unit volume). This was likely the first time that science **was used in a trial,** contributing to the sad fate of the goldsmith.

This discovery was related to more basic and important discoveries, presented in his treatise ***On Floating Bodies,*** now known as **Archimedes' principles [3]**. These consist of the following **relationships:**

(a) If a solid floats in a liquid, it will be immersed so far that the weight of the solid equals the weight of the displaced fluid.

Fig. 1.3.
"There goes Archimedes with his confounded lever again." From the fertile imagination of Sid Harris [5].

(b) If a solid submerges in a fluid, its weight in the fluid (as measured by a suspending scale) will be less than its true weight by the weight of the displaced fluid. (1.2.2)

(c) If the solid is **forcibly immersed,** the solid will be acted on by an **upward force** (as measured by a stabilizing weight) equal to the difference between its weight and weight of the displaced fluid.

Note that the shape of the solid is not important, nor is the type of fluid (e.g., water, oil, or mercury). From these results, **Galileo** (around 2000 years later!) obtained the concept of a **"force"** by reading Archimedes discussion of the force on a light solid when submerged. Galileo wrote **[3, p. 321]**, "Those who read his works realize only too clearly how inferior are all minds compared with Archimedes and what small hope is left of discovering things similar to what he discovered."

Archimedes never held a high regard for his many "practical" discoveries, and focused much more on his mathematical results. One that should be mentioned, since it played a significant role in the future of **Kepler's and Newton's dynamic understanding,** was his method for establishing the area of a circle. It used a method of approximately covering irregular bodies with many smaller regular bodies, whose areas (or volumes) are known, such as triangles, squares, etc. (or cubes, etc.). This idea was invented by Eudoxus of Cnidus (c. 390–340 BCE), known as "the method of exhaustion." To this, Archimedes added the development of a **systematic limiting approximation [3],** which ultimately recurred in **Newton's and Leibniz's development of calculus.** The application of this limiting process to dynamics is outlined in **Section 4.1.**

Following Archimedes, the subject of dynamics was dominated by **Ptolemy** (2nd century CE), who applied Aristotle's metaphysical dynamic teachings to the planetary motions about a fixed Earth. Aristotle had argued the only the "perfect forms" of circular motion at uniform speeds could exist in the heavenly domain of the Gods. Ptolemy then attempted to reproduce the observed motions of the five known planets about the Earth, using only combinations of such circular motions with uniform speeds. As a mathematical exercise, his development has been highly praised by some mathematicians, but it gave **no scientific understanding,** since no common relationship between the planets and the earth, nor between each other, was even considered, much less any constraint on the number of circular motions.

1.2 The Early Scientific Studies of Dynamics (300 BCE–1800)

The compound-circular construction of planetary orbits was then applied by **Copernicus (1473–1543)** to the ancient concept of heliocentric (Sun-centered) motion of the planets. However the circular constraints could only approximate the observed data if the center of the circular motion of the Earth lies at a point well outside of the sun **(a nonheliocentric "theory")**. [6] Indeed, Copernicus introduced **no spirit of scientific inquiry [6, 7]**, despite frequent claims to the contrary. Apparently, he even required more epicycles (circles moving along circles) than Ptolmey's construction **[7, p. 195; 16, p. 65]**, thereby not even simplifying the dynamic representation. Moreover his noncentral status of the Sun shows that he had no conception of an interaction with the Sun. Worse yet, Copernicus' views corrupted all of Galileo's concepts of celestial dynamics, despite Kepler's tutoring **[6, 7],** which he never was able to reconcile with his basic terrestrial discoveries and understandings. There is a very important lesson in detailing these events:

Lesson: Both Ptolemy's and Copernicus' theories, and Galileo's acceptance, were based on the **cultural belief** that the **Gods' celestial dynamics** must involve only uniform circular motions. Their subsequent efforts to produce ("explain") the observed planetary motions, based on this belief, was a logical process and in this sense (alone) does not differ from modern science. It was not irrational, but based on a faulty, **culture-induced "theory."** However, this "theory" fails to explain anything in the modern sense, because it could always be made to fit any planetary motion simply by using enough epicycles, and thereby established no general relationships between any physical factors. A more modern example of culturally misguided theories in the field of biological evolution is discussed in **[8]**. As will be discussed in **Sections 2.1 and 2.2,** such **culturally misguided beliefs continue to exist** even in "hard" sciences, but will be replaced by the developing understanding and appreciation of complex dynamic character of natural phenomena.

Following Archimedes, Europe lapsed into approximately 2000 years of intellectual stagnation, and indeed intellectual regression, losing touch with much of the ancient Greek knowledge. As A. N. Whitehead noted **[18], "In the year 1500 Europe knew less than Archimedes who died in the year 212 B.C."** One notable exception in this gloomy era was the varied and brilliant conceptions of the genius **Leonardo da Vinci (1452–1519).** Among his less-noted (and documented) accomplishments may have been his discovery of the first relationship in **fluid flows. [9;** unknown original reference]. The story goes that by using bits of grass and seeds, he noted that water in a stream of **uniform depth** flows proportionately faster the narrower the stream; presently this is described as

velocity × width = constant   ("**law of continuity**")   (1.2.3)

This "law of continuity" is one of the most basic understandings of fluid flows.

Much later, **a new form of understanding** was the **functional significance** of blood flow, recognized by **William Harvey (1578–1657)**. He discovered the **circulation of blood,** from the heart, through the veins and arteries, and then to the lungs, accomplishing the **"feeding" of the body,** thereby sustaining life. During this discovery, he noted in his book that if an artery was punctured (during human vivisections!), a jet of blood projected to different heights during the cycle of the heart beat, being the greatest when the heart contracts. **Daniel Bernoulli (1700–1782),** beginning with this observation, went on to discover how to measure the **pressure of a flowing fluid,** and then by **generalizing** Leonardo's law of continuity **(1.2.3)** to flow in a pipe, he discovered the (Daniel) **Bernoulli relationship** between the speed of the fluid flow in a pipe and its pressures on the walls **[9]**. This illustrates nicely the fact that one needs to look at information from a variety of physical sources when seeking inspiration for a discovery.

The earlier return to Thales' philosophy of scientific observations for celestial dynamics was begun by **Tycho Brahe (1546–1601) [6, 7]**, referred to as Tycho, as was the custom of the time. This colorful and contentious individual **[19]** made a number of basic discoveries in celestial dynamics for which he is rarely recognized **[20]**. He established the fact that dynamics occur in Aristotle's "fixed celestial sphere" of the stars, by careful observations of the "new star" (nova), which appeared and then faded from view in roughly a year (and therefore was dynamic). His observations were accurate enough to establish (by the **lack of parallax motion [21]**) that the nova's location was well outside the dynamic planetary region. He thereby **extended dynamics to include all celestial regions.** Second, he established by **the same parallax method** that the motion of comets occurred outside of the Earth's atmosphere (refuting Aristotle's terrestrial claim), and hence that **all celestial dynamics is not circular** (now refuting Aristotle's mythical circular motion dictum, basic to Ptolemy and Copernicus). Thus **Tycho was responsible** for establishing that **all celestial regions are dynamic,** and moreover these involved **noncircular dynamics** (which he did not thoroughly appreciate **[6]**).

During the period 1576–1597, Tycho made very frequent observations of planetary motions in the world's finest observatory, on the island of Hveen, near Copenhagen, Denmark. Both the island and the observatory were given to Tycho by an admiring and grateful King Fredrerick II. **[19]** With this wonderful observatory, in which Tycho had constructed a quadrant some fourteen feet in diameter (telescopes were first constructed in 1608), he improved the existing observational accuracy from ten minutes of arc to two minutes of arc. **[6]**

In order to derive some understanding from this wealth of precise observational data, Tycho fortunately had the good sense to invite the mathematician **Johannes Kepler (1571–1630)** to join him in Prague, where Tycho had moved after the King's death, and disagreements with the new King. It was also fortunate that Tycho did not wait too long, because he died suddenly (1601), only eighteen months after Kepler arrived. Their direct interactions were infrequent, due to trips and numerous arguments between these two strong personalities. Fortunately Kepler did not entrust the safekeeping of Tycho's to data Tycho's heirs, and as Kepler recounted, "I confess that when Tycho died, I quickly took advantage of the absence, or lack of circumspection, of the heirs, by taking the observations under my care, or perhaps usurping them" **[10]**. It is hard to imagine the future course of science had Tycho not rapidly contacted Kepler, and Kepler had not also absconded with Tycho's data! This is certainly an important example of human dynamics! **[7, p. 350; 10]**

Kepler's analysis of Tycho's data was based on the premise that the planetary motion was due to an **interaction with the Sun**, thus developing **the first truly heliocentric theory**. He had a remarkable insight into this attractive interaction, anticipating Newton's action–reaction law, as he noted in the introduction of his *Astronomia Nova* **(1609;** *The New Astronomy***)**:

> It is therefore clear that the traditional doctrine about gravity is erroneous.... Gravity is the mutual bodily tendency between cognate bodies towards unity or contact (of which kind the magnetic force also is) so that the earth draws a stone much more than the stone draws the earth....
>
> If two stones were placed anywhere in space near to each other, and outside the reach of force of a third cognate body, then they would come together, after the manner of magnetic bodies, at an intermediate point, *each approaching the other in proportion to the other's mass.*" [7, p. 342, emphasis added]

He likewise understood the influence of the Moon's attraction on the Earth's ocean, reasoning (but not explaining) that this should cause **two** high tides each day (whereas Galileo championed a single real tide, in order to be consistent with Copernicus' ideas, rather than Kepler's, about which he had been informed).

Kepler then spent seventeen years making sense of Tycho's data (among a variety of both scientific and mystical activities), always with an appreciation that the planetary motions came from some type of interaction with the sun. To make sense of Tycho's observational data, Kepler had to first estimate the earth's motion about the sun. He showed that it was necessary to discard both the ancient concepts of the uniform speed and circular forms of motion.

The end product of this dedicated effort was his **three remarkable relationships** (he never used the terminology "law"), quoted in the order of discovery, with the first two published in his *Astronomia Nova:*

> **K1.** The line joining the Sun and a planet sweeps through equal areas in equal times.

Three more years were required to discover that the "oval" orbit of Mars is an ellipse (having explicitly rejected this possibility earlier, he reflected "Ah, what a foolish bird I have been!" **[7, 10]**). Then followed his second relationship:

> **K2.** The planets orbit the sun on ellipses, with the sun at one focus.

In 1619, Kepler published in *Harmonices Mundi* (Harmony of the World) the third **relationship, between the period and size of the orbits of all planets,** an association that he had been seeking for over twenty years:

> **K3.** The period of revolution of a planet, and the semi-major axis of its orbit (half the longest dimension of the ellipse), are related to each other such that the square of the period is proportional to the cube of the semi-major axis of the orbit.

Note that these relationships are **not** simply between **"observable" quantities,** such as positions, population numbers, or amounts of money, etc. Rather, they involve what will be referred to as **"holistic concepts."**

> **Holistic Concepts:** These are concepts that our minds somehow invent and develop, based on some type of **correlations over intervals of space and**

time between isolated observed events (such as the concepts of **periods, orbits, ellipses, a line connected between a fixed and a moving point, the area swept out by this line in a given period of time, etc.**). (1.2.4)

Holistic Understanding: Involves finding some relationship between holistic concepts, as exemplified by **Kepler's relationships [21]**. (1.2.5)

The use of holistic concepts to establish a holistic understanding of dynamics was lost from science for roughly three hundred years, until it was **redeveloped by Poincaré.** This was an opportunity **missed by Darwin,** as will be discussed in **Section 2.1.**

Think 1.2B

What **holistic concepts** were used by **Pythagoras** and **Archimedes?** Develop some ideas about the nature of such concepts. Can you think of (many) examples of holistic concepts that we use in **everyday life?** But now comes a **challenge!** What do you think is required to make some of these concepts "scientific" and not just personal?

Kepler's accomplishment is a wonderful example of the scientific process of using facts (from observations), coupled with the imaginative **development of concepts,** to generate understandings of the dynamics of a system, namely the above relationships. Which one of us would have ever sought for the relationship K3? (Kepler said that "On 8 March of this present year 1618, if precise dates are wanted, [the solution] turned up in my head.") This was one of Kepler's great Eurekas! **[10]**. Without the full three relationships, Newton would not have been able to establish his great law of gravitational attraction **[6, 7, 17]**.

We have here an outstanding example of the systematic development of a scientific understanding of dynamics:

Facts (observations) ⇒ **Imagination (concepts, Eureka!)** ⇒
↑   **Understanding (relationships)?**
                      ↓   (1.2.6)
Further Observations for generalized confirmation

Following Kepler, **Galileo Galilei (1564–1642)** first established his fame by discovering the four moons of Jupiter, the structures on the moon, the phases of Venus, and numerous new stars, all with the help of the newly discovered telescope, which he improved, and turned from a toy into a scientific instrument [11]. Galileo choose not to publicly acknowledge the findings that Kepler had sent to him, and wasted a large portion of his life defending until his death in 1642 the dynamically obsolete Copernican system, alienating his many admirers in the Church in the process [7; on the influence of the patronage system (funding ?), see **16,** chapter 10]. At the age of seventy, after his devastating scientific and psychological experience during the Church trial, he returned to the subject of the science of dynamics, which he had abandoned a quarter of a century earlier, and wrote the book ***Dialogues Concerning Two New Sciences* (1638),** on which his immortal fame rests.

One of Galileo's great contributions was to dispose of Aristotle's claim that a force must act on a body to sustain its uniform motion. It was this Aristotelian concept that crippled Kepler's attempt to fully understand the nature of interaction between the sun and planets. **Galileo, for the first time, introduced time intervals as a variable** in describing observed physical relationships in his *Two New Sciences*. He also gave definitions for both uniform velocities and accelerations, and used the clever method of studying motions down an inclined plane in order to slow down the acceleration for his water clocks. He thus experimentally established that the dynamics can be understood as a combination of these two motions, a uniform horizontal and accelerated vertical motion. He did not introduce a time variable in a mathematical (e.g., algebraic) context, but rather gave an equivalent **verbal description** of the **spatial–temporal relationships** of the dynamics.

It was a mixed bag of concepts developed by Galileo, only for terrestrial dynamics, and Kepler's celestial relationships (interspersed with Gilbert's theory of the earth's magnetic force, which Kepler recognized as another example of action at a distance) from which **Issac Newton (1642–1727)** had to select and generalize in order to form a **coherent theory of both celestial and terrestrial dynamics.** Newton developed his laws of motion and the theory of gravitational attraction, which he used to mathematically analyze planetary motions, in his book ***Philosophiae Naturalis Principia Mathematica* (1687)** (*Mathematical Principles of Natural Philosophy*). This is now frequently referred to simply as ***Principia,*** which loses all the significance of the original title.

Newton's analysis in this publication was based on a geometric analysis, in which the dynamics are broken into equal time intervals, as is illustrated in **Fig. 1.4**. Here the smooth planetary orbit

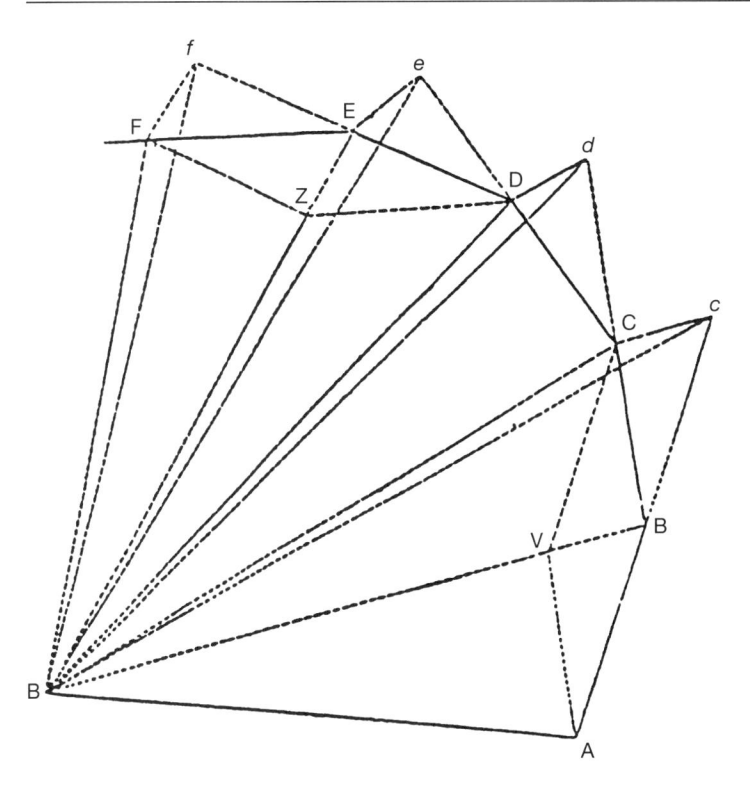

Fig. 1.4.
The figure used by Newton in the proof of his Proposition II. Theorem II in *Principia* **[15]**.

is first approximated by a series of straight-line segments, between the points A, B, C, D, E, and F. At these points Newton imagined that an impulsive force acts on the planet, directed toward the Sun (S) (called a **central force**), causing a change in the direction and speed of its motion. Hence, in traveling **from A to B,** it is given this impulsive force at B such that it acquires an **additional** component of the **velocity BV.** Thus, instead of going from **B to c** in the next interval of time, the additional velocity component causes it to move from **B to C**. At C, it obtains another impulse towards S, whose magnitude depends on the distance from S, and so on. Using this construction, Newton could prove that **Kepler's relationship K1** is satisfied for **any** such central force. However, to derive all of Kepler's relationships, he showed that the magnitude of **the impulsive force had to decrease as the square of the distance from S,** thereby establishing his **"universal law of gravitation,"** as it is often referred to these days.

Such **discrete-time characterizations** of dynamics is often quite natural, as will be shown in the next section, and in modern time it has become a fundamental method of understanding complicated dynamics [see Chapter 3, and Sections 5.2, 8.2, and 8.4].

To the present day, many find an interest in unraveling or replacing this method of Newton's analysis [12, 13, 14]. Later, this analysis was replaced by the **differential equations of calculus,** which is based on the idea of **taking smaller and smaller time steps in the above figure,** until the broken curve becomes a **smooth, continuous curve.** This subsequently became the dominant formulation of dynamics, until the latter half of the twentieth century. The details of the relationship between the discrete and "continuous" mathematical representations of dynamics will be discussed in **Section 4.1.**

Newton showed that not only could he derive Kepler's relationships [17], but also many other dynamics associated with terrestrial activities (such as the flight of projectiles through the air, and the first explanation of the two daily high tides, from which he deduced the mass of the earth). Thus was born the realization that **celestial** and **terrestrial** phenomena can have a **common explanation.** This fact had a tremendous philosophical impact, since it joined together "Heaven and Earth," which had always been viewed as being governed by entirely different influences. Ever since then, science has continued to seek a unified understanding of phenomena in nature. These, and other influences of Newton's discoveries on the twentieth century philosophy of, and **culture within,** science will be discussed in **Section 2.1.**

Next we will begin the process of understanding how to **represent dynamic processes,** and to see some features that **distinguish linear dynamics** from the **real world of nonlinear dynamics.**

Review of Concepts

- **Natural philosophy** (Thales) was essential for the birth of **Western science**

- different **understandings** of phenomena based on the discovery of **different types of empirical relationships** between observable properties

- **cultural influences** on science

- dynamic relationships such as: **stationary algebraic** (Archimedes, Leonardo, Bernoulli); **holistic** (Pythagoras, Kepler); **celestial parallax** (Tycho); **temporal** (Tycho, Galileo); **algebraic with time variable** (Galileo); **differential equation of motion** (Newton); Newton's **"universal law"** of gravity, **unifying celestial and terrestrial dynamics**

References and Notes

1. For a useful brief account of Thales of Mileitus, with helpful additional references, see J. A. Moore, *Science as a Way of Knowing: The Foundations of Modern Biology,* Harvard University Press, Cambridge, MA, 1993; also see B. Russell, *A History of Western Philosophy*, Simon & Schuster, New York, 1945, "Philosophy begins with Thales . . .," p. 3.

2. J. G. Roederer, *The Physics and Psychophysics of Music, an Introduction* (3rd ed. Springer-Verlag, Berlin, 1995.

3. A. M. Wilson aptly describes Archimedes as "The Divine" in his book, *The Infinite in the Finite*, Oxford University Press, 1995. Even though his discussion of the problem of King Hieron's crown is unduly complicated, he gives a fine detailed account of Archimedes' development of a mathematical limiting process that predated modern integral calculus by over two thousand years. Both Kepler and Galileo acknowledged their indebtedness to Archimedes' ideas in their dynamic research.

4. *Encyclopedia Britannica*. William Benton, Chicago, 1961.

5. See the enjoyable and insightful book, *There Goes Archimedes, Cartoons by S. Harris* (edited and annotated by S. N. Arseculeratne). Unigraphics (Pte) Ltd., Colombo, Sri Lanka, 1995. For more of Sid Harris' wonderful insights into science, see: *Chalk Up Another One, the Best of Sidney Harris* (foreword by Leon M. Lederman), AAAS Press, 1992.

6. A fine source of the historical development of astronomy is contained in A. Pannekock, *A History of Astronomy,* Interscience New York, 1961. For a quotation about Copernicus' ambivalence about a moving sun, read pp. 197–198, and the background for his assessment, "Nowhere the breath of a new era, nowhere the proud daring of a renovator, nowhere the symptoms of a new spirit of scientific research!" This is echoed in [7].

7. For a carefully documented history of both the scientific concepts and personal characteristics of Copernicus, Tycho, Kepler, and Galileo, read A. Koestler, *The Sleepwalkers: A History of Man's Changing Vision of the Universe,* Arkana, Penguin Books, London, 1989; an amusing example is Kepler's dating the time of his conception as 16 May A.D. 1572, at 4. a.m. (p. 227).

8. S. J. Gould, This View of Life: The Rule of Five, *Natural History*, 93(10), 14–23 (1984).

9. The Bernoulli relationship is $P + (1/2) \rho v^2$ = constant, where P is the pressure, ρ is the density of the fluid, and v is its velocity. For a nontechnical account of the developments of these understandings of fluid flows, see M. Guillen, *Five Equations That Changed the World*, Hyperion, New York, 1995, pp. 86–103.

10. For a sensitive discussion of the discovery process in science, read S. Chandrasekhar, *Truth and Beauty: Aesthetics and Motivations in Science* (University of Chicago Press, Chicago, 1996).

11. For a discussion of Galileo's many other experimental and theoretical works, see Raymond J. Seeger, *Galileo Galilei, His Life and His Works,*

Pergamon Press, Oxford, 1966; on dynamics [16, pp. 74–86]; for his personal interactions, see [7]; for his method of measuring short intervals of time, see his *Two New Sciences*, MacMillan, New York, 1914; translated by H. Crew and A. De Salvio; Second New Science; Treating Motion, Third Day, p. 179; for other references, see [16, p. 86]; For an extensive critical assessment of Galileo, see [7].

12. D. L. Goodstein and J. R. Goodstein, *Feynman's Lost Lecture*, W.W. Norton, New York, 1996.
13. J. B. Brackenridge, *The Key to Newton's Dynamics,* University of California Press, 1995.
14. S. K. Stein, Exactly how did Newton deal with his planets?, *Math. Intelligencer 18* (#2), 6–11 (1996).
15. *The Principia; translated by A. Motte,* Prometheus Books, Amherst, NY, 1995, p. 41, or see [12], p. 84, or [13], p. 9.
16. J. T. Cushing, *Philosophical Concepts in Physics: The historical relation between philosophy and scientific theories* (Cambridge University Press, New York, 1998) This gives a rare examination of this subject by a thoughtful physics professor, which extends over much of the history of scientific theories (and not so scientific ideas) that relate to the inanimate world.
17. For a discussion of the logical connection between Newton's three laws of motion and Kepler's three relationships for planetary motion, see [16, Chpt. 9].
18. A. N. Whitehead, *Science and the Modern World,* The Free Press, New York, 2000 (originally published 1925).
19. J. R. Christianson, *On Tycho's Island: Tycho Brahe and His Assistants,* Cambridge University Press, New York, 2000.
20. O. Gingerich, Tycho and the Ton of Gold, *Nature, 403,* 20 Jan 2000, p. 251.
21. For a detailed appreciation of Kepler's achievements see: A. Einstein, *Ideas and Opinions,* Bonanza Books, New York, 1954, pp. 262–266; *On the Occasion of the Three Hundredth Anniversary of Kepler's Death.* Published in Frankfurter Zeitung (Germany), Nov. 9, 1930.

1.3 AN EXAMPLE OF LINEAR, NONLINEAR, AND UNCERTAIN DYNAMICS: MONEY IN TWO VERY DIFFERENT BANKS!

Now we will begin the process of learning one way of representing dynamic phenomena in a mathematical manner. To do this it is helpful to start by seeing how we can represent some dynamics that we all are familiar with, like putting money in a bank. (As will be seen, **many very general and important lessons** can be learned **from this "simple" example.**)

1.3 An Example of Linear, Nonlinear, and Uncertain Dynamics

We begin with an account in the **Safe Bank,** which gives a very conservative interest of **5%,** compounded once a year. Let us represent the amount we initially deposit by $A(0)$ (so 0 simply represents "time zero," when we start our clocks; $A(0)$ might be $100, or whatever). Then next year we will have an amount

$$A(1) = A(0) + 0.05\,A(0) = 1.05\,A(0)$$

in the account [so, if $A(0)$ $100, then $A(1)$ $105]. After two years we will have

$$A(2) = 1.05\,A(1) \quad \text{(or \$110.25, in the above case)}$$

and so on in succeeding years.

To represent this dynamics for any year, y, we introduce a **dynamic equation** that tells us how $A(y)$ changes from year to year. As we just saw, it can be written in the form

$$A(y+1) = 1.05\,A(y) \quad (y = 0,1,2,3,\ldots = \text{years})$$

where now the amount $A(y)$ has been written so as to show that it depends on y (A is called a **"function" of y**), and $A(y)$ is called a **dynamic variable,** which depends on the **independent variable, y** (because it can be selected to by **any** of the above integers).

We can also write this more generally, to represent the case when any percentage interest rate is given, which we will represent by p ($p = 0.05$ in the above example). Since the amount next year equals what is already in the bank plus the percentage increase $[p \cdot A(y)]$,

$$A(y+1) = A(y) + p \cdot A(y) = (1+p)A(y)$$

and the dynamic equation involves the multiplicative factor $C = (1+p)$. Therefore, we can write the more general equation (for any p) in the form

$$A(y+1) = C \cdot A(y) \quad (C = 1+p; \text{ "Safe Bank," } p = 0.05) \quad (1.3.1)$$

A relationship of this type, where the value of dynamic variable at the **next time step** is given in terms of its value at the **present time,** is known as a **recursive relation**, because the same relation, (1.3.1), is applied recursively (repeatedly) for all increasing values of $y = 0, 1, 2, \ldots$. Since the time of **Newton, recursive relations have been one of the most common ways of representing dynamic processes**. Other types of representations of dynamic processes will first be discussed in Sections **1.2** and **2.2.**

In the case of (1.3.1), the recursive relation is a typical example of what is called a **linear dynamic equation.** It is called **linear** because it has the property that if the initial deposit, $A(0)$, is increased by some factor, say to a new initial deposit of $A^*(0) = 3\,A(0)$, then the new $A^*(y)$ will likewise be increased by the same factor, $A^*(y) = 3\,A(y)$, for all values of y. This simple **proportional relation** is called a **linear relation**. Therefore **(1.3.1) is a linear recursive relation.**

Because of this proportionality feature, such linear equations can be easily **"solved,"** which means that we can **obtain $A(y)$ for any value of y,** in terms of C and the initial deposit $A(0)$. To do this, we note that by taking $y = 0, 1, 2 \ldots$ in equation (1.3.1), we obtain **successively**

$$A(1) = C \cdot A(0)$$
$$A(2) = C \cdot A(1) = C[C \cdot A(0)] = C^2 \cdot A(0)$$
$$A(3) = C \cdot A(2) = C^3 \cdot A(0)$$

Therefore, it is not difficult to see that, for **any year, y,**

$$A(y) = A(0)C^y \qquad (y = 0, 1, 2, \ldots) \tag{1.3.2}$$

This is called the **solution of (1.3.1)** for all values of y. It "solves" the problem of directly **determining $A(y)$,** for any y, in **terms of C** and the initial deposit, $A(0)$.

▓ Think 1.3A

Now determine the relation between $A(y)$ and another case, $A^*(y)$, in which the initial deposit is $A^*(0) = 5\,A(0)$. This illustrates the **linear (proportional) relation** between $A(y)$ and $A^*(y)$, for **any value of C.** ▪

From (1.3.2) you can **predict** the amount in the account in any year, **provided,** of course, that the bank keeps paying this interest (an **assumption** of this model). This type of predictability has historically (since the time of Newton) always been considered to be of great importance, and indeed a necessary requirement for any science (here a "banking science"?). With the study of more complicated systems, the **simplistic idea** that all phenomena have **predictable outcomes** (for **all** future times) has undergone a dramatic reevaluation, as will be discussed later. This bank, hopefully, is an example of a reliable (predictable) source of income, at least for our lifetime. But even here **one needs to be careful,** as we will soon see.

Because the variable y in (1.3.2) appears as **an exponent of a constant, C, the growth of $A(y)$** is called **an exponential growth.** It is a faster rate of increase than the case of a **linear**

1.3 An Example of Linear, Nonlinear, and Uncertain Dynamics

growth, for example. In a **linear growth** the increase in $A(y)$ is only proportional to y, such as

$$A(y) = A(0)(1 + p \cdot y) \qquad (1.3.3)$$

This is the amount one gets from a bank that only pays a percentage p on the **initial** deposit, not on what is presently in the account (so that the money is **"compounded" at each iteration**). Thus (1.3.3) is the solution of another type of recursive relation

$$A(y + 1) = A(y) + p \cdot A(0) \qquad (1.3.4)$$

In contrast to the linear recursive relation (1.3.1), **(1.3.4) involves repeatedly adding a fixed amount each iteration.** This type of recursive action will be referred to as an **additive recursive relation.** No bank would actually survive with this offering, but the **total earnings** from weekly or monthly **wages,** are described by an **additive recursive relation, (1.4.1),** which you can investigate with a **computer program** in the **next section.**

▪ **Think 1.3B**

Can you show that (1.3.3) is indeed the solution of (1.3.4)? Try a substitution of (1.3.3) into (1.3.4). Referring to the discussion following (1.3.1), is (1.3.4) a linear dynamic equation? ▪

We can make a graph of these two solutions, (1.3.2) and (1.3.3), in **Fig. 1.5,** and see the rate that $A(y)$ increases only becomes large for

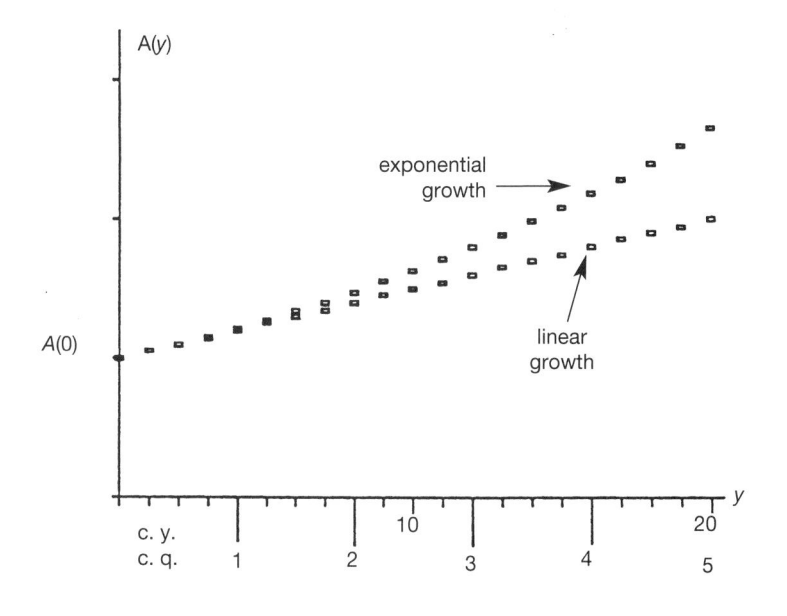

Fig. 1.5.
Comparison of the **linear growth** of $A(y)$ versus y, (1.3.3), with that of the **exponential growth,** (1.3.2), **both** with $p = 0.05$ (5%), when compounded yearly **(c. y.),** and quarterly **(c. q.)** (which is **much less conservative**).

long times, when the interest is compounded yearly (the **c. y.** time line). However if these interest rates are **compounded every quarter** (three months), then their difference becomes larger in a shorter period of time (the lower **c. q.** time line).

Now let us consider **another possible banking establishment,** which **claims** to be a **great deal less conservative.** Assume that you saw an advertisement by the **Swift Bank,** owned by Mr. and Mrs. P. D. Swift (not to be confused with any Swiss Bank). Fortuitously, the owners' name fits well with their claim that they will **double** your deposit every quarter (three months). In other words, $p = 1$, and this is done every **quarter, q.** Since $C = (1 + p) = 2$ in this case, the **solution (1.3.2)** shows that the amount the Swift Bank says you will have after the **quarter q** (the new time interval for adding the interest) is

$$A(q) = A(0) \cdot 2^q \quad \text{(the Swift claim!)} \quad (1.3.5)$$

If you put in $100, then one year later you are told you will have

$$A(4) = 100(2^4) = \$1600$$

and after two years, it presumably will be

$$A(8) = 100(2^8) = 100(256) = \$25,600.$$

In a mere four years you will supposedly be a multimillionaire (~$6.55 million!). Wow! Not bad (if true)!

So some people rush to make deposits in the Swift Bank, making Mr. and Mrs. P. D. Swift very wealthy, very swiftly. The depositors may rationalize that this rate is reasonable because the Swifts are very clever and know how to use these deposits in some very profitable investments. This is, of course, is an **extreme example,** in order to clearly illustrate a **general and important real issue,** which is often less obvious because of the **time scales involved** (as in Fig. 1.5).

▌ Think 1.3C

Both (1.3.2) and (1.3.5) are exponential growths. Why is (1.3.5) so much faster? Comparing (1.3.5) with (1.3.2), which has the dynamic equation (1.3.1). Obtain the dynamic equation for the Swift bank as a function of the years, $A(y + 1) = S \cdot A(y)$ (where S is related to the Swift interest each year). ▌

But alas, we know that we live in a **real world,** and it is clear that something is wrong with the Swift claim, (1.3.5). After all, the above rationale is not substantiated by any known facts. We might

1.3 An Example of Linear, Nonlinear, and Uncertain Dynamics

(or certainly should) begin to wonder how one can realistically expect the Swift Bank to be able to sustain this rate of return. Where is it getting its money? Of course, initially they may be taking in much more money from deposits than they are supposedly paying in interest, but this illusion can only last as long as the depositors do not start drawing out their "deposits." Even if Mr. and Mrs. Swift have other reasonably good trading activities, or loans, these financial resources are finite.

Important Lesson: Finite resources is a basic fact of nature.

So, even though the Swift Bank claims that its dynamics is governed by the equation

$$A(q+1) = C \cdot A(q) = 2 A(q) \quad \text{(the Swift promotional equation)} \quad (1.3.6)$$

we can be quite sure that **C must decrease** from the value of 2 as **A(q) increases** because the Bank is going to run out of monetary resources sometime soon. Therefore, the exponential growth rate, shown in **Fig. 1.5,** must decrease, so the curve for $A(y)$ must "bend over" if $A(y)$ becomes "large."

The **more general lesson,** of which the Swift phenomena is only a more trivial example (except to the depositors), is illustrated in **Fig. 1.6** (with the **uncertain future** indicated). This all means that a "realistic" recursive equation for the Swift Bank must be of the form (at least until the **?? factors** come into play)

$$A(q+1) = C[A(q)]A(q) \quad \text{(the "realistic" Swift equation)} \quad (1.3.7)$$

where the function $C(A)$ must decrease as A becomes large, to either near the value $C = 1$, or possibly lower, all depending on **what type of people** the Swifts happen to be (this is where **an assumption** is made, and **why?? generally applies** in **Fig. 1.6**). Whatever are the details of this case, $A(q+1)$ is no longer simply proportional to $A(q)$ [**no longer linear in $A(q)$**]. Equation **(1.3.7)** is referred to as a **nonlinear equation,** or more specifically, a **nonlinear recursive relation.**

In contrast with the ease of obtaining the solution (1.3.5) of the linear equation (1.3.6), the **mathematical solutions of nonlinear equations,** such as (1.3.7), **are generally unknown.** So, in order to study the dynamics of such nonlinear equations, it is generally necessary to make use of **computer programs.** We will therefore **defer the possible fate** of the Swift Bank until **Section 1.5,** after we have discussed **simple computer programs in Sec-**

"Understanding" Nature's Dynamics

Fig. 1.6.
The general influence of **limited resources** on something that initially grows exponentially, requiring a **resource $R(I)$** at each iteration of its recursive relation, I. **One possible nonlinear response** to this limitation is illustrated by the dark rectangles. This system has an **uncertain future (??),** depending on the **character of future actions.**

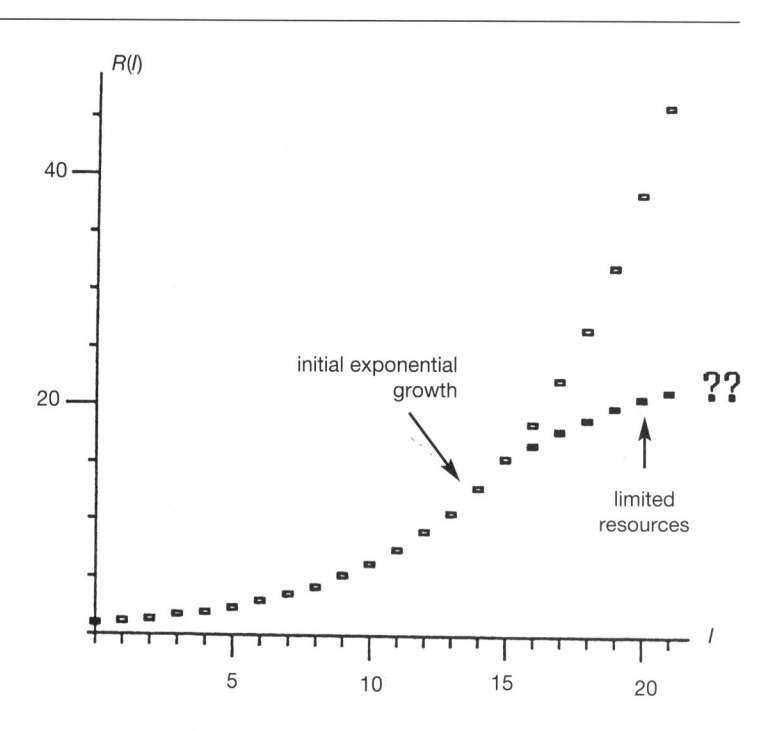

tion 1.4. At that time, we will look into some ways of representing **uncertain (nondeterministic) aspects of dynamic systems.**

Thoughts about Dynamic Models, Simplicity, and Assumptions

The **dynamic equations (1.3.1), (1.3.4), (1.3.5), and (1.3.6)** are all referred to as **dynamic models.** The term **models** emphasizes the fact that they are not real-world dynamics, but rather the dynamics that would occur if the **assumptions** that have been used in constructing them really are "reasonably" valid, **for some period of time.** Since we do not know all the factors that influence phenomena over long periods of time, **all models are** to varying degrees only **approximations of real phenomena.** In science, we seek to find the **simplest dynamic model** that approximates the phenomena of interest, to the **desired accuracy,** over **some time span.**

This concept of **simplicity,** which is one of the **basic principles of science,** dates back to **William of Occam (or Ockham, 1290–1350).** Presently he is primarily known for a maxim that is not found in his writings, but nonetheless is called **Occam's razor:**

Entities are not to be multiplied without necessity

However he did write [1]

"It is vain to do with more what can be done with fewer"
(1.3.8)

which in a sense is a more profound statement. Indeed, this leads quite naturally into another way of expressing Occam's idea in the present dynamic context:

Keep dynamic models as simple as possible to represent a dynamic phenomena (within specified accuracy and time spans) (1.3.9)

It was this concept of Occam's razor that was not part of the thinking of either **Ptolemy or Copernicus (Section 1.2).** They simply joined together any number of circular motions in order to "explain" the planetary motions. They only dealt with a mathematical puzzle, and did not resolve any **physical issue** of real planetary motion, nor even the heliocentric versus the geocentric hypotheses [2]. This is one way in which their efforts are distinguished from modern science.

Think 1.3E

This is a good time to think about **what aspect of reality** has been included in the Swift Bank dynamics that **was not considered** in the **"Safe Bank's"** dynamics. That is, how realistic is the "Safe Bank" model (1.3.1), for what **values of p,** and for how long a **period of time?** Thus, even in the "Safe Bank" circumstances, it may **not always** be **obvious** how to apply Occam's razor.

Review of Dynamic Concepts

- dynamic variables
- independent variables
- dynamic equations as recursive relations linear and nonlinear dynamic equations
- solutions of dynamic equations;
- linear and exponential growth with time
- the finite resources in nature produce nonlinear effects
- Occam's razor;
- implicit assumptions are part of any dynamic model

References

1. B. Russell, *A History of Western Philosophy,* p. 472. Simon & Schuster, New York, 1945.
2. A. Pannekoek, *A History of Astronomy.* Interscience, New York, 1961. Also see [6] in Section 1.2.

1.4 A FIRST LOOK AT EXPLORING WITH BASIC PROGRAMS

The computer is an indispensable tool for studying nonlinear dynamics. You might have noted that in the case of the Swift Bank in the last section only the general character of the **nonlinear dynamics, (1.3.7),** was suggested. No details of this dynamics were obtained, because we do not know how to obtain their mathematical solutions! This is common for nearly all nonlinear equations, so we need to write some simple computer programs to understand what dynamically occurs from these recursive relations.

More than **60 Basic programs,** which illustrate a wide variety of natural dynamic phenomena, are available from the Wiley ftp site or the disk at the back of this book. You can, in fact, run these programs, learn about **new dynamic ideas,** and do **some explorations** of suggested ideas and questions, with **no knowledge** about **Basic programming.** However, unless you learn a few rudimentary aspects of the Basic language, you will miss the **great fun and stimulation of being able to try out your own ideas,** and not just look at the ideas of some teacher or program writer (like myself). This is a good chance to **expand your imagination,** and fortunately you will soon find that **it is not difficult.** But, like all worthwhile activities, it takes **some study and thought.**

In the back of the book you will find the **Computer Appendixes** (see the Table of Contents for the layout of the Appendixes). Regardless of your initial objectives, the **first thing** to take care of is the mechanics of **accessing, running,** and **transferring** between the programs. Although all these and other processes will be done in **MS-DOS,** it is assumed that you have access to **Microsoft Windows,** and a simple method of transferring between the Windows page and **running programs** will be discussed. For those unfamiliar with such matters, they are explained in detail in **Appendix C0. Other resources** for running **Basic programs on DOS and MacIntosh** computers can be found in references [1] and [2].

The **second mechanical issue** concerns setting up **your**

1.4 A First Look at Exploring with Basic Programs

"**working space,**" where you can **copy programs** from the ftp site to a disk or your C drive, **modify** them, make **new programs,** and **save them,** giving them **any name.** These matters are explained in **Appendix C1,** and should also be read carefully, since they will be used shortly.

Once you have **understood** the contents of Appendixes **C0 and C1,** and are **at A:\>** on the disk, you can get an **initial experience** of the Basic programs by **<QB FIRST>**. The dynamics of this program concerns the **total wages earned** at the **end** of each week t, $E(t)$, over the course of one year, when **you specify** the weekly wage, *WAGE*. The numerical values of $E(t)$ are given by the simple **additive recursive relation** [like (1.3.4)]

$$E(t+1) = WAGE + E(t) \qquad [t = 0, 1, 2, \ldots, 51; E(0) = 0] \qquad (1.4.1)$$

You could obviously figure this out for yourself, but that is not the point, and it would soon get tedious. The point here is to learn how the numbers from such recursive relationships are generated by a program. It is **another example** of a **linear growth rate,** seen before in the **last section.** So start with:

Explore FIRST

As discussed in Appendix C0, to run the program, go to **Run** and then **Start.** Here the dynamics is given by the **additive recursive relation (1.4.1).** The program generates the **total amount earned** at the **end of each week,** over the course of **one year,** once you have specified the **weekly wage,** *WAGE*. Where is this entered in the program? After running the program you should to read all of the program comments, and begin the process of understanding the program structure and commands. You can also refer at any time to **Appendix C2** for further programming information. If you want to try modifying this program, using you knowledge from **Appendix C1,** the safest thing is to **make a copy,** by exiting the program to **A:\>,** and then use **<copy FIRST.bas mod1st.bas>** (with two spaces). You can now run the modified program **mod1st.bas,** and make any modifications, while safely retaining FIRST. To change anything in the new program, put the cursor at the location and either use the **delete key** or type in whatever you want. If you are doubtful about some remarks, **make a small change in the program** (e.g., change a number, remove an apostrophe, change a colon to a comma, etc.) and run the program again, noticing the change in the behavior of the dynamics.

There are several **specific changes** that you can try to make:

1. The challenge here is to insert a new line into the program, which reads **GOTO A**. This tells the computer to return to the symbol **A:** at the beginning of some line (note the colon). In the program FIRST there is such a line, beginning with **A:**, where one can input a new selection for the *WAGE*. Note that this makes use of an INPUT command in the program (locate it). Try out your idea of where GOTO A should be inserted **after** all the calculations are completed, and **adjust it** until it **takes the computer back to the line A:.** (There is a **substantial** hint in the program)

2. Another approach to this same idea is given in the program, which now allows you to choose **from the keyboard,** between returning to **A:,** or exiting the program. This has been made **inactive** by the **apostrophe** at the beginning of the line that starts with INPUT. To make this **active,** simply **remove this apostrophe,** by locating the cursor at this point and using the delete key. This shows **one way to exit a program,** involving a command **END** in the program. **Another way** (discussed in C0) that is sometimes very helpful, is to **[CTRL+PAUSE],** which means to **press both of these keys simultaneously.** If you would like **more programming details** at this point you can look into **Appendix C2.**

Now you should have a **your own new program,** modlst.bas. But it is not **really yours** until you give it a **name** you like, and **store** it someplace of your own. It is easy to give it any name, say ####, by starting at **A:\>** and **<move mod 1st.bas ####.bas>,** where there is a space after "move" and before #### (hopefully you have thought of a more imaginative name!). This removes the program **modlst.bas** and **generates** the program **####.bas,** on the floppy disk. But now you need somewhere to put this program in a **"working space" of your own.** So this is the time to use the information in **Appendix C1,** which shows how to **make your own directory** on the computer (e.g. "MINE"), and either **move or copy programs copies** from the disk into MINE, and how to **store new and modified programs in MINE.** Also, to **protect you** from a future disk failure, you will find an easy way to **make a copy** of the **book's disk,** both in MINE, and onto **your own disk.** So there are a number of very useful mechanical operations which need to be **practiced** at this point.

Before going into more details with the programming, look into the program SECOND.

1.4 A First Look at Exploring with Basic Programs

Explore SECOND

In this program most of the descriptive remarks have been removed from FIRST, so this is **closer** to (but not entirely) a simple "working program," in which nearly every line is read by the computer. In addition to the linear growth rate of **wage earnings,** the second recursive dynamics of **compound bank interest** has been added. Using the **keyboard selection method,** illustrated at the end of FIRST, you will now be able to **choose** either the additive recursive dynamics **(1.4.1), or** the linear recursive relation [as was explained following (1.3.1)]

$$D(t + 1) = RATE * D(t) \qquad (1.4.2)$$

Here **D(t)** is the **amount deposited** in a bank at the end of week, t, and *RATE* stands for **(1 + the *weekly* interest rate).** So if the weekly interest rate is **1%,** then *RATE* = **1.01.** Of course, one also needs to specify the **initial deposit, D(O),** which is requested by another **INPUT** command.

The difference between the numbers generated by **(1.4.1)** and **(1.42)** was illustrated in **Fig. 1.5 in Section 1.3.** This **graphical representation** of dynamic information is very important, and this quantitative graphical information will now be illustrated in the program **THIRD.**

Explore THIRD

Go to **program THIRD.** This is a modified form of the program SECOND, in which the data is no longer printed out, but instead is used to construct a **graphical representation of the dynamics (1.4.1) and (1.4.2).** Here you will find how a simple coordinate system is set up, and how **data points,** such as D(t), can be graphically represented as a **function of t.** You will also find an understandable notation for the **color selections** for **Lines and Circles.** In all **future programs these graphical details will be given,** but you should become familiar with what is involved, so that you will be able to **develop your own ideas.** So study the comments carefully, and explore some modifications to understand them better. More descriptions of these matters can be found in **Appendix C3.**

Explore FOURTH

Finally, the **program FOURTH** will illustrate how the discrete data output of a computer can be turned into a series of connected lines. This is similar in spirit to **what Newton did in Fig. 1.4.**

When these data points are close to each other (e.g., when the **time steps are small**), the resulting curve looks **"continuous."** This is the beginning of Newton's introduction of calculus into science, and will be discussed further in **Chapter 4**. The dynamics are the same as in **program THIRD,** but now the line connections between data points hopefully give greater clarity. Note again that **the graphical process will always be set up in the future programs,** but the more you know, the greater will be your range of opportunities to invent new approaches in the future. When you examine this program, you should look at the further details in **Appendix C3**.

The following programs give more detailed information about the construction and use of Basic programming, which you can look over at least briefly, and then **return to for more details later on.** The program DYNAMICS will graphically illustrate several **more dramatic** types of dynamics, both as a function of time, and in a **two-dimensional** example. The program DETAILS is just that, containing a discussion of a number of graphic and programming details, which can be usefully referred to at any time.

The **program DYNAMICS** illustrates how numbers that are produced by **some given functions** of time can be turned into a **graphical representation.** This is similar to what was done in THIRD and FOURTH, except those numbers were generated by the recursive dynamics (1.4.1) and (1.4.2). Two types of examples will be illustrated:

(I) The first example will develop a graph of a **"continuous" curve** (related to the remarks in FOURTH), but now of a more interesting, but rather **strange function $X(t)$** ver-

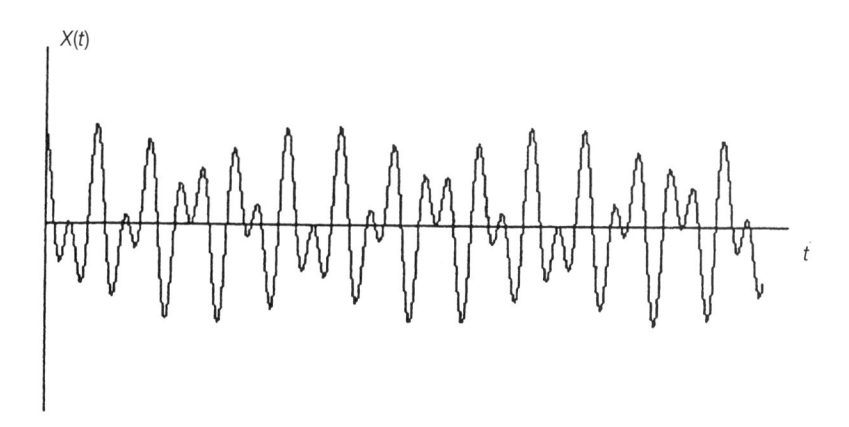

Fig. 1.7.
The graph of $X(t)$ versus t, for an arbitrarily selected function $X(t)$.

sus t. The static characterization of this dynamics is illustrated in **Fig. 1.7.**

(II) The second example will be the graph of a "continuous" curve in a **two-dimensional space** defined by the values of **two functions, $X(t)$** and **$Y(t)$,** which have been selected to give some complicated dynamics. This is **(partially)** illustrated in **Fig. 1.8.** Here it is essential that you use the **clear screen,** `<C>` (capital C!), in order to see what is going on (in particular, the **orientation of the orbits.** Since computers are now very fast, it is nice to be able to **slow down the programs** so that we humans can see what is going on in the dynamics. Therefore a **Delay option** is introduced, so that you can make the dynamics go **slower** or **faster** while the program is running, by using `<S>` or `<F>`.

■ **Think**

Where is the time recorded, as it is in **Fig. 1.7,** when one uses this phase-space" representation of the dynamics in **Fig. 1.8?** Later on, read about Poincaré's introduction of this representation of dynamics.

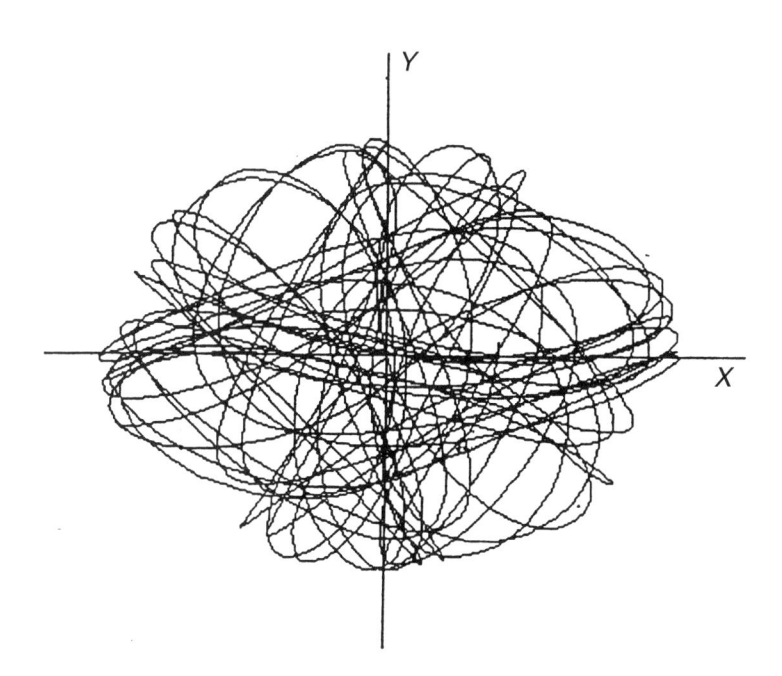

Fig. 1.8.
The graph of $X(t)$ and $Y(t)$ in a two-dimensional **"phase space,"** for a short period of time. This requires **clearing the screen** periodically in order to follow the dynamics. **What is the orientation?**

Explore DYNAMICS

Now go to the **program DYNAMICS** to see the **actual dynamics,** statically characterized in **Figs. 1.7 and 1.8.** In addition to seeing some more interesting forms of dynamics, this program is intended to be used as a **review and extension** of the contents of the previous programs in this section, and can be studied in conjunction with **Appendixes C1–C3.** More specifically, it contains: **(i)** more on coordinates systems; **(ii) automatic** clearing of screen **and** a subroutine to refresh coordinates, etc.; **(iii)** more use of **colored-line connections to form "continuous" curves; (iv)** further **interactions with the keyboard** to **clear the screen** and **slow down the dynamics** in this era of **very fast computers,** and to **introduce new data/conditions;** produce **beeps** and **tones** with selected **frequencies and duration** (see Appendix **C7); (v)** review of the iteration loop, with **temporal count; (vi)** more on the **two-dimensional "phase–space" graphics,** [$X(t),Y(t)$]. There are many things here for your creativity!

Finally, an explanatory **nondynamic program** is offered, both to review some of the concepts in the above programs and for further detailed information:

Explore DETAILS

The short program DETAILS contains no dynamics, but a great deal of information about its programming structure. Much of it involves textual setups, which should be looked over to get an idea of what the commands and operations are all about. The best way to proceed is to make a copy of this program **into your directory,** as discussed in **Appendix C2,** and then use this program to make changes, and see what effects are produced (retaining the original on the floppy). Also it is useful to make **a printout of the program,** by getting the program on the screen, then go to `File`, and down to `Print`, and strike `Enter`. **This will be useful for making exploratory notes. Again see the details of the use of INKEYS** at `A:`, `B:`, and `C:`, which allows **access from the keyboard.** At each of these locations, if **no key is struck,** indicated by **" ",** the computer keeps returning to `A:`, `B:`, and `C:` (i.e., **causing a pause of any desired length).**

Following all of this background, we now return to the **uncertain and nonlinear recursive dynamics** of the banking world, and some issues of the **uncertain** and very nonlinear dynamics of **human population growth.**

References and Notes

1. While I know of no resources for **studying dynamics** using the Basic language, one resource for an **introduction to Basic programming** is the collection of books by: David A. Lien (all published by CompuSoft Publishing, San Diego, CA, 1987):
   1) *MS-DOS: The Basic Course*. (Contains extensive details about the mechanics of dealing with Basic programs on MS-DOS.)
   2) *Learning Basic for Tandy Computers.* (**A very useful general** programming reference, regardless of the DOS computer!)
   3) ***Learning Microsoft Basic for the MacIntosh:*** *New 2.0 Version, and 2.1 Version.* (This might well be a **very valuable resource for new types of dynamic innovations and explorations,** which is the **raison d'être of this book.** Unfortunately I never learned about it in time for dynamic explorations.)
2. A resource for more advanced applications of Basic to **mathematical topics,** requiring Numerical Recipes Software, can be found in J. C. Sprott, *Numerical Recipes: Routines and Examples in Basic*, Companion manual to *Numerical Recipes: The Art of Scientific Computing*. Cambridge University Press, New York, 1998.

1.5 IMPLICIT ASSUMPTIONS IN MODELS: "REAL" SWIFT BANK DYNAMICS (AND HUMAN POPULATIONS?)

In **Section 1.3,** we saw that several basic issues arose in trying to understand the Swift Bank's promotional claim, which was that the amount of your money in their bank each quarter, $A(q)$, would be determined by

$$A(q + 1) = C \cdot A(q) = 2 A(q) \qquad (1.3.6)$$

This would yield the unbelievable amount

$$A(q) = A(0)2^q$$

or, on a yearly basis, the even more absurd amount, discussed in Section 1.3

$$A(y) = A(0)16^y$$

Because of the finite resources of the Swifts (and of any natural process), we are sure that (1.3.6) must be replaced by **some form** of nonlinear dynamics

$$A(q + 1) = C[A(q)]A(q) \qquad (1.5.1)$$

which would limit the growth in $A(q)$. This type of limitation was schematically illustrated in **Fig. 1.6, Section 1.3**. Note however there is a **question mark (?)** in this figure, because it is **not clear** what is the **precise behavior of $C[A(q)]$** as $A(q)$ becomes large. It only **must limit** the magnitude of $A(q)$, because of finite resources. But there are **many ways** that this might happen, which **depend** on the **Swift's moral and philosophical character!** Thus, there is a great deal of **uncertainty** in the dynamics, which involves the mysterious world of **cognitive processes.** This situation exemplifies a **general basic fact about modeling:**

> **Any model is only as good as the assumptions** (1.5.2)
> **used in constructing that model**

So **what assumptions** might be made about the Swifts?

Think 1.5A

Here is a good place to use **your imagination** to extend this strange list, and try to program that idea at some time.

1. One possibility is that they are **strangely philanthropic,** simply wishing to distribute a sizable portion of their wealth to a number of people by this strange method. This is not likely, of course, but what dynamic model might characterize this behavior?

2. Another possibility is that they are planning to attract a number of depositors who will put in large deposits, let the money accrue interest from their limited investment resources, and then, when they think the time is right, simply leave town with all the money. **They are crooks!** But **when** they decide that the time is right to leave town is **not predictable,** probably even by them. It just may "seem right" for a variety of personal reasons. So there is an uncertainty (**nondeterministic**) character to this **dynamics.** How might we model that feature?

3. Another possibility is that they used this **extravagant promotion** in order to attract depositors, but then **plan to decrease $C(A)$ as A increases,** hoping the depositors continue to leave their money in the bank. On a much less drastic

level, some banks and credit card companies use similar promotional methods, with the subsequent decreased interest rate information hidden in small print.

Therefore, **behind any model** of the form (1.5.1) are some **implicit assumptions** about such possible behavior. Normally, one would not take the trouble to model all three of these possibilities, but this will be done, **very crudely,** just to **illustrate different forms of nonlinear modeling.** This requires **assuming** some other **possible** representative **functions,** or **nondeterministic** forms of $C(A)$. In the present modeling, these three possibilities will be characterized by:

1. In this case, it will be assumed that the Swifts want to limit their gift to any depositor by an amount *LIMIT*. Thus they will indeed set $C[A(q)] = 2$ if $A(q) < LIMIT$, but if $A(q) > LIMIT$, they will go back to the last quarter and set $C[A(q-1)] = 1$. In other words, the *LIMIT* can never be exceeded, and it may not even be reached by each depositor. Finally, they require that $A(0) < LIMIT$ (the Swifts are not interested in giving money to the wealthy).

Think 1.5B

What happens during the time that $C(A) = 1$? Given the action described in (1), what determines the **actual** limiting value of a **specific** deposit $A(0)$? Can you think of ways to modify this effect?

2. In this case, **C drops to zero,** at a time when they think it is best to take all the money and get out of town. So the **implicit assumption** that C is a **deterministic** function of A **is incorrect.** One can model the Swift's uncertain behavior on some probability basis, which is of course **another assumption.** A probability can be introduced using the random number generator, **RND,** which generates numbers between **0 and 1** with equal probability. Thus $A + B*RND$ will generate numbers between A and $A + B$ with equal probability (See Appendix C2).

Think 1.5C

A **statistical description** is needed when knowledge of all known factors, such as A, does not describe the resulting dynamics, which however **may** be accounted for by some probable range of outcomes that is based on an assumed "probability distribution" (in this case, *RND*). What are some possible factors that might influence the Swifts in this way? Check **Program 1-5A** for the present **toy modeling** of this probability distribution. See if you can modify this modeling, based on **your own ideas.**

3. Here $C(A)$ must rapidly decrease to **slightly above $C = 1$,** which pays a small interest. Recall that the safe bank offers $C = 1.05$, so very likely C will end up between $C = 1$ and $C = 1.05$. See Program 1-5A for one method to recover the promotional expenses. There are many ways this could be accomplished, and the present model is **pretty crude,** so polish it up!

Think 1.5D

What are the issues involving the maximum value of $C(A)$, the time required to decrease to the lower value, and the future duration required to recover the promotional cost? **Is this model really realistic?** What would make it more realistic?

Of course one should not take any of this very seriously. These **toy models** are simply constructed to illustrate some very different forms of nonlinear and statistical dynamics. Hopefully, however, this will make you aware of this issue of implicit **assumptions to all models.**

These models of the three possibilities [including the nondeterministic case (2)] are schematically illustrated in **Fig. 1.9.**

Explore 1.5A

You can explore these nonlinear models by using **Program 1-5A.** Use this program to develop your ideas in the **above Think questions,** and also to learn some further techniques for programming these **very nonlinear and diverse forms of dynamics.**

Relating Dynamic Ideas in Different Areas

When **dynamic ideas are learned in one area,** such as in these banking examples, it is important to see if some of these **ideas can apply** in other very **different dynamic physical contexts,** which is the ongoing search for **generalities of dynamic processes.** One might think that there is no more different area of

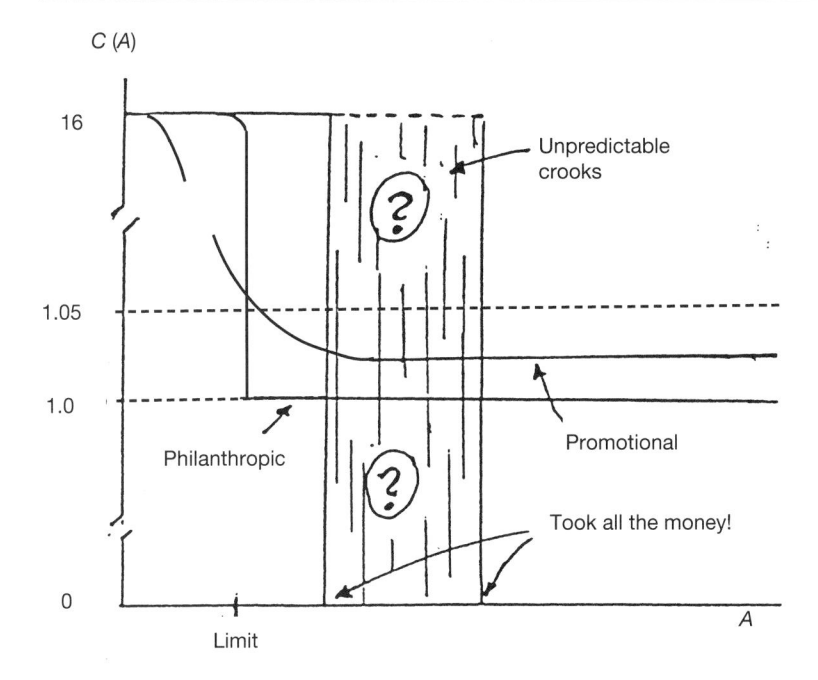

Fig. 1.9.
Two deterministic, and one nondeterministic, possible nonlinear behaviors of $C(A)$, due to finite resources and the character of the Swifts.

concern, and of dynamics, than the possible future development of the **human population,** either as a **global issue** or in **specific localities.** The **latter issue** is most likely to be what would motivate our **individual concern and attention,** as caricatured in **Fig. 1.10.**

Think 1.5E

Perhaps Fig. 1.10 may stimulate you to think about how the banking lessons may be allegorically related to **populations in contrasting worlds.** If one associates the amount, $A(y)$ and the yearly **regional** population, $P(y)$, along with nonlinear (real) percentage interest rate, $p(y)$, and the population yearly growth rate, $g(y)$, and the "Two Banks" as **"Two Regions"** (or, more ambitiously, "Two Worlds"), with **different factors influencing the growth rate,** it is not much of a stretch of the imagination to see that some lessons learned in the banking example may be transcribed to this **human ecological concern.** Beginning at a **simplistic level,** write down a **recursive relationship for $P(y)$,** which seems like a "reasonable" dynamic model, similar to (1.3.7). Think about what changing factors could cause different magnitudes of the yearly growth rate, $g(y)$. What must the long-term value of $g(y)$ be in order

"Understanding" Nature's Dynamics

Fig. 1.10. **Look who has come to stay!**

to avoid some problems **(what problems? how "long-termed"?).** What are the **specific implicit assumptions** associated with some problems? You will soon find some help in such matters, but first think about it yourself.

At present, the only modeling that is widely referred to are the **simple linear models,** which you may have constructed in Think 1-5E. Such models yield **exponential growths** at different rates, depending on **regional policies** and **resources.** For **short periods of time,** these are useful for comparisons of different regions, as will be **illustrated in Explore 1-5B,** below. However, because of **limited resources,** we know that such modeling can only apply for a **limited duration of time.** For periods of a century or less, the influences of **human polices** in different regions are becoming increasingly important, and even critical. Such policy making is obviously **very uncertain,** and only a **variety of models** can be offered, depending on what **assumptions** are made, some of which can only give **probabilistic predictions.** Then, of course, there are many natural events (e.g., plagues, massive volcanic eruptions, etc.) that **no assumptions** are likely to cover.

1.5 Implicit Assumptions in Models

By way of general **background information** on these issues, you would benefit by referring to references [1] and [2]. For example, for even **a single country (Egypt),** over **any 200 year period** in its nearly 3000 year known history, the estimated population is a wildly unpredictable phenomenon **[2, p. 39]**. Over **very long periods** of time, it has proved to be **impossible** so far to obtain **any model** (understanding). For example, see the estimates of the **world's population** over the **past million years [2, p. 96:** "I know of no published **mathematical model** to explain (these results)"].

So our **present "understanding"** of the world's **regional** population growth is simply assumed to be characterized over a **"short period of time"** by linear recursive relations, $P(y + 1) = C \cdot P(y)$. However, due to the rapidly changing **interconnectedness** of the world, in terms of people, diseases, commerce, and political policies, this **"short period"** is itself becoming **rapidly shorter.** With this forewarning, it is still useful to explore some of the **present** information concerning such **exponential population growths.**

Explore 1-5B

In **Program 1-5B** there are illustrated some of the data of the **human population in the world,** for various cities and countries [1]. Using this **you** can select **an initial population,** and an **exponential growth rate,** and see when it will **bypass** the present populations of various cities and countries. The references should be referred to for more details, particularly concerning the great variability of the exponential rates around the world. You should pay attention to such **data** in the future, for **it is within your children's future—See Fig. 1.10!**

Think 1.5F

Following this linear excursion, you should make a list of at least **5 to 10 factors** that can strongly **influence** (for better or worse) this exponential growth over say **50 years.** To enlighten you about these matters, you will find references **[2–4]** to be valuable quantitative, conceptual, and policy making resources. In reference **[2]** you will find fine graphical representations of the influences on past populations of **conflicts** and **diseases,** and projections for future world populations depending on the control of **fertility** at different stages in the future. Reference **[3]** contains short discussions of a list of **19 factors** that can influence the human population: (1) grain production, (2) fresh water, (3) biodiversity, (4) energy, (5) ocean fish catch, (6) jobs, (7) infectious diseases, (8) cropland, (9) forests, (10) housing, (11) climate change, (12) materials, (13) urbanization, (14) protected natural areas, (15) education, (16) waste, (17) conflict, (18) meat production, and (19) income.

Although numbered 15 in this list, **education,** in its multitudinous forms, surely ranks high in influencing most of these factors. How recent **economic policy issues** have influenced **such social factors** and their repercussions in a **number of countries** have been **documented** and **"understood"** to a considerable degree by employing **dynamic modeling [5].** This is the most ambitious and complex scientific approach that I have seen into this area of macrocognitive dynamics.

At the bottom of many of the other factors influencing the future human condition is the **biological diversity** of the world. The fragile character of the world's biological diversity "is not just a source of aesthetic pleasure for *Homo sapiens*—**it is the reason for our being,"** as is beautifully discussed in reference **[6].** This contains discussions of a number of **ecological models,** both pro and con. In addition there is now a **fine collection** of resources for raising your awareness of some of these **ecological and biological factors,** among them being references **[6–11].**

Now let us turn to other examples in **the world of natural sciences,** which are **simple enough** to model reasonably realistically, and explore some of the **startling new dynamic features** that have been uncovered in even these **simple modern dynamic models.**

References

1. *The World Almanac, 1998*, p. 838. K-III Technical Corp., Mahwah, NJ, 1997. *Population of World's Largest Cities*. United Nations, Dept. for Economic and Social Information and Policy Analysis, New York, 1996. *Current Population and Projections for all Countries: 1997, 2020, and 2050*. Bureau of Census, U.S. Dept. of Commerce, Washington, DC, 1996.
2. J. E. Cohen,, *How Many People Can The Earth Support?* W. W. Norton, New York, 1996.
3. L. R. Brown, G. Gardner, and B. Halweil, *Beyond Malthus: Nineteen Dimensions of the Population Challenge*, W. W. Norton, New York, 1999.

4. A. Gore, *Earth in the Balance: Ecology and the Human Spirit*, Houghton Mifflin, Boston, 1992.
5. W. W. McMahon, *Education and Development: Measuring the Social Benefits*, Oxford University Press, New York, 1999.
6. S. Levin, *Fragile Dominion: Complexity and the Commons*. Helix Books, Perseus, Cambridge, MA, 1999.
7. Biodiversity, The Fragile Web, *National Geographic, 195*(#2), Feb. 1999.
8. S. L. Pimm, *The Balance of Nature? Ecological Issues in the Conservation of Species and Communities*. University of Chicago Press, Chicago, 1991.
9. E. O. Wilson, *The Diversity of Life*. W. W. Norton, New York, 1992.
10. R. E. Ulanowicz, *Ecology, The Ascendent Perspective*. Columbia University Press, New York, 1997.
11. L. Garrett, *The Coming Plague: Newly Emerging Diseases in a World Out of Balance*. Penguin Books, New York, 1995.

1.6 SOME BASIC DYNAMIC INSIGHTS

While the dynamics considered in this chapter have generally been easy to model (given our assumptions), and to program, they have illustrated some **very important general issues** concerning nature and our assumptions about it. These are worth collecting for future reference and reflections:

1. **Linear recursive relationships,** such as $x(t + 1) = A \cdot x(t)$, produce exponential growth rates, $x(t) = x(0)A^t$, which become arbitrarily large as t increases. This **can only represent a natural process for a limited period of time, because ...**

2. In dynamic processes there are **always limited resources,** and this requires that such **recursive relationships must be nonlinear,** $x(t + 1) = A[x(t)]x(t)$, where $A(x)$ **decreases as x increases (Fig. 1.6).**

3. All dynamic **models are based on some assumptions** about how nature behaves, which are **often implicit** (i.e., not explicitly noted or recognized). All models are **only as good as the assumptions used in constructing them.** Some basic assumptions that need to be explicitly recognized are the **applicable duration of time, extent in space, degree of observational reproducibility, and**

the environmental conditions for which a model is believed to be valid.

4. **Keep a model as simple as possible** in order to represent a dynamic phenomenon, **involving a specified range of initial conditions,** and **assumptions noted in (3);** this is a dynamic variation of "Occam's razor."

5. Not all **dynamic phenomena** can be represented by **deterministic dynamic models.** We saw above several examples that are controlled by unpredictable **cognitive processes.** Other phenomena are influenced by unpredictable environmental actions, and by **inherent internal dynamic sensitivities.** These can severely limit the **time duration** of predictability, or require a **probabilistic description** of the dynamic phenomena. These concepts will be illustrated in more **detail in Section 2.1.**

2

Dynamic Aspects of Modern Science

*2.1 THE NEWTON-TO-1950 STAGES OF DYNAMIC DEVELOPMENTS

As noted at the end of **Section *1.2,** Newton was the first to use the **recursive relation** as a description of dynamics. He did this in his analysis of planetary motion, using discrete time steps, as illustrated in **Fig. 1.4.** A modern discussion of Newton's analysis, using this figure, can be found in reference **[1, p. 84].** Briefly put, the dynamics is broken up into **equal time intervals,** represented by the planet being located at the points A, B, C, ... The planet is first taken to be traveling in the direction A to B, at which point (B) it receives an impulsive force toward the Sun (labeled S), giving it an additional velocity component B–V. Thus, instead of traveling on to the point c in the next time interval, it goes to C, which is a vertex of the parallelogram (V, B, c, C).

This same analysis is then repeated, using the impulse received at C, to change the velocity B–C, so that it travels to D (the parallelogram is not always shown). The essential feature to note is this is a **recursive analysis,** where the same analysis is applied at each set of the succeeding points. Then, instead of describing the dynamics in these large equal discrete time steps, Newton envisioned reducing these time intervals to **"infinitesimal"** time dif-

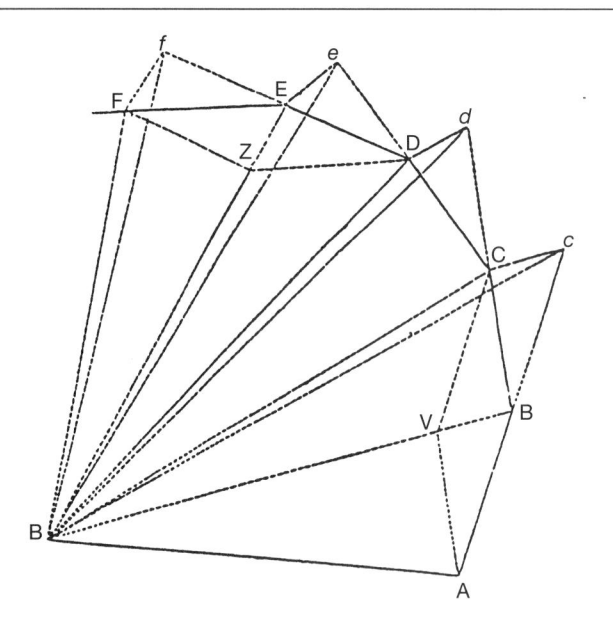

Fig. 1.4.
This figure is from Newton's *Principia*, showing his **recursive analysis** of a planet's motion about the sun, employed at discrete time steps.

ferences, thereby approaching a **"continuous-time"** form of recursive relationships (as hinted at in program **FOURTH, Section 1.4**). This major change in the character of recursive relationships was developed both by **Isaac Newton** and **Gottfried Wilhelm Leibniz (1646–1716),** which led to years of bitter disputes concerning the priority of developments [2]. Such "infinitesimal" recursive relationships led to the **calculus of "ordinary differential equations" (ODEs).** The **full (relevant!) course** concerning the development of ordinary differential equations, and their solutions, is given in **Section, 4.1.** You can look at this if you like, but it will not be used until Chapter 4. It turns out that it is generally easier to analyze the solutions of ODEs than it is of the discrete recursive relationships, as will be made amply clear in Chapter 3.

The dynamics of a physical system is generally represented by a collection of ODEs, which can be written in the form

$$\frac{dA_1}{dt} = F_1(A), \frac{dA_2}{dt} = F_2(A), \ldots, \frac{dA_N}{dt} = F_N(A)$$

or more compactly in the "vector" form

$$\frac{dA}{dt} = F(A) \qquad (2.1.1)$$

Here $A(t) = \{A_1, A_2, \ldots, A_N\}$ is a vector, representing a group of N observable properties (such as the total population of different animals, or species; the concentrations of different chemicals or vegetation; the positions and velocities of various particles; or voltages, currents, etc.). Likewise, $F(A)$ is a collection of N functions of these variables. The details of such equations are not important at this point. Equation (2.1.1) is shown only to give some **sense of the notation** that replaces the discrete form of recursive relations encountered in Chapter 1.

In addition to the application of these differential equations to gravitational forces, Newton developed equations of motion to apply to **any type of "force"** (for example, due to Archimedes' buoyancy force, or electric, magnetic, and frictional effects). This led him to his **"universal form"** of equations of motion for particles, his famous

$$\textbf{Force} = \textbf{Mass} \times \textbf{Acceleration} \qquad (2.1.2)$$

which greatly exaggerated the popular impression about the ability of science to predict the phenomena that occur in the world.

The following historical stages in the study of dynamics (up to 1800) were based solely on Newton's ordinary differential equations, and other differential equations that were developed to describe the dynamics of **sound waves, vibrating strings** (back to Pythagoras, Section 1.2), and **fluids.** (For an authoritative, nontechnical discussion of this history of mathematics applied to science, read Chapter 3 in [3].) These new types of equations described the dynamics of **continuous systems,** which are **spatially extended,** so their dynamic variables, such as the flow velocity of water (or blood!), $V(x,t)$, generally requires both space and time to specify the state of the system. (As discussed in Section 1.2, Archimedes, Leonardo, Harvey and D. Bernoulli all made basic advances in the understanding of fluids, but they did not mathematically describe time-dependent effects.) This new form of dynamics led to the so-called **partial differential equations (PDEs),** the details of which are not of concern here. However, their introduction was essential in the development of some of the most important advances in physics of the 19th and 20th centuries, most notably **James Clerk Maxwell's (1831–1879),** remarkable theory of the propagation of **electromagnetic radiation** (e.g., light), and as the basis of **Erwin Schrödinger's (1887–1961)** theory of the **quantum mechanical properties of systems,** whose philosophical impact will be discussed below. Both of these equations, as well as Boltzmann's equation, introduced a fundamental change in the character of dynamic equations, in that these equations **do not in-**

volve observable variables. They are abstracted away from observation but deal with functions that can be used to account for some "expected" observable properties.

Now it is important to point out several **general features** of these differential equations, because they strongly influenced the **culture of science,** particularly in the 20th century. A differential equation, such as (2.1.1), is useful only if it is possible to obtain a **"solution,"** $A(t)$, which **"satisfies this equation."** An **example of a solution** of the recursive relation (1.3.1) was given in **equation (1.3.2).** Here the meaning is much the same, but now a "solution" is a **specific function,** $A(t)$, such that the differential operation on the left side (represented by dA/dt) produces a **new function** that is identical to the function $F(A)$ on the right side, when this particular $A(t)$ is inserted into $F(A)$. In this sense, it satisfies this equation." The details of this will be made clear in **Section 4.1.** This all sounds pretty complicated; and in general it is indeed very complicated, so complicated that such explicit functions can only be found for limited types of "linear equations" (similar to the **linearity of section 1.3,** and discussed below).

But there is one very **general statement** that can be made about **any such solution,** even if we can not obtain these explicit functions. If the functions $F(A)$ **satisfy several simple and common properties** (such as changing their values smoothly as the value A changes, and they do not increase too rapidly when A increases), then the following properties are true:

Existence and Uniqueness of Solutions

Given arbitrary initial conditions, $A(0)$ at $t = 0$, there is a unique function (solution), $A(t)$, which satisfies the equations (2.1.1) and the prescribed initial conditions, and "exits" for all $t > 0$ (which means that it remains finite for all finite values of t). (2.1.3)

This result ensures that, for any initial state of the system (at $t = 0$) that we may be interested in, **such specific solutions of (2.1.1) exist** (even if we cannot find them), and moreover, they specify finite values for $A(t)$ for **all future times.** Furthermore, such solutions are **unique.** This means that **if $A(t)$ and $B(t)$ are two solutions** of (2.1.1) and if they have the same initial conditions, $A(0) = B(0)$, they will also have identical values **for all $t > 0$, $A(t) = A(t)$.**

▓ **Think 2.1A**

Notice that the time, t, does not appear explicitly in (2.1.1), so that "$t = 0$" can equally well really refer to any time on the clock, say $t = 22$ minutes, 43 seconds on your watch. Given this insight, explain **what must be true** if two solutions of (2.1.1), **"intersect"** each other in the sense that $A(t_0) = B(t_0)$ **at some other time** $t = t_0$. This is an important basic feature of uniqueness, which will be used after Chapter 3. ▬

The Mechanistic and Deterministic View of Nature

Due to these **mathematical** features of **existence and uniqueness,** the era of mathematical developments from the time of Newton to around 1800, and the examples of numerous solutions related to **simple** physical phenomena, all conspired to generate a sense of assurance among many scientists, philosophers, and the general public that such mathematical equations as (2.1.1) could be used to uniquely **predict the future behavior** of **any** physical phenomenon that can be described by $A(t)$, with **arbitrary precision** and for **any duration of time into the future.** This viewpoint is often referred to as a **mechanistic and deterministic view of nature.** It is **"deterministic,"** because once the initial state of a system is prescribed, **all future states** of that system **are determined.** That is, given the solution $A(t)$, one should be able predict the future **with arbitrary precision.** Thus, from this viewpoint, **nature is simply ticking along in a mechanical fashion.** It is simply a Newtonian **"clockwork universe."** This viewpoint predictably had a profound influence on both the social culture and the technical aspects of the **scientific culture, particularly in the 20th century.** The impact of such **cultural beliefs** can have an **ossifying influence** on the development of science, as attested to by the two thousand year **scientific hiatus of the Aristotle–Ptolemy–Copernicus** culture, recounted in **Section *1.2.** There are **modern science-ossifying influences,** some proclaimed by eminent scientists, that are **no less dangerous,** as will be discussed below.

▓ **Think 2.2B**

It is none too soon to begin to **think more carefully** about the above mechanistic and deterministic viewpoint. Note that (2.1.3) is entirely a mathematical statement (theorem) about a mathematical equation. You might begin by thinking about **the differences** between **mathematical functions,** $A(t)$, and their **associated physical properties,** which we observe in nature. Make a list of several fundamentally different features of these two concepts.

Then, in the **19th century,** new forms of physical dynamics were investigated. In the areas of chemistry and population ecology, **nonlinear** ODEs were found to be required, but these still were described by deterministic equations of the form of (2.1.1). On the other hand, the writings of **Robert Malthus (1766–1834),** which contained no mathematical equations, were based on the **three social dynamic propositions: (1)** populations cannot increase without the means of substance, **(2)** the population invariably increases if the substance is available, **(3)** the superior power of population growth cannot be checked without producing misery and vice. In other words **wars, pestilence, and famines** are the main checks to population increases. This was Malthus' response to the questions in **Think 1-5D,** and it not only had a great impact on social policies, such as opposition to the Poor Laws of the day, but also to **Darwin's "survival of the fittest"** concepts later in the century (see below). For some references on these matters, see references **[4, 5, 6]**. The point here is to again note the impact that **cultural views** can have on science, which in this case supported the "deterministic" and "inevitable" viewpoint of Newtonian dynamics.

Entirely different was the discovery in the 19th century of the need for **nondeterministic forms of dynamics.** The first was in the area of biology, where **Gregor Mendel (1822–1884),** an Augustinian monk, published a paper in **1865** that contained a set of **rules to determine the probability** that sets of "alternative factors" carried by each parent (e.g., in the case of peas: yellow or green; wrinkled or round) will be transmitted to their offspring (e.g., chapter 9 of reference **[6]**; chapter 14 of reference **[7]**). If one applies these rules to successive generations, one has an entirely new form of probabilistic recursive dynamics. However, as in the case of Kepler's mode of holistic understanding, Mendel proceeded to propose four general principles of hereditary phenomena **[8]**. Mendel's results were ignored until they were rediscovered around 1910, ushering in the ongoing and evolving understanding of genetic dynamics **[6, 7]**.

Later, **Ludwig Boltzmann (1844–1906)** proposed an entirely new (and very difficult) dynamic equation to describe the **probabilistic behavior** of an **"ensemble"** (collection) of macroscopically identical **nonequilibrium gas systems.** Their average observable properties can be obtained from a "probability distribution function," which is the function in his equation. This was the first dynamic characterization of a **probability function.** It met fierce opposition from believers in the deterministic Newtonian theories, but is understood today (or should be) as a more basic representation of our ability to understand nature than the scientific culture could then accept (à la Poincaré, to follow) **[9]**.

Then came something else entirely new in the consideration of nature's dynamics, and which remains an evolving theory. **Charles Darwin (1809–1982)** developed the concept that species evolve by a process of **"natural selection,"** governed by a concept of the **"survival of the fittest," based on the dynamic competitions between populations in the living world,** a concept which he developed after reading Malthus **[4]**. His insights into these matters are contained in his masterpiece, *The Origin of Species* **[10]**. Darwin's vision did not involve any specific dynamic processes, much less equations, but he was apparently strongly influenced by the Newtonian paradigm, as was Malthus. Darwin's dynamic concerns were primarily directed toward understanding such concepts as "survival of the fittest" and "natural selection" of different species within various environmental conditions. The environmental factor, which is so critical to much of Darwin's ideas, was very non-Newtonian in character, so the picture is by no means simple. The dynamic processes that are basic to the "natural selection" process are even **presently** largely undefined, even though they are fundamental in making the characterization "survival of the fittest" a nontautological concept.

The implications of Darwin's vision have been discussed by innumerable authors. The reader should be warned that the topic of biological evolution is one of great controversy and acrimony. This is not easy to untangle in a balanced fashion, as is attested to by a sample selection of readings **[7, 11–20]**. Indeed, this area of understanding is very much in an **evolutionary stage,** due in considerable part to the **evolving genetic technology and knowledge.** It is **worth appreciating** that what is going on here is an **evolution of science** itself, which will be discussed in **Section 2.2.**

Toward the end of the 19th century, **Henri Poincaré (1854–1912)** made an astonishing discovery involving **relationships between all solutions** of certain **deterministic equations,** (2.1.1). What made this all the more astonishing was that it was made in the heartland of what had come to be viewed as Newton's **"clockwork universe,"** namely planetary dynamics.

The history of this discovery is a fascinating backdrop to the amazing new perceptions that he developed. In 1889, Poincaré, then thirty-five and unknown, submitted an entry to an international competition, sponsored by Oscar II, King of Sweden and Norway, related to the question of the stability of the planetary dynamics in the solar system. Due in part to Poincaré's entirely new (and uncomprehended) mathematical approach to this problem, his entry won the gold medal and was published in the prestigious journal, *Acta Mathematica,* which made him instantly famous. However, it was soon discovered that his results contained a deep and fundamental error, causing a traumatic period for all concerned. The few distributed copies of the journal were hastily recovered, and a very different publication appeared only after several years **[21]**.

All of this is by way of the background to Poincaré's invention of an entirely different method for the investigation of dynamic processes (which the award committee did not fully understand). He introduced into mathematics a variety of **qualitative and holistic forms** of understanding properties of the collection of **all solutions** (not used since **Kepler**), thereby giving birth to the modern field of **topology.** It is important to note that these qualitative and holistic concepts are established within a logically precise framework, and not as a matter of philosophy, thereby making them suitable for the further development of science.

These methods exposed an entirely different world of dynamic phenomena than had even been dreamed of up to that time. Even Poincaré doubted the implications of his new methods of analysis at first **[21]**, and it required the mathematical talents of George Birkhoff to prove rigorously that Poincaré's conclusions were correct **[22]**.

However, these are mathematical results, and the **empirical implications** of Poincaré's mathematical conclusions to the field of science were apparently initially appreciated only by the French physicist Pierre Duhem **[23]**. It was only much later that other physicists focused on the empirical impact of Poincaré's results **[24,25]**. Paraphrasing Duhem's careful discussion of Poincaré's insights, one comes to the following very basic relationship between the solutions of mathematical models and empirical science:

Duhem's Lesson

Deterministic dynamic models of natural phenomena do not imply that these phenomena are empirically deterministic. (2.1.4)

To expand on this statement a little, by a **"deterministic dynamic model"** is meant one that satisfies the existence and uniqueness conditions, (2.1.3), where the variables, $A(t)$, are related to observable properties of the phenomenon of interest. The concept of **"empirical determinism"** means that the more accurately one can determine the initial conditions, the longer one can predict the future outcome of the system (to the same accuracy). The **erroneous idea** that the first implies the second remains a broadly held belief even today, despite Duhem's analysis and the rigorous **Poincaré–Birkhoff theorem.**

Thus, **at the end of the 19th century** there were numerous examples of the fact that there exist **entirely new forms of dynamics,** distinct from those described by (2.1.1). A very important addition to this were Poincaré's **new topological methods for understanding** the holistic characteristics of **all** solutions of (2.1.1)

In the first half of the 20th century, the new form of dynamics that is most relevant to terrestrial phenomena (as contrasted with Albert Einstein's theory of general relativity), was **Erwin Schrödinger's (1887–1961)** introduction of a **"quantum mechanical"** partial differential equation that gave the space–time dynamics of a probabilisitc **"wave function,"** $\Psi(x, t)$. Using this (unobservable) wave function, a mathematical procedure is then prescribed to predict the **expected values** of certain observable properties at the **atomic level.** That is, the dynamics of any physical system is only described probabilistically, and once an observation is made, giving a specific number to the observable, the standard "Copenhagen interpretation" is that the wave function has "collapsed" to a new state determined somehow by the observation process. Since it is thereby impossible to observe a particle in a succession of two locations (without interfering in some unknown fashion), this formulation of particle dynamics, with the standard "Copenhagen interpretation," is **incapable of describing any particle dynamics.** This is clearly a serious limitation of this theory. See reference **[26]** for a helpful introduction to this issue, and for more details of **David Bohm's resolution** to this inadequacy see reference **[27]**.

Overview

From these discoveries, Poincaré showed that buried within Newton's equations for the motion of planets is a wonderland of complicated dynamics, which today is called **deterministic chaos,** or more popularly, simply **"chaos."** In more recent times, we have discovered that even simpler-looking nonlinear dynamic equations can also possess this remarkably complicated "chaotic" behavior.

This type of dynamics is always related to the dynamic sensitivity of systems, meaning that small disturbances can cause very different types of dynamic behavior. This discovery has profound implications on our ability to "accurately" predict the behavior of most physical systems in nature over "long" periods of time. On the positive side, there are many indications that the sensitivity of this type of motion may also be responsible for many healthy capabilities of our hearts and mind to respond to changes in our activities and in our environmental situations. We obviously need to begin to learn more about this type of dynamics, and this will therefore be explored in what follows, in a variety of diverse systems. Because we do not know how to solve most nonlinear equations, many of these discoveries have come to light because of the development of the digital computer, which is a unique resource for exploring ideas, as we will do throughout the following studies.

References and Notes

1. D. L. Goodstein and J. R. Goodstein. *Feynman's Lost Lecture*. W.W. Norton, New York, 1996.
2. H. Hellman. *Great Feuds in Science: Ten of the Liveliest Disputes Ever*. Wiley, New York, 1998. Chapter 3: Newton vs. Leibniz: A Clash of Titans.
3. M. Kline. *Mathematics: The Loss of Certainty*. Oxford University Press, 1980.
4. D. Winch. *Malthus*. Oxford University Press, Oxford, 1987.
5. L. R. Brown, G. Gardner, and B. Halweil. *Beyond Malthus: Nineteen Dimensions of the Population Challenge*. W. W. Norton, New York, 1999.
6. D. J. Depew and B. H. Weber. *Darwinism Evolving: Systems Dynamics and Genealogy of Natural Selection*. MIT Press, Cambridge, MA, 1997.
7. J. A. Moore. *Science as a Way of Knowing: The Foundations of Modern Biology*. Harvard University Press, 1993.
8. For a very readable and authoritative survey of biological ideas, see E. Mayr, *This is Biology: The Science of the Living World,* Harvard University Press, 1997.
9. C. Cercignani. *Ludwig Boltzmann: The Man Who Trusted Atoms*. Oxford University Press, Oxford, 1998.
10. C. R. Darwin. *The Origin of Species*, John Murray, London, 1959 (originally published 1959); *The Portable Darwin,* D. M. Porter and P. W. Graham, Eds., Penguin Books, New York, 1993.
11. For an authoritative viewpoint, with a sensitive vision of the history of science, see [7].

12. For broad introductory discussions of evolutionary dynamics, see E. Mayr, *Toward a New Philosophy of Biology: Observations of an Evolutionist,* Harvard University Press, 1988.

13. For one view of population dynamic issues see S. J. Gould, *Full House: The spread of excellence from Plato to Darwin*, Three Rivers Press, New York, 1997; for his variety of evolutionary insights, read his *Ever Since Darwin: Reflections in Natural History*, W. W. Norton, New York, 1992.

14. One critical assessment of the evolutionary ideas of S. J. Gould has been given by R. Wright, The Accidental Creationist: Why Stephen Jay Gould is Bad for Evolution, pp. 56–65, *The New Yorker,* Dec. 13, 1999.

15. A. Brown. *The Darwin Wars: How Stupid Genes Become Selfish Gods*. (Simon & Schuster, London, 1999); a review by M. Ruse, Stores from the Front, *Science, 284,* May 7, 1999, p. 923.

16. R. Dawkins. *The Selfish Gene,* 2nd Ed.; Oxford University Press, Oxford, 1989.

17. For a contrary vision of evolution see B. Goodwin, *How the Leopard Changed Its Spots: The Evolution of Complexity,* Scribner's, New York, 1994.

18. E. Mayr. *One Long Argument: Charles Darwin and the Genesis of Modern Evolutionary Thought.* Harvard University Press, Cambridge, MA, 1991.

19. E. Mayr. Darwin's Influence on Modern Thought, *Scientific American,* July, 2000, pp. 79–83.

20. D. J. Depew and B. H. Weber. *Darwinism Evolving: Systems Dynamics and the Genealogy of Natural Selection*. MIT Press, Cambridge, MA, 1993.

21. The remarkable history of this development can be found in three books (in order of increasing technical content): I. Peterson, *Newton's Clock: Chaos in the Solar System,* Freeman, New York, 1993, Chapter 7; F. Diacu and P. Holmes, *Celestial Encounters: The Origins of Chaos and Stability*, Princeton University Press, Princeton, 1996, ff. 27, ff. 45; J. Barron-Green, *Poincaré and the Three Body Problem*, American Mathematical Society, Providence, RI, London Mathematical Society, London, 1997, Chapter 4 (largely nontechnical discussion).

22. For a relatively accessible survey of Birkhoff's imaginative works, with statements of the present Poincaré–Birkhoff theorem and references to it higher-dimensional generalizations, see M. Morse, George David Birkhoff and His Mathematical Work, *Bull. Amer. Math. Soc.,* May, 1946, Vol. 52 (#5, part 1) pp. 357–391, specifically, p. 378. Also read his preface to *George David Birkhoff: Collected Mathematical Papers*, American Mathematical Society, Providence, RI, 1950. For a **fine modern mathematical survey of dynamic systems,** with many references to **scientific interests** (despite its forbidding title!), see: M. W. Hirsch, The Dynamical Systems Approach to Differential Equations, *Bull (New Series) Amer. Math. Soc., 11*(#1) July, 1984, pp. 1–64; the Poincaré–Birkhoff connections are found around p. 37.

23. P. Duhem, *The Aim and Structure of Physical Theory*. Princeton University Press, Princeton, 1991. (Translated from *La Théorie Physique: Son Objet, Sa Structure*, Marcel Rivière & Cie, Paris, 1914.) Of particular scientific importance was his appreciation, in Section 3.4, of the distinction between the mathematical interest in the stability of the gravitational three-body problem, and the quite distinct interest of astronomers in the planetary stability issue. **Duhem recognized,** as Poincaré apparently did not [e.g., see E. Lorenz, *The Essence of Chaos* (University of Washington Press, Seattle, 1993) p. 121], that Poincaré's conjecture [15] meant that, although the **problem of stability** of the solar system is quite **meaningful to mathematicians,** "for the initial positions and velocities of the bodies are for him elements known with mathematical precision, for **astronomers** these elements are determined only by physical procedures involving **errors** that will gradually be reduced by improvements in the instruments and methods of observation, but **will never be eliminated.** Consequently it might be the case that the problem of the stability of the solar system should be **for the astronomer a question devoid of meaning,** [for] ... among this data there are some (solutions) that would eternally maintain all heavenly bodies at a finite distance from one another, whereas others would throw some one of the bodies into the vastness of space. ..."

24. L. Brillouin. *Scientific Uncertainty and Information*. Academic Press, New York, 1964; Chapters II, IV, and VI, Sect. 1 (for starts!).

25. I. Prigogine. *The End of Certainty: Time, Chaos, and the New Laws of Nature*. The Free Press, New York, 1997.

26. J. T. Cushing. *Philosophical Concepts in Physics: The Historical Relation between Philosophy and Scientific Theories*. Cambridge University Press, Cambridge, 1998.

27. D. Bohm and B. J. Hiley. *The Undivided Universe: On Ontological Interpretation of Quantum Theory*. Routledge, London, 1993.

2.2 THE CHANGING SCIENCE OF NATURE'S DYNAMICS (AFTER 1950)

In the decade of the 1950s, an entirely **new source of information** was introduced into the study of dynamics, namely the **digital computer.** No longer would we be left in total ignorance about the features of those unattainable analytical solutions of the **nonlinear** dynamic equations [1], which **must** be used to represent **most of the dynamics in nature.** The sense of dominance of nonlinear dynamics in nature was delightfully characterized by the mathematician Stanislaw Ulam [2] (paraphrased): **"The study of** nonlinear dynamics in nature is like the study of nonelephant mammals in zoology."

So now scientists could really begin to **seriously study the dynamics of nature,** rather than just the mathematically dictated simple cases (like the elephants in zoology).

The first use of a digital computer to study the nonlinear dynamics of a physical system was due to **Enrico Fermi, John Pasta, and Stanislaw Ulam,** using the brand-new primitive computer of the day (MANIAC) at Los Alamos, New Mexico. They were interested in the dynamics of a chain of particles ("atoms"), connected by nonlinear springs and fixed at both ends (similar to a discrete form of an elastic band). Their original objective was to see how some energy, which was initially distributed in a single long wavelength form (like one musical tone), would become uniformly distributed over tones corresponding to higher harmonics. Physically, this latter state would correspond to a very complicated and varying shape of the elastic band. The distribution of energy equally over many modes is referred to as an "equipartitioning of energy," and would represent how this system tends to **thermal equilibrium.** The question they were interested in was simply **how long it would take** for this distribution of energy to occur, and to see how it would compare with the prediction of Fermi's famous and widely used **"golden rule"** application of perturbation theory (which is **not nonlinearly valid, [1]**). They were therefore quite surprised to discover that this widespread distribution of the initial energy **did not occur,** but rather the energy initially became distributed over only a few harmonics, and then returned to nearly the initial wave form, and **continued** to do this periodically. This surprising result is now known as the Fermi–Pasta–Ulam **(FPU) phenomenon [3,4,5].** Fermi later expressed to Ulam that this was indeed **"a minor discovery."** Unfortunately, Fermi died soon after this discovery, before he could further reflect on this phenomenon **[6],** and these results remained cloistered in a Los Alamos report **[3],** and so were generally unknown. The first theory that accounted for their results did not occur until 1963 **[7],** requiring the theoretical incorporation of the **nonlinear properties discussed in Section 4.3.**

Thus, in its first application to physical dynamics, the computer showed that it could uncover surprising and fundamentally new dynamic features **[8]**. This discovery stimulated computational and mathematical research that led to the discovery of **solitons** (localized packets of energy that retain their identity even when they collide) **[9],** and this has further led to important practical applications.

Other important early studies done with computers, **prior to 1980** were:

1. **Hénon and Heiles** applied the **Poincaré map** method (see **Sections 7.3 and 8.2**) to a simple-looking dynamic model, to study how **chaotic dynamics** arise when the energy is sufficiently large (**Section 6.11 of [4]**).

2. **Edward Lorenz,** a meteorologist at MIT, discovered the phenomena of "sensitivity to initial conditions" and a "strange (chaotic) attractor," while studying a simplified model of atmospheric dynamics **(see Section 8.2)**.

3. **Robert May** discussed the chaotic behavior of certain models of the **population ecology** of insects. These details are in **Chapter 3**.

The discovery of Hénon and Heiles is closely related to an important **mathematical study,** referred to as the **KAM theorem** (for Kolmogorov–Arnold–Moser). Their results are in turn closely connected to the **chaotic dynamics** established in the **Poincaré-Birkhoff theorem,** referred to in **Section 2.1**. (For more details and references, refer to **KAM in reference [4]**.) However, this theory is not able to establish the **quantitative** details of the chaotic structure found in the special example of Hénon and Heiles, illustrating some specialized benefits of the computer explorations of dynamics (to particular systems).

Since 1980, there has been an explosive use of computer simulations of dynamic systems, which will be selectively discussed in the following chapters. For those interested in a fine **general mathematical survey** of modern aspects **of dynamic systems,** with many references to concerns of **scientific interest,** see reference **[10]** (despite its forbidding title!)

These examples, and many others, have now established the fact that computers give us access to dynamic information that is unobtainable in any other manner. This means that the informational basis of science has been fundamentally enlarged, complementing both empirical information mathematical information. This is not simply an issue of computational capabilities, but the fact that we can now explore much more involved and **novel forms of dynamics,** some of which may be basic to the **future understanding** of the more **complex dynamic systems in nature.**

Now it should be clear that the very **foundations of science have been changed,** so that it is necessary to spend at least a little time considering the basic challenges of modern science in general. Foremost, it is a time to keep in mind the simple adage:

**You know ** what you know ** when you know **
how you know**

which means giving careful thought to the **assumptions and constraints** involved in the approaches that are applied to **"understanding"** nature, as has already been emphasized in **Section 1.5.**

A more expansive set of goals, concerning what science should **ideally** be about, was expressed once by **Richard Feynman [11]:**

> **Science is a way to teach how something gets to be known, what is not known, to what extent things are known (for nothing is known absolutely), how to handle doubt and uncertainty, what the rules of evidence are, how to think about things so that judgments can be made, how to distinguish truth from fraud, and from show.**

In keeping with what has been noted above, it is now important to consider **more specifically the changes** that are occurring in **"how something gets to be known,"** and most importantly **"to what extent things are known,"** specifically in **the use of new empirical, mathematical and computational sources of information.** The issues of **"rules of evidence"** and **"how to think about things"** are also in need of **reevaluation,** particularly in the **areas of dynamic phenomena.** So let us briefly consider some of these issues.

In science, **"how something comes to be known"** is built on the combination of **(1) empirical information** ("facts"); **(2)** the imaginative development of a possible **"understanding"** from these facts (Eureka! or otherwise), and **(3)** some form of **logical validation** of these proposed understandings, by making consistent associations with **physical observations.** These ideas were already discussed in **Section 1.2,** where this relationship also appeared:

$$\begin{array}{c} \textbf{Facts (observations)} \Rightarrow \textbf{Imagination (concepts, Eureka!)} \Rightarrow \\ \uparrow \qquad \textbf{Understanding (relationships)?} \\ | \qquad \qquad \downarrow \\ \textbf{Further Observations} \\ \textbf{for generalized confirmation} \end{array} \qquad (2.2.1)$$

One profound impact of **technology** on science is that there are now **three sources of information** to facilitate **"how something comes to be known"**:

1. **Natural Phenomena**—empirical sources of physical "facts" (involving communally agreed upon "physical features")

2. **Mathematical Models**—logically precise characterizations of information, often with limited temporal inferences

3. **Computer Explorations**—a uniquely general source of constrained temporal information, using models that are limited only by an algorithmic ability to formulate imaginative dynamic ideas

When obtaining information from each of these sources, it is important to remember that the information is subject to various **limitations,** which all go to the issue of **"to what extent things are known."** This relates, for example, to such empirical informational limitations over a specific duration of time or spatial extent, and empirical accuracy, all of which are subject to **technological changes in the future.** Moreover, the **environmental conditions** that can influence observations always needs to be kept in mind.

The first and essential source of information is **natural phenomena.** Once we have collected an assortment of facts from observations of some phenomena, we need to derive some understanding from these facts. The distinction between a **collection of facts** and **scientific understanding** is nicely illustrated by Poincaré's comparison [12]: **"Science is built up of facts, as a house is built of stones; but an accumulation of facts is no more a science than a heap of stones is a house."**

What turns a heap of stones into a house is obviously the relationship they have to each other, and a talented builder of houses incorporates both the practical needs and esthetic desires of the owners in this relationship.

Similarly, what is missing from a collection of facts is the discovery of **some** relationship between **some** of these facts, which gives a scientist **some** type of understanding of the physical world. Thus, an "understanding" is achieved by the **discovery of some relationship** between some facts, which appears to hold over a **variety of circumstances,** within some **prescribed accuracy, temporal limitations,** and **environmental influences.** This

2.2 The Changing Science of Nature's Dynamics (After 1950)

yields a working assumption, or **assumed understanding,** to be checked and generalized as in **(2.2.1).**

One of the essential new features about the modern study of dynamics is that we now have the enormous assistance of the **digital computer,** as **indicated above.** This has opened up the entirely new area of **"computer explorations,"** in which **new complicated dynamic features,** for even apparently **"simple systems,"** can be uncovered, as will be demonstrated in **Chapter 3** and beyond. One emphasis of this book will be to see how computer explorations can be used both to confirm conjectured understandings (relationships), and to also **stimulate discoveries,** which may lead to unsuspected new relationships. So be prepared in what follows to **explore your ideas,** with the aid of computer programs. Some starting suggestions will be given in the **"Explore"** exercises throughout this book, but you should **keep asking "What if . . . ?"**

Another essential aspect of science to address the issues of **what the rules of evidence are and how to think about things so that judgments can be made.** One approach to such issues is to try to establish a **"consistent association"** (noncontradictory relationship) **between** a hypothetical **relationship** and the **predictions** of a mathematical model, or an algorithmic model, concerning some phenomenon. This **"validation process,"** which is quite **different** from any precise form of **reductive analysis [13],** constitutes part of one vague modern vision of validation, to replace the even more vague historic **"scientific method" [14]. It does not directly address many of the challenging issues raised by Feynman.** But it should be emphasized that the results of any such process does not prove anything absolutely, for **"nothing is known absolutely."** The limitations of both the information and the models or computer explorations are fundamental facts of scientific life, and must be kept in mind, in order to avoid **". . . a misconception of the scientific attitude: it is not *what* the man of science believes that distinguishes him, but *how* and *why* he believes it. His beliefs are tentative, not dogmatic; they are based on evidence, not on authority or intuition" [15].**

These issues are of **increasing importance** in the future development of the science of dynamic systems. We will refer back to the ideas in this brief outline from time to time in what follows.

Concepts: **There are now three (rather than two) sources of scientific information:**

1. **Natural Phenomena**—empirically constrained sources of physical "facts"

2. **Mathematical Models**—logical sources of limited dynamic inferences

3. **Computer Explorations**—a unique logical source of temporal information, using algorithmic models

These are each **logically distinct** types of information, so their comparisons are limited to **consistent associations.** These are the first elements in a **proposed general revision [13] of the historic "scientific method,"** starting from the time of Newton. This revision incorporates the inclusion of;

A) **Modern sources of scientific information**

B) **Many modes of understanding dynamics, requiring more imagination**

C) **Mathematical generalizations, needed to represent natural dynamics**

D) **Developing methods for consistent associations between logically distinct representations of natural dynamics; compatibility checks**

References and Notes

1. The nonexistence of uniform analytic solutions to the class of "canonical" equations, considered to be of fundamental importance in physics, was established by H. Poincaré in his Méthodes *Nouvelles de la Mécanique Céleste,* 3 vols., Gauthier-Villars, Paris, 1892–1899; reprinted Dover, New York, 1957. This limitation in physical theories was emphasized in Chapter IX of L. Brillouin, *Scientific Uncertainty and Information*, Academic Press, New York, 1964.

2. Stanislaw Ulam, 1909–1984. *Los Alamos Science*, No. 15, Special Issue, 1987.

3. E. Fermi, J. Pasta, and S. Ulam. Studies of Nonlinear Problems. *Los Alamos Scientific Laboratory Report LA-1940*, May, 1955; also in Fermi's *Collected Works*, Vol. 2, 978–988, University of Chicago Press, 1965.

4. E. A. Jackson. *Perspectives of Nonlinear Dynamics*, Vol. 2, Section 8.6. Cambridge University Press, Cambridge, 1991.

5. J. Ford. The Fermi–Pasta–Ulam Problem: Paradox Turns Discovery. *Physics Reports 213*, 271–310, (1992).

6. This quotation and other remembrances of this period can be found in S. M. Ulam, *Adventures of a Mathematician*, Chapter 12, Scribner's, New York, 1983.

7. E. A. Jackson. Nonlinear Coupled Oscillators: I. Perturbation Theory; Ergodic Problem. *J. Math. Phys. 4,* 552 (1963). Nonlinear Coupled Oscillators: II. Comparison of Theory with Computer Solutions, *J. Math. Phys. 4,* 686 (1963).

8. T. P. Weisser. *The Genesis of Simulation in Dynamics: Pursuing the Fermi–Pasta–Ulam Problem.* Springer-Verlag, New York, 1997.

9. For some pictures and discussion of the early research of N. J. Zabusky and M. D. Kruskal in ref. [4], see pp. 368, 369, 381, and 384.

10. M. W. Hirsch. The Dynamical Systems Approach to Differential Equations. *Bull. (New Series) Amer. Math. Soc., 11*(1), 1–64, 1984.

11. Quoted in J. Gleick, *Genius: The Life and Science of Richard Feynman*, Pantheon Books, New York, 1992, p. 285.

12. J. H. Poincaré. *La Science et l'Hypothèse.* Flammarion, Paris, 1902; p. 141 in *Science and Hypothesis*, Dover, New York, 1952.

13. E. A. Jackson. Unbounded Vistas of Science: Evolutionary Limitations. *Complexity*, 5(5), 35–44, 2000.

14. H. H. Bauer. *Scientific Literacy and the Myth of the Scientific Method.* University of Illinois Press, Urbana and Chicago, 1992.

15. B. Russell. *A History of Western Philosophy*, p. 527. Simon & Schuster, New York, 1945.

2.3 THE MODERN USES OF MODELS IN THE SCIENCE OF DYNAMICS

Some Background Aspects about "Models"

The introduction of the digital computer as one of the basic sources of scientific information, particularly in dynamic situations, has raised entirely new issues about what types of "models" can be used to obtain scientific "understandings" of natural phenomena. There is presently no consensus on this issue, in large part because the horizons of science are rapidly expanding into unknown areas, where the mechanisms are likewise unknown. What is clear is that the historic analytic methods of "deducing" some natural phenomenon, based on "fundamental physical laws," and frequently perturbation methods, need to be extended into much more **versatile methods of exploration,** which are made possible by computers.

One issue that then arises is whether a particular algorithmic **model** is relevant (related) to some natural process. Here much hinges on what one means by **"relevant,"** and again there is no consensus. So in this brief excursion into this morass, only a few rough distinctions will be offered for thought, and illustrated in this book.

For example, as was discussed in the last section, the first application of the digital computer in this context was in the study of a **"nonlinear lattice" dynamics,** and the discovery of the **Fermi–Pasta–Ulam (FPU) phenomenon.** The model used in this case was intended to **"simulate"** the dynamics of a one-dimensional nonlinear lattice, based on a model involving "particles" connected by nonlinear springs (a very classical treatment of interacting atoms). This type of treatment is not uncommon, since it is generally felt that such treatments capture some "essential aspect" of the real dynamics, even at the atomic level (e.g., Boltzmann's equation, Einstein's theory of Brownian motion, etc.). The details of the FPU phenomenon were **not subjected to empirical tests,** but it certainly challenged the "physical understanding" (i.e., assumptions) of the Nobel Laureate physicist, Enrico Fermi, concerning a physical "insight" for which he was famous. Was that model scientifically "relevant"? It obviously was, if "relevance" of a model includes its stimulation of **new insights into physical processes.** There is no suggestion that an algorithm that is used to explore dynamic behaviors is **generally** more relevant than that.

Physical Models

Some models attempt to generate dynamics of some "physical features," which can be directly compared with some physical observations, or at least used to predict these observations. In this case, there is an attempt to find those **physical features** that are important to account for some dynamic phenomenon. This involves showing that a variety of examples of some phenomenon are **consistent** with a particular relationship between these features. A **rich resource for dynamic phenomena,** which will be used throughout this book, are the **recursive dynamic models.** But, as recounted in **Section 1.2,** there have been many modes of understanding **dynamics** in the past, associated with a variety of **"physical features"** such as: physical observables, and many "conceptuals," including the holistic physical features of Kepler's planetary motions, Mendel's genetics, Boltzman's atomic interactions, Poincaré's topological concepts, Maxwell's "fields," or Schrödinger's wave functions. The common feature of these models is that they attempt to characterize, in some fashion, a **specific physical phenomenon** in a system, which in fact is **limited in accuracy** and **duration of time,** and may be probabilistic. Thus, any physical model is based on a number of implicit factors, not the least of which is our personal or cultural bias and sensory and technological limitations that we bring to any set of observations of nature.

To emphasize this generally neglected topic, consider the more general fact that **all of our "models" of nature,** whether mathematical, computational, or paintings and sculptures, are but the product of our mental **"projected impressions"** of nature. As Brillouin phrased this in science, **"Physical models are as different from the world as a geographical map is from the surface of the earth"** ([3] p. 52). In the world of art, the content of this issue was captured in a series of **thought-provoking** paintings by the Belgian surrealist artist, René Magritte. A nice example of our **"projective"** representations of nature was illustrated by Magritte in a painting, a poor caricature of which is in **Fig. 2.1.**

It should be abundantly clear why any model representation in science of a natural phenomena should more appropriately be referred to as a **"humble, limited, tentative model."** Although the biological areas of science have long recognized this fact, it has been appreciated for some time by only a small group of sensitive physicists **[1–6]. A final constraint** on such a **physical model** is that it not be in logical conflict with the principles used in constructing other "established models" (or "theories",) as they may have become designated), unless it intends to empirically challenge these models.

Exploratory Models

There can be a **second purpose** of developing a **dynamic recursive relationship,** which is simply to **explore some dynamic idea.** These ideas may be abstractly related to known or suspected physical behaviors, or they may be even more "wild and crazy" forms of dynamics, which might contain features that could stimulate our physical imagination, by broadening our "dynamic hori-

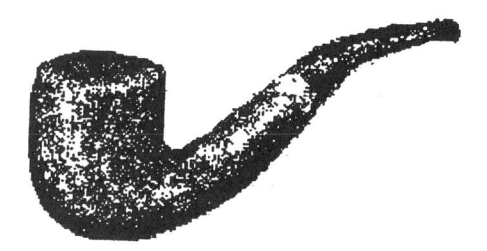

Ceci n'est pas une pipe

Fig. 2.1.
A distinction between reality and our projective representations, as captured by René Magritte.

zons." These types of models might be referred to as **exploratory models,** or something like that. Some dynamic models in the literature are simply "fun and games," with no suggested connection with nature. Exploratory models, on the other hand, retain a concern with ultimately understanding some natural dynamic process, possibly at a much **higher level of complexity** than has yet been dealt with by science. This not only refers to many forms of traditional biological dynamics, but also various levels of **cognitive, social, and cultural dynamics (to name but a few).**

An important aspect of exploratory models is that they can be **entirely personal,** without regard to others' opinions **(at least initially).** If a model is going to contribute to science, it will need to ultimately lead to some understanding that can be communally accepted. But initially the variables in the recursive relationship may be only your metaphors (abstract "likenesses") for observable properties of systems, and the model represents some relationship to a highly constrained behavior you envision in the real world. It may have different purposes, ranging from some **conceptual aspects** of dynamics to some class of **physically motivated** processes. In either case, it should help us **explore** some dynamic ideas, possibly giving some new insights, and, who knows, stimulating a **real discovery of something!** A **Eureka** event! That is, exploratory models might stimulate new ideas, which in turn may influence the **development of some physical model in the future.** Sometimes, exploratory models only reveal their value at a later time, when one tries to understand some more complex aspects of nature. Of course, sometimes they may never be of present scientific value, but who knows how the mind may process this information in the future.

So we now have some rough concepts of both **physical and exploratory models,** to help us understand and explore our dynamic world. We will consider aspects of both types of models in the following chapters. In either case, always keep in mind the **limitations of the knowledge** about the real world that we can gain **from any model.**

Before turning to more established forms of science, we will first look into **developing models that relate to an "exploratory idea."**

References

1. P. Duhem. *The Aim and Structure of Physical Theory.* Princeton University Press, Princeton, NJ, 1991. [Originally published 1914; discussed in **Section 2.1,** with the lesson **(2.1.4).**]

2. D. Bohm. *Causality and Chance in Modern Physics*. Van Nostrand, Princeton, NJ, 1957.
3. Leon Brillouin. *Scientific Uncertainty and Information*. Academic Press, New York, 1964. Also see Chapter IV, this volume.
4. P. Anderson. More is Different. *Physics Today*, 177(4047), 393–396, 1972.
5. D. Bohm and F. D. Peat. *Science, Order, and Creativity*. Bantam Books, Toronto, 1987.
6. A. J. Leggett. *Problems of Physics*. Oxford University Press, Oxford, 1987. (See, particularly, Chapter 4.)

2.4 A "RANDOM PLUS RULE" DYNAMICS THAT EXHIBITS HOLISTIC FEATURES

As has been discussed in **Section 2.1,** there were a number of non-Newtonian types of natural dynamics that were discovered in the 19th century, among them being **Mendel's probabilistic-reproduction** dynamics and **Darwin's ill-defined evolutionary dynamics.** As discussed in that section, the understanding of what is involved in this type of dynamics is still in the process of evolution, in part due to the evolving technology that uncovers increasingly detailed molecular and genetic information. It is now clear that part of this evolutionary dynamics involves **nondeterministic genetic processes,** whereas the interactions between or within species appears to often be understandable by **deterministic dynamic rules** (e.g., Chapters 3, 5, and 8). Whatever is the future understanding of these details, shortly after **Charles Darwin** published his *Origin of Species,* he referred to one of his detractors, and said "[the astronomer **Sir John**] **Herschel** says **my book is 'the law of higgledy-piggledy'"** [1]. In recent times, John Archibald Wheeler, a well-known physicist, has used this expression to characterize a type of dynamics that has **both a random and a deterministic** component (a set of rules) in its recursive relation **[2].**

Here we can take advantage of the freedom granted by the computer to illustrate, with a well-known example, **a surprising result** that can occur in this unusual type of dynamics. One might reasonably think that if a recursive relationship has **one step that is random** in character, then the end result of this dynamics will likewise be random, even if there is **another step that is deterministic.** In **one sense** this expectation is **correct. But,** as was illustrated by Kepler's (Section *1.2) and Poincaré's (Section *2.1) **holistic forms of understanding dynamics,** sometimes you

Dynamic Aspects of Modern Science

need to think about **dynamics "in the large,"** that is, how it develops over time and space, rather than individual steps.

To proceed, in order to study some **nondeterministic forms of dynamics,** one needs to use the generator of **random numbers, RND,** in the computer program, which was already illustrated in **Explore 1.5A, case (2).** To illustrate this again in a spatially extended setting to be used shortly, begin with Explore 2.4A.

Explore 2.4A

Look in Program **2-4A** and find how the random points (x, y) are generated, with the help **of the uniform random number generator, RND.** This means that if one sets $x = RND$, then each time this is encountered it will **generate values of x between zero and one with equal probability** (in other words, any value of x between zero and one is equally likely to occur). Note how the values of (x, y) are confined within the triangle. But the **pattern is not very interesting.** Can you make a more interesting pattern? To try, **see the location noted in the program** where you can make changes.

Now, with this background we can turn to **a exploratory model** involving both **a random and a deterministic aspect** of the dynamics. While it might seem that this exploratory model is very far from nature, there are some **scientific ideas** around (for how long?) that are related to this type of dynamics [2]. In any case, see what you think about it.

This exploratory model is based on the following dynamics. Begin with an equilateral triangle, with sides of length L and vertexes labeled 1, 2, 3, as illustrated in **Fig. 2.2.** The location of the vertexes on the (x, y) plane are at

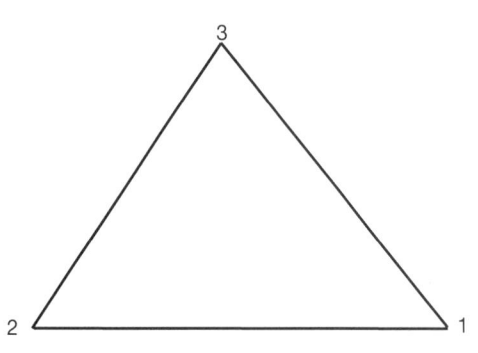

Fig. 2.2.
An equilateral triangular region in the (x, y) plane.

2.4 A "Random plus Rule" Dynamics that Exhibits Holistic Features

$$[x(1) = L/2, y(1) = 0]; [x(2) = -L/2, y(2) = 0]; [x(3) = 0, y(3) = \sqrt{(3)}L/2]$$
(2.4.1a)

The dynamics begins by taking an initial point $[x(0), y(0)]$ **anywhere** within this triangular region (or anyplace nearby, for that matter).

The **dynamic rules** are: (2.4.1b)

1. pick one of the numbers (1, 2, 3) **at random;** call it V

2. move the point (x, y) **one-half the distance toward the vertex V**

3. return to (1) (this generates the **recursive dynamics**)

You will note that this dynamics has two distinct features: **step(1) is totally random,** whereas **step (2) is totally deterministic** (V is given). This dynamics will be characterized as **an example of**

*** **Random-plus-Rule Dynamics** ***   (2.4.2)

Whether this type of dynamics occurs anywhere in nature is unclear, but the **holistic consequence** of this dynamics is certainly interesting. It may stimulate you to ask some questions concerning its generality. But in any case, let us discover what this dynamics produces.

Explore 2.4B

In **Program 2-4B** there is a "slow motion" of this dynamics, showing its unexpected consequence. Basically, the dilemma resulting from this is how does **something orderly** come out of **a dynamics that involves random (nondeterministic) choices** at each time step (of course, don't forget the deterministic factor!)?

Think 2.4A

See if you can explain why the above dynamic rules give the orderly structure found in the **Program 2-4B**. [**Hint:** select any value for V and draw a line between the midpoints of the two sides of the triangle that connect with V. Thus, inside the triangle there is now a large and a small enclosed region. Where do **all** the points in the largest enclosed region go in the next step? Repeat this idea for a few iterations.] Proceed **only** after you have thought this over carefully.

Explore 2.4C

Having thought about **Think 2.4A** carefully, proceed to **Program 2-4C** for a further examination of these ideas. If you successfully grasp what is involved in this Explore, you will particularly benefit from the following **Think.**

Think 2.4B

A Maxim: This is a good place to learn an **important lesson.** When you discover such a new form of dynamics, look for ways that you can establish the **generality of this dynamic result,** by "slightly" modifying the spatial domain and/or the dynamic rules. Is this a singular case? **What are the requirements for this holistic feature "to work"?** [4]

References and Notes

1. Charles Hershaw Ward. *Charles Darwin: The Man and His Warfare*. Bobbs-Merrill, Indianapolis, IN, 1927, p. 829. A further exploration of these ideas can be found in Section 6.2.

2. John Archibald Wheeler. On Recognizing "Law without Law," Oersted Medal Response at the joint APS-AAPT Meeting, New York, 25 January 1983, *Am. J. Phys., 51*(5), 398–404, 1983. *Frontiers of Time*. North-Holland for the Società Italiana Di Fisica, Amsterdam, 1979; Chapter 1: Law without Law; "Are the laws of physics eternal and immutable? or are these laws, like species, mutable and of a "higgledy-piggledy" origin?" The quotations are from J. A. Wheeler, *Geons, Black Holes, and Quantum Foam: A Life in Physics* (with Kenneth Ford) p. 351 W. W. Norton, New York, 1998. Drawn from Charles Hershaw Ward, *Charles Darwin: the Man and his Warfare*. Bobbs-Merrill, Indianapolis, IN, 1927, p. 829.

3. E. Mayr. Darwin's Influence on Modem Thought. *Scientific American*, July 2000, p. 79.

4. In following sections of this book you will find many examples of **holistic forms of dynamic understanding,** both in physical space and in Poincaré's "phase space"; e.g., Chapters 4 and 5; Sections 6.2, 7.2, 7.3; Chapter 8.

3
Different Types of Population Dynamics

So far we have studied the dynamic examples of money in two banks, and a little concerning the human population, which involved both **exponential growths** and the nonlinear effects of **finite resources.** Now we will look at some complex physical systems, which can have some of the same dynamic features, but also many more dramatic activities. To illustrate this, we will be interested in this chapter in the dynamics of populations of living organisms, the study of which dates back to the 19th century.

In the 1840s, there was particular interest in how the human population might develop in time, and how this might lead to overcrowding and various social, economic, and political difficulties [1] (not dissimilar from present concerns [2–8]). Also of interest is the fact that, after reading **Malthus'** ideas of competition in ***Population,*** **Darwin** conceived of the struggle for survival as the agent for the formation of new species [9]. Such concerns about increasing populations led to relatively simple models of the dynamic behavior of populations, the most famous of which became known as the **Verhulst differential equation,** which will be discussed in **Section 4.1.**

In the latter part of the 20th century, the study of the population dynamics of living systems developed into the field called **pop-**

ulation ecology [10]. In this field, studies could be made of insects that reproduced on rather short time scales, which is much more suitable for experimentation, and from which much has been learned about nonlinear influences.

This naturally leads to considering two extreme cases of reproduction, which have quite different types of dynamic models:

I. **The case in which there is nearly continual reproduction within the population.** This is the situation with us humans, and also those prolific rabbits, to name but two examples. Since the reproduction does not occur at the same time for all members of the population, this type of population dynamics can be approximately represented as a continuously changing density of animals.

II. **The case in which there is effectively no overlap between successive generations, so the populations are made up of a single generation.** Examples of this extend over a large range of time periods. Many smaller insects reproduce in approximately this fashion in periods of a few days, whereas in temperate zones there are many arthropod species that have short-lived adult generations each year. At the other extreme are the cicadas, whose adults emerge periodically to reproduce every 7, 13, or 17 years.

In this chapter, we are only going to consider **type II reproduction,** leaving the **continuous type (I)** dynamics for consideration in **Chapter 4.**

3.1 DISCRETE-TIME MODELS OF POPULATION DYNAMICS

In type II dynamics, the natural unit of the "time" is the discrete generation, so $t = 0, 1, 2, \ldots$ is the generation count, regardless of how much actual time passes between each generation. If $N(t)$ represents the number density of a population at the generation t, one might expect that if the density is not too large, the population number would increase linearly in each generation (like the "Safe Bank" example of Section 1.3), so that the dynamic equation for $N(t)$ would be

$$N(t + 1) = R \cdot N(t) \qquad (3.1.1)$$

Here R is some "reproduction rate," and the population only grows if $R > 1$, otherwise it would die out. However, as in the banking case, there is only a **finite amount of resources** to sustain this growth in population, so the dynamics must involve **nonlinear effects,** which limit this growth when the density becomes large. Therefore the dynamic equation for $N(t)$ must have a more general form, which we can represent as

$$N(t + 1) = F[N(t)] \quad (3.1.2)$$

where $F(N)$ **is a nonlinear function of** N, which depends in some manner on the species in question. This discrete form of dynamics, (3.1.2), is often referred to as a **map dynamics,** and the mapping action is sometimes represented by the notation

$$F{:}N(t) \to N(t + 1) \quad (3.1.3)$$

A few examples (models) of such functions that have been used in population ecology (to represent different species) are listed here for future explorations [11]:

	$F(N)$	$(N > 0)$
I)	$R \cdot N(1 - N/N^*)$	(and $N < N^*$)
II)	$R \cdot N \exp(-a \cdot N)$	
III)	$R \cdot N/(1 + a \cdot N)^b$	

(3.14)

All of these functions reduce to $(R \cdot N)$ if N is small, in which case (3.1.2) reduces to the linear dynamics (3.1.1). The **different forms of nonlinearity** in (3.1.4) represent the different characterizations of the constraints on different species, which are **due to finite-resource effects** [hence **all $F(N)$ decrease when N is larger than some amount,** depending on the model]. These models will be discussed further in **Section 3.7.**

In **Fig. 3.1** there is a graphical representation of functions that are similar to those in (3.1.4). They differ only in changing some constants for the purpose of an interesting display.

▩ **Think 3.1:**
Decide which of the above figures have the type of nonlinearity found in the functions in (3.1.4). Does one of these models seem unreasonable? ▩

Explore 3.1
In computer **Program 3-1** you can see how the map (3.1.2) [also represented by **(3.1.3)**] appears graphically for three func-

Different Types of Population Dynamics

$$N(T+1) = F[N(T)]$$

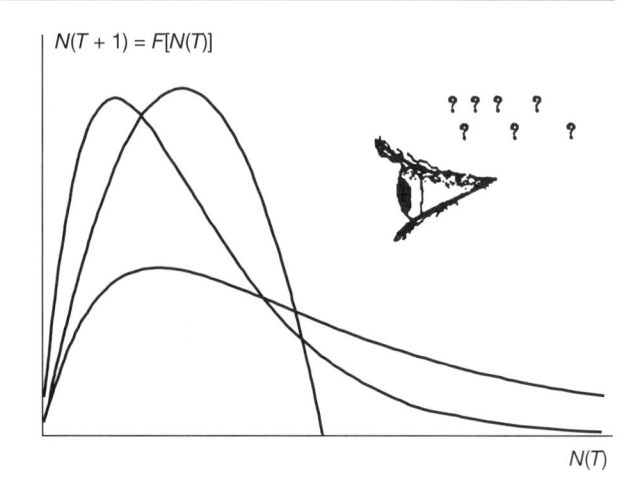

Fig. 3.1. The graphical representation of functions similar to those in (3.1.4).

$N(T)$

tions like those in Fig. 3.1. What is shown is the **one-step map** for each function, from the same initial state $N(0)$ to the state $N(1)$, that is

$$N(1) = F[N(0)], \quad \text{or} \quad F(0) \to F(1) \qquad (3.1.5)$$

You can **make changes in the program** to explore new initial values, $N(0)$, or to change constants in the three functions. This one-step dynamics, $N(0) \to N(1)$, for the all three models in Eq. (3.1.4), is illustrated in **Fig. 3.2.**

In this chapter we will primarily focus on **model I**, which is known as the **logistic map.** This is done both because of historical reasons, which will be made clear in **Section 4.1,** and also because it has been widely used, due to its algebraic simplicity. This makes life easier for some of the following mathematical analyses, and at the same time it does not change **some** essential features that are common to all the maps in (3.1.4). Nonetheless, it is **important to explore** some of the features of the **physically more realistic models II and III.** These explorations will be done in **Section 3.7.**

Initially, however, let us begin to **explore the remarkable dynamics** that the "simple" **logistic-map dynamics (model I)** can produce.

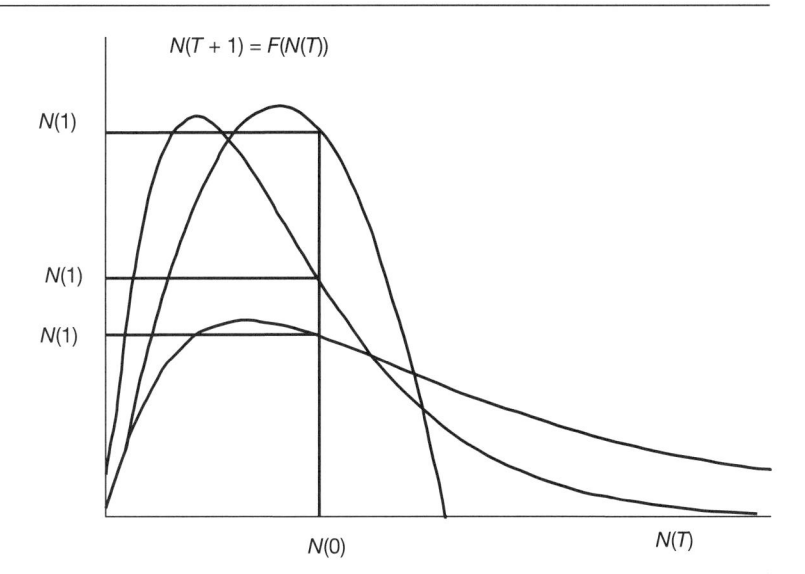

3.1 Discrete-Time Models of Population Dynamics

Fig. 3.2.
The one-step dynamics, Eq. (3.1.3) ($t = 0$), for all three models in Eq. (3.1.4).

References

1. D. Winch, *Malthus*. Oxford University Press, Oxford, 1987.
2. J. E. Cohen, *How Many People Can the Earth Support?* W.W. Norton, New York, 1995.
3. L. R. Brown, C. Gardner, and B. Halweil, *Beyond Malthus: Nineteen Dimensions of the Population Challenge*. W. W. Norton, New York, 1999.
4. A. Gore, *Earth in the Balance: Ecology, and the Human Spirit*. Houghton Mifflin, Boston, 1992.
5. E. Ayensu, et al., International Ecosystem Assessment. *Science*, 286, 22 Oct. 1999, p. 685.
6. *National Geographic*, Oct. 1998, pp. 36–75 (Millennium in Maps: Population).
7. F. B. M. de Waal, F. Aureli, and P. G. Judge, Coping With CROWDING. *Scientific American*, May, 2000, pp. 76–81.
8. R. E. Ulanowicz, *Ecology, The Ascendent Perspective*. Columbia University Press, New York, 1997.
9. See the discussion and quotation in S. J. Gould, *Ever Since Darwin: Reflections in Natural History* (W. W. Norton, New York, 1977), pp. 21–22.
10. S. E. Kingsland, *Modeling Nature: Episodes in the History of Population Ecology*. University of Chicago Press, 1995.
11. R. M. May (Ed.), *Theoretical Ecology: Principles and Applications*. Blackwell, Oxford, 2nd Ed., 1977. (Section 2.3.2)

3.2 THE LOGISTIC MAP AND ITS FIXED POINTS

We now consider the so-called **"logistic map" dynamics,** as given by the form (I) in (3.1.4):

$$N(t + 1) = R \cdot N(t)(1 - N/N^*) \quad (0 < N < N^*) \quad (3.1.4\text{:I})$$

The historical origin of this strange name is discussed in [1].

In this expression, the constant N^* represents the **maximum possible number density** of this species. This is a physical drawback of this model, which is not present in the other models of (3.1.4), for it strictly limits the number density, which is not realistic. However (3.1.4:1) is used because it is analytically simple, which helps in some of the following analysis; and at the same time, the resulting dynamics is qualitatively the same as the dynamics of the other models in (3.1.4), differing only in quantitative details.

Next we put this equation into a simpler form, by setting $x(t) = N/N^*$, so $x(t)$ is now **normalized population density,** and we denote the **linear reproduction rate** by $c = R$. These changes produce a standard form of the **logistic map**

$$x(t + 1) = c \cdot x(t)[1 - x(t)] \quad [1 > x(t) > 0, c > 0] \quad (3.2.1)$$

Introducing the function $F(x, c) = c \cdot x(1 - x)$, the logistic map can also be written in the form

$$x(t + 1) = F[x(t), c] = c \cdot x(t)[1 - x(t)] \quad (3.2.2)$$

where the parameter c is explicitly noted. Such a parameter is often referred to as a **"control parameter,"** and it will be seen that it does "control" the character of the dynamics in a dramatic fashion. It is also sometimes referred to as an **"endogenous parameter,"** because it represents the reproduction property of the system itself, and not something that is controlled from the outside environment (an "exogenous parameter").

Since c is simply a multiplicative factor in this logistic map, when it is changed, it has the simple influence that is illustrated in **Fig. 3.3.** It can be seen that $F(x, c)$ has a **maximum value,** for any value of c, at the value $x = 1/2$, which means that $F(1/2, c)$ is the maximum (normalized) population density for that value of c. Now, on the other hand, because of the normalization, $x = N/N^*$, and the requirement that $N < N^*$, $x(t)$ must always remain less than 1. The condition that $N < N^*$, or $x < 1$, is required by the fact that if $x(t) > 1$ then (3.2.1) **yields $x(t + 1) < 0$,** which makes no sense for a population density (unless, perhaps, we interpret this to mean extinction!). This relates to its unrealistic feature, noted above.

Given the above result for the maximum population, we see that in order to ensure that $x(t + 1)$ will remain less than 1 requires that $1 > F(1/2, c)$. But, according to (3.2.2), $F(1/2, c) = c/4$, hence the value of c can not be larger than 4. With the other condition of **$c > 0$, in (3.2.1),** we find that c must lie in the range **$4 > c > 0$,** so (3.2,2) now becomes, in its final form:

$$x(t + 1) = c \cdot x(t)[1 - x(t)] \quad (1 > x(t) > 0, 4 > c > 0) \quad (3.2.3)$$

Let us begin by considering the **simplest dynamics** of (3.2.3). Clearly this is the case when **$x(t)$ does not change with time,** so that

$$x(t + 1) = x(t) = x^* \quad (3.2.4)$$

Such a value x^* is called a **fixed point,** or an **equilibrium point,** of the system. If we substitute (3.2.4) into (3.2.3), we find that there are two possible fixed points:

$$x^{**} = 0 \quad \text{(for any } 4 > c > 0\text{)} \quad (3.2.5a)$$

$$x^* = 1 - (1/c) \quad \text{(only if } c > 1\text{)} \quad (3.2.5b)$$

Physically, these equations show how the steady-state populations (fixed points) depend on the parameter c. The fixed point x^{**} is sometimes referred to as the **trivial case.** It is not a physically interesting case, since it corresponds to zero population. The condi-

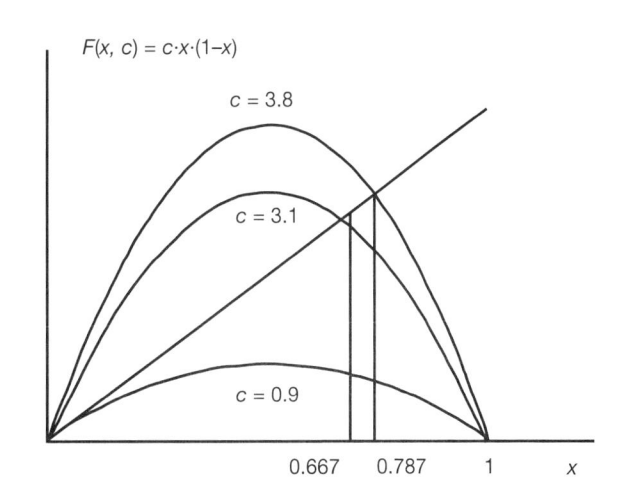

Fig. 3.3.
The **intersection of the 45 degree line,** $F[x(t), c] = x(t)$, corresponds to the **fixed points, x^*,** of the dynamics. Note that the "trivial" fixed point, $x^{**} = 0$, also intersects this line at the origin.

tion $c > 1$ in **(3.2.5b)** is required to ensure that x^* **is a positive number.**

The value of any **fixed point** x^* is made clear by returning to Fig. 3.2, which shows how $x(t + 1) = F[x(t), c]$ depends on $x(t)$. It can be seen that the only values of $x(t)$ that equal $x(t + 1)$ are those for which $F[x(t), c]$ **intersects the line** $x(t) = x(t + 1)$**,** or in other words the straight line that **bisects the angle between the axes** $x(t)$ **and** $x(t + 1)$**.** In Fig. 3.3, this line has been included, and it will prove to be very useful for understanding the dynamic gases as well, as we will see in the next section.

Since (3.2.4) is satisfied for a fixed point, the value of x repeats every iteration, and therefore this is a simple periodic motion, with a **period of one iteration (a "period-one" solution).** Hence, the value of x in (3.2.5) is also referred to as a **period-one point.** This is the first of many periodic solutions that we will soon discover.

Reference

1. S. E. Kingsland, *Modeling Nature: Episodes in the History of Population Ecology*. University of Chicago Press, Chicago, 1995, p. 2.

3.3 GRAPHICAL REPRESENTATION OF MAP DYNAMICS: FIXED-POINT STABILITY

In this section we will look at several graphical representations of map dynamics. First, we will see how the graphical Fig. 3.3 can be used to easily understand the dynamics defined by **any map**

$$x(t + 1) = F[x(t)] \qquad (3.3.1)$$

In this chapter we will be primarily concerned with the **logistic map,**

$$F(x, c) = c \cdot x(1 - x), \qquad (0 < x < 1, 0 < c < 4) \qquad (3.3.2)$$

but **the method applies equally as well to any map,** such as those presented in **(3.1.4).**

A graph of $F(x, c)$ versus x for the logistic map, (3.3.2), was shown **Fig. 3.3,** for several of the allowed values of c. Now we will **take $c = 2.8$** (for reasons that will soon become clear), and describe how we can **extend the one-step result,** shown in **Fig. 3.2,** to **more steps,** by using a simple graphical device.

We begin by introducing the straight line $x(t + 1) = x(t)$ into the

3.3 Graphical Representation of Map Dynamics

figure of $F(x, 2.8)$ versus x, as shown in **Fig. 3.4.** We select for an **initial condition $x(0) = 0.9$**. In Fig. 3.4, we draw a vertical line upward from this point, until it **intersects $F[x(0)]$**. Then, according to (3.3.1), this intersection is the new value, $x(1)$. Therefore, using a **horizontal line** from this intersection, it intersects the vertical axis at $x(1)$, as in **Fig. 3.2.**

Next, consider the 45 degree line, $x(t+1) = x(t)$ that we introduced in the last section, and note that the horizontal line intersects this 45 degree line **directly above this new value of $x(1)$** on the **horizontal axis.** This is illustrated in **Fig. 3.4.** For the next time step, we can now draw a vertical line upward from that intersection with the 45 degree line, and see where it intersects the curve $F(x)$. That will give $x(2) = F[x(1)]$ on the vertical axis. Now we use the same trick to draw a horizontal line over to the 45 degree line, and that will now lie directly above $x(3)$ on the horizontal axis. But now we find that in order to draw a vertical line to $F(x)$, it will need to **go downward,** but that does not change the concept, as illustrated in **Fig. 3.4.** One can proceed in this manner as long as one likes.

It can be seen in Fig. 3.4 that the solution $x(t)$ tends toward the fixed point $x^* = F(x^*)$.

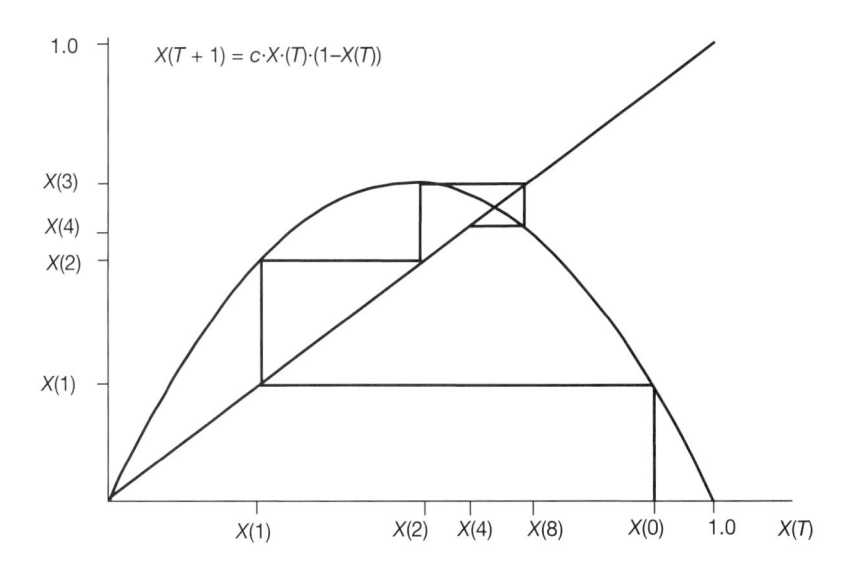

Fig. 3.4.
The **step-by-step** graphical representation of the iterates $x(0)$, $x(1)$, $x(2)$, $x(3)$, and $x(4)$, given by (3.3.1) and (3.3.2). The values $x(0) = 0.9$, and $c = 2.8$ were selected.

Different Types of Population Dynamics

▪ **Think 3.3**
Show that this fixed point is $x^* \cong 0.643$, using previous results.
▪

Since x tends to x^* this **fixed point is called an attractor.** Dynamic attractors are very important in nature, because they represent the **stable forms of behavior** (dynamics). We will find a variety of such attractors in what follows.

Explore 3.3A
To see a few examples of **different attractors,** use **Program 3-3A,** which shows where some given initial states tend to go, for the values of **$c = 2.8$** (extending the results in Fig. 3.4), or **$c = 0.95$, and $c = 3.2$**. Are all of these attractors **fixed points?** For each value of c, how many fixed points are there? For the case $c = 3.2$, how do solutions near the fixed point behave? **Is it still an attractor?**

Explore 3.3B
The dynamics shown in Fig. 3.4 can also be viewed in a different way, namely as **$x(t)$ versus t.** In this case, the behavior for large values of the time is particularly clear. Using the **Program 3-3B,** try this for **the values of c in Explore 3.3A**. One example is shown in **Fig. 3.5.**

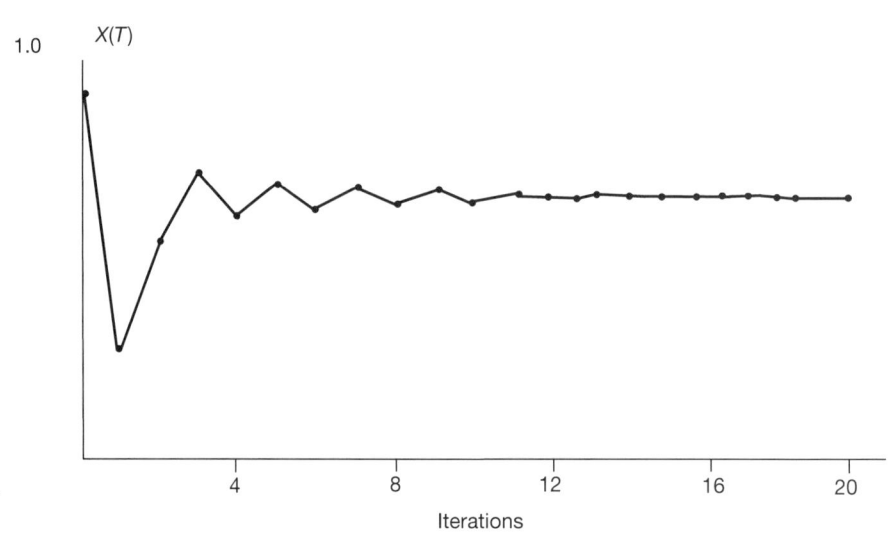

Fig. 3.5.
This shows $x(t)$ versus t for the case **$c = 2.8$**

3.4 THE BIFURCATIONS OF THE FIXED-POINT ATTRACTORS OF THE LOGISTIC MAP

We have seen that all models, such as those in (3.1.4), contain a number of parameters that remain constant during the dynamic process. Moreover, we have seen that the qualitative character of the dynamics can be changed if the value of these parameters is changed. For example, in the case of the logistic map, (3.2.1),

$$x(t + 1) = c \cdot x(t)[1 - x(t)]$$

we found in (3.2.5) that there is always a fixed point (period-one dynamics) at $x = x^{**} \equiv 0$ for any value of c in the allowed range $0 < c < 4$. Moreover, there is a second fixed point at $x = x^*(c) = 1 - (1/c)$, provided that $c > 1$, which is necessary since x must be positive.

Now any dynamic system that has two fixed points has qualitatively different dynamics than a system that has only one fixed point. To illustrate this fact for the logistic map, consider the dynamics that is illustrated in Fig. 3.4. Here there is an attracting fixed point at $x^* = 1 - (1/c)$, which is called an **fixed point attractor.** On the other hand, there is also a fixed point at $x^{**} = 0$, but since all nearby points move away from x^{**} it is called a **fixed point repellor.**

If you studied Explore 3.2 (as you should have!), you would have found that the dynamics for the case $c = 0.95$ is qualitatively different from that in Fig. 3.4.

What does **"qualitatively different"** mean? One way to give this a precise meaning is to see if we can find a method of "relating" (or "associating") the dynamic steps (i.e., for all initial states) for two systems with different values of c. To relate the steps of $x(t, c_1)$ to those of $x(t, c_2)$ for all values of $t = 0, 1, 2, \ldots$, when they start out at any point (associate in a smooth fashion), each point of the set $\{x(t, c_1)\}$ associates with the point at the same time in the set $\{x(t, c_2)\}$. The idea of a "smooth association" is simply that we want there to be a functional relationship (making things smooth) between these two sets of points. If this is the case, then there is a functional relationship of the form:

$$x(t, c_1) = S[x(t, c_2)], \quad t = 0, 1, 2, \ldots, \quad \text{for all } x(0, c_1) = x(0, c_2) \tag{3.4.1}$$

If there is a relationship (3.4.1), the c_1 and c_2 dynamics are said to be **qualitatively similar.** The function $S(x)$ is sometimes called the similarity map between these two dynamics. If no such relationship exists, then the two dynamics are said to be **qualitatively different**. In this case, it is said that **the dynamics has undergone a bifurcation** as c changes from c_1 to c_2. When this happens, there must be some value of $c = c^*$, which lies between c_1 and c_2, such that for any

$$c_1 < c^* < c_2$$

there does not exist the relationship (3.4.1). This value of c^* is **called a bifurcation point** (it is a point, of course, along an axis of c values).

Stated again, c^* **is a bifurcation point** if the dynamics when $c < c^*$ is qualitatively different from the dynamics when $c > c^*$ [there exists no association (3.1.4) between these two dynamics]. If the two dynamics are **qualitatively similar,** they are also called **topologically equivalent,** and **otherwise topologically distinct.** This terminology is quite common and will be made more relevant in Chapter 4. Fortunately, we do not have to find an explicit expression for $S(x)$, but only establish whether it can exist or not, and this can be done with simple pictures. Moreover, we generally need a computer to find the different bifurcation points, c^*, but not always.

To illustrate this idea of a bifurcation, and the bifurcation point c^*, consider again the logistic map when $c < 1$. All of the initial points map to smaller values, and tend to $x = x^{**} = 0$. If $c_1 < 1$ and $c_2 < 1$, then both behave in the same fashion, and we can characterize the general behavior of all points $0 < x < 1$ as "flowing" towards $x = 0$, as illustrated in Fig. 3.6a.

Therefore, it is rather clear that there is some relationship between two dynamics with this same flow characteristic that would satisfy (3.1.4), even if we do not want to take the trouble to find $S(x)$ explicitly. In other words, we know that $S(x)$ exists, and that is all that counts.

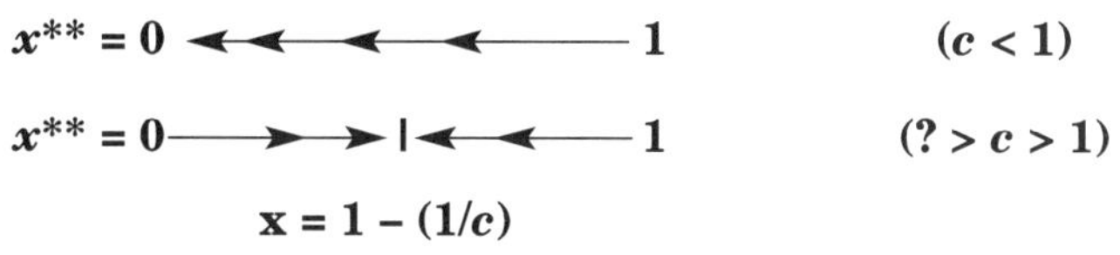

Fig. 3.6.
(a) The "flow" of mapped points, when $c < 1$, is to the attractor $x^{**} = 0$. (b) Now x^{**} is a repellor. The flow of mapped points is to the attractor, $x = 1 - (1/c)$, provided that $? > c > 1$.

$x^{**} = 0 \longleftarrow\longleftarrow\longleftarrow 1$ $(c < 1)$

$x^{**} = 0 \longrightarrow\longrightarrow | \longleftarrow\longleftarrow 1$ $(? > c > 1)$
$\qquad\qquad\quad x = 1 - (1/c)$

3.4 The Bifurcations of the Fixed-Point Attractors of the Logistic Map

However, if $c_1 < 1$ and $c_2 > 1$, the situation is different, for if $c > 1$ there are two fixed points. The one that is the same as above, $x = x^{**} = 0$, and now another at $x(0) = 1 - (1/c)$ (if $c > 1$). Now the fixed point x^{**} is a **repellor,** whereas $x = 1 - (1/c)$ is an attractor (provided that $c < ?$, as we will see below). The "flow" of points the x axis is shown in Fig. 3.6b. With some thought, it should be clear that there is no smooth way that we can associate the mapped points of the flow in Fig. 3.6b with those in the flow in Fig. 3.6a. Thus there must be a bifurcation point for some value c^* between $c_1 < 1$ and $c_2 > 1$. Since the flows in Figs 3.4 hold for all $c < 1$ and $? > c > 1$, the **bifurcation point is $c^* = 1$.**

Think 3.4

In fact, this can be seen fairly easily by focusing your attention on one x region of both figures, and showing that the $S(x)$ that would be needed to satisfy condition (3.1.4) for this region cannot be smoothly related to an $S(x)$ that would be needed to satisfy (3.1.4) in another x region. That is, they would differ radically for all initial conditions in these regions. What are these two x regions?

This idea of comparing the dynamic behavior of systems that differ by changing a parameter in the equations of motion is an idea that was invented by Poincaré in the 1890s. It is a different and very interesting way of studying the properties of dynamic systems. These parameters are often called **"control parameters"** (because we can change them), or in areas such as economics they are called **"exogenous variables** (because they are variables that are controlled "outside" the system). They are both the same idea, and we will refer to them as control parameters.

Let us now **return to the ? mark in Fig. 3.6b.** What happens as c is increased beyond the value $c = 1$? We saw above that as c is increased from $c = 0$, the fixed point $x = x^{**}$ **changes from an attractor to a repellor** as c passes the **bifurcation point $c = c^* = 1$.**

An attractor fixed point is called a **stable fixed point,** since all nearby initial states tend to this fixed point as t increases. Similarly, a **repellor** fixed point is called and **unstable fixed point.** The stability of fixed points can be analyzed mathematically, using the method of linear analysis of the dynamics, which is applicable if x is very near to the fixed point x^*. This method of **stability analysis** is discussed in some detail in the **Mathematics Appendix M2.** Those who would like to learn this useful insight into dynamics

Different Types of Population Dynamics

should now refer to that appendix, where the answer to the above **?** will be precisely determined mathematically.

An alternative approach to such problems, which however can only give answers of limited precision, is to make use of computers. On the other hand, the computer can give information that cannot be obtained from the linear methods used in the stability analysis. In particular, it can give information about where $x(t)$ goes when it moves far away from the fixed point. Moreover, it is not generally possible to develop nonlinear analytical methods to answer such questions, so the computer becomes a unique source of information in these cases (and in nonlinear dynamics in general). The nature of this information is illustrated in **Fig. 3.7**.

Explore 3.4

You can begin to explore this **bifurcation concept** by using **Program 3-4**. This shows the dynamics that leads to the large time ("asymptotic") behavior of $x(t)$ for various values of c. Slowly vary the value of c between the values $c = 0.5$ and $c = 3.2$, and determine the approximate values of the bifurcation points, c^*. Identify the meaning of the vertical axis, and the numerical scale on the horizontal axis in **Fig. 3.7**.

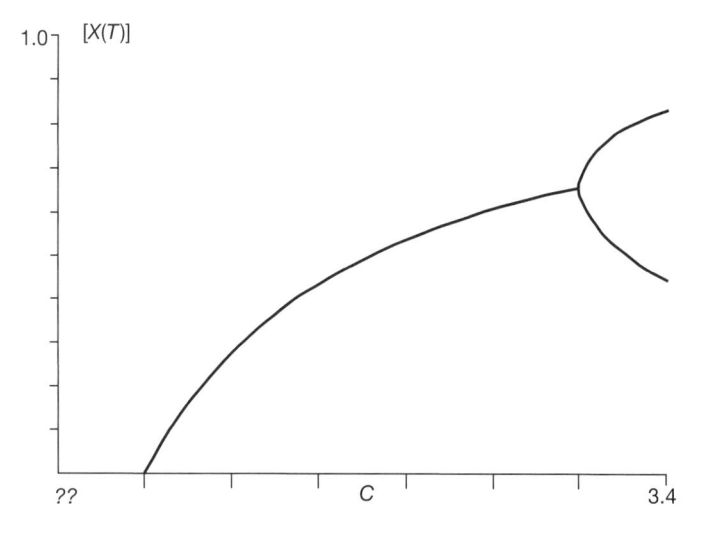

Fig. 3.7.
Here is some dynamic information obtained by a computer, for the logistic dynamics in the control parameter range ?? $< c <$ 3.4. What this information is, and what the scale is on the horizontal axis, can be discovered using Explore 3.4.

We will now proceed to see how the asymptotic behavior of the logistic map dynamics varies as the value of c is increased up to $c = 4$. **Hang on to your hat, because there is a dynamic jungle out there to be explored!**

3.5 THE LOGISTIC-DYNAMICS JUNGLE!!

In the last section we found that as c is increased from zero there is a bifurcation at $c = 1$, changing the dynamics from one **fixed-point attractor at $x^{**} = 0$** to one **fixed point repellor at $x^{**} = 0$** and a **fixed-point attractor at $x^* = 1 - (1/c)$**. Then, if you did an honest bit of research and learning, either with the Mathematics Appendix **M2,** and/or **Explore 3.4,** you found that there is a **second bifurcation at $c = 3$**. At this value of c, the **fixed point at $x^* = 1 - (1/c)$ becomes unstable** and turns into a **fixed point repellor.** But now the bifurcation involves a new feature, because it also generates a **period-two attractor,** rather than another fixed-point attractor. This period-two attractor is represented in the **"pitchfork" portion of Fig. 3.7.** As you found from Explore 3.4, for these values of c, **all initial states,** $x(0)$, aside from $x(0) = 0$, **tend toward** the two states for that c, but **jump back and forth** between them, yielding a **dynamics with a period of two.**

We will see that this development of **new periodic attractors** is a common feature as the value of c is increased over a **limited range of c values;** namely, periodic attractors will become unstable, becoming repellors, and at the same time generate new periodic attractors.

To explore this more general situation, note that these bifurcations begin (for small values of c) with the loss of stability of fixed-point attractors (period-one attractors). The results of Appendix M2 showed that bifurcation of a fixed-point attractor of a map occurs when the fixed point, x^*, becomes **unstable** once c has a value above $c = c^*$, the so-called **bifurcation point.** Moreover, this instability occurs for **any map**

$$x(t + 1) = F[x(t), c] \quad (3.5.1)$$

when the value of c is such that it has a slope of magnitude one at the fixed point x^*. In the case of the logistic map, we found that when $x = x^{**} = 0$, the bifurcation point is $c^* = c_0 = 1$, and when $x^* = 1 - (1/c)$, the bifurcation point is $c^* = c_1 = 3$. Here we have **introduced a subscript** to these values of c, writing the **bifurcation points as c_n, $n = 0, 1$**. We can likewise denote the corresponding fixed points as x_n, $n = 0, 1$. So, for **any map (3.5.1), bifurcations of fixed points** occur at

$$|dF(x, c)/dx| = 1 \quad \text{(evaluated at } x_n \text{ and } c_n) \quad (3.5.2a)$$

and (the fixed-point statement)

$$x_n = F(x_n, c_n) \quad (3.5.2b)$$

For the logistic map, when c **is a little larger than** c_1 there is a **stable period-two attractor,** which no longer is a fixed point of the logistic function $F(x, c) = c \cdot x (1 - x)$. However, in order to use the results (3.5.2), we can only use the fixed points of maps. So we **need to discover a new map** that has a fixed-point dynamics when the logistic map has this stable period-two dynamics.

This turns out to be easy to do. If $F(x, c)$ is a map that has a stable period-two dynamics, then this dynamics has the following property:

$$x(1) = F[x(0), c] \quad \text{and} \quad x(2) = F[x(1), c] \quad (3.5.3a)$$

and $x(2) = x(0)$ because it is period-two dynamics. Using a few substitutions, it follows from this that

$$x(0) = F[F(x(0), c] \equiv F^2[x(0), c] \quad (3.5.3b)$$

Note that $F^2[x(0), c]$ is an **entirely new function,** and does not represent the square of the function $F(x, c)$, which would be written as $F(x, c)^2$. Moreover, **(3.5.3) applies to any map function $F(x, c)$,** so that any period-two point, $x(0)$, of $F(x, c)$ is a fixed point of the new map, which is now denoted by $F^2[x(0), c]$.

Specifically, for the logistic map, $F(x, c) = c \cdot x (1 - x)$, we find that this new function is

$$F^2(x, c) = c^2 x(1 - x)[1 - c \cdot x(1 - x)] \quad (3.5.4)$$

A **fixed point of** $F^2(x, c)$ is any value of x that satisfies the equation

$$x = c^2 x(1 - x)[1 - c \cdot x(1 - x)] \quad (3.5.5)$$

As we know, and as is obvious, $x = 0$ is always one fixed point, for any value of c. For any nonzero fixed point, we can divide (3.5.5) by x and obtain the equation for the other **fixed points of period two**

$$1 - c^2(1 - x)[1 - c \cdot x(1 - x)] = 0 \quad (3.5.6)$$

However, this is a "messy" third-order polynomial equation, so it might look like we have a big problem on our hands. Not so! The important observation to note is that the fixed points of period one, namely $x = 1 - (1/c)$, clearly also repeat their value every two steps, and hence must be fixed points of period two! That means that $x = 1 - (1/c)$ must be a solution of (3.5.6). That means, in turn, that we must be able to write (3.5.6) in the form

$$[x - 1 + (1/c)]G(x, c) = 0 \qquad (3.5.7)$$

and the problem is to determine the function $G(x, c)$, in order to obtain the **new period-two fixed points.** There will be two of them, and they will correspond to the **two branches** that are shown below in **Fig. 3.8**.

▪ Think 3.5A

So here is a little old-fashioned algebra! Factor the third-order polynomial (3.5.6) into the first-order and second-order, $G(x, c)$, polynomials in (3.5.7). Once you **obtain $G(x, c)$,** solve the polynomial equation $G(x, c) = 0$ to **obtain the two new period-two fixed points.** Show that the two roots are given by

$$x = [1 + (1/c)]/2 \pm [1 + (1/c)]/2\sqrt{[1 - (4/(1 + c))]} \qquad (3.5.8)$$

As c is increased above the value $c_1 = 3$, a value of c is reached where the period-two dynamic attractor becomes unstable. To determine this value of c_2 one uses the fixed points (3.5.8), together with function $F^2(x, c)$ in (3.5.4). These are substituted into the condition for the new bifurcation point c_2, given by Eq. (3.5.2), where F is now the function $F^2(x, c)$. ▪

▪ Think 3.5B

The result of this is a rather complicated equation for determining the value of c. However, once again, one knows that this equation has the two roots, $c_0 = 1$ and $c_1 = 3$, so $(1 - c)$ and $(3 - c)$ can be factored out of the equation. See if you can show that the new bifurcation point is given by $c_2 = 1 + \sqrt{6} \cong 3.45$. (Since this is the solution of a quadratic equation, why isn't there a second root?) ▪

When c is above the value c_2, the period-two dynamics becomes unstable, and the new attractor has a period of four. This **period-four dynamics** then becomes unstable when c increases past a value $c_3 \cong 3.54$, and a **period-eight dynamics** then becomes the attractor. Note that all of the attractors have a **period 2^n, for some value ($n = 0, 1, 2, \ldots \infty$)**.

Explore 3.5A

To understand the **dynamic attractors** that result at each of these bifurcations, see **Program 3-5A**. In the first ten iterations, note the ***x*-ordering** of the dynamics. The values of c have been limited by $c < 3.45$ (see **Table 3.1** for this significance). Once you have understood these results, you can modify the program by replacing 3.45 by a new value (**everywhere!** To do this use the **"search"** and **"change"** commands, e.g., **find 3.45, change to 3.57**). Note that the asymptotic dynamics takes longer to be achieved.

Explore 3.5B

Rather than follow the dynamics, as in 3-5A, one can produce the **asymptotic values** of the dynamics as a function of c, as in Fig. 3.7. This is done in **Program 3-5B, which can** be used to automatically scan through all the values of **cmin** < c < **cmax** of this infinite set of period-2^n bifurcation points ($n = 0, 1, 2, \ldots, \infty$). To do this, first select **cmin = 3** and **cmax = 3.57.** You will discover that there are **difficulties** in displaying all this information. **Why?** Try "zooming in" by using some smaller c range, just below $c = 3.57$. **Does this help? Do you have other ideas?**

When $c > c_\infty$, there is an amazing diversity of dynamics, a veritable **"jungle"** of dynamic varieties; all of this from the very simple and innocent looking logistic map! This is but one of many examples of the fact that **recursive relations can hide a wild variety of dynamic behaviors,** which are totally unexpected when the recursive relations are initially proposed. In this section, we will be able to explore only a limited amount of dynamic details within this jungle, leaving much for you to uncover in the future.

Table 3.1. Approximate values of some bifurcation points, $c = c_n$. For $c > c_n$ the period-2^n dynamics is a repellor (unstable), whereas the period-2^{n+1} dynamics is an attractor (stable).

n	c_n	n	c_n
0	1.0	5	3.568759
1	3.0	6	3.569692
2	3.449499	7	3.569934
3	3.544090	8	3.569934
4	3.564407	∞	3.569946

3.5 The Logistic-Dynamics Jungle!!

Briefly, we will see that there are **periodic motions that do not have the periods 2^n, which we found so far, various types of nonperiodic, "chaotic" types of motion,** and motions that **seem to be periodic,** but then **give "spurts" of chaotic behavior,** all accompanied by **new types of birfucation processes.** Needless to say, the details of this type of dynamics can only be explored with the help of a computer. So let us begin to explore.

Explore 3.5C

The **Program 3-5B** lets you find the **asymptotic dynamics** of the logistic map for **any range of c values** you choose. By **"asymptotic dynamics"** is meant the dynamics that **only occurs for large values of the time, neglecting the early stages of the dynamics.** You should check the program to see **how much dynamics is being neglected.** Here you might begin by getting an overview of all the types of dynamics that is contained in the allowed range, $0 < c < 4$. When you do this, you will see how a very famous figure, similar to **Fig. 3.8,** is generated dynamically.

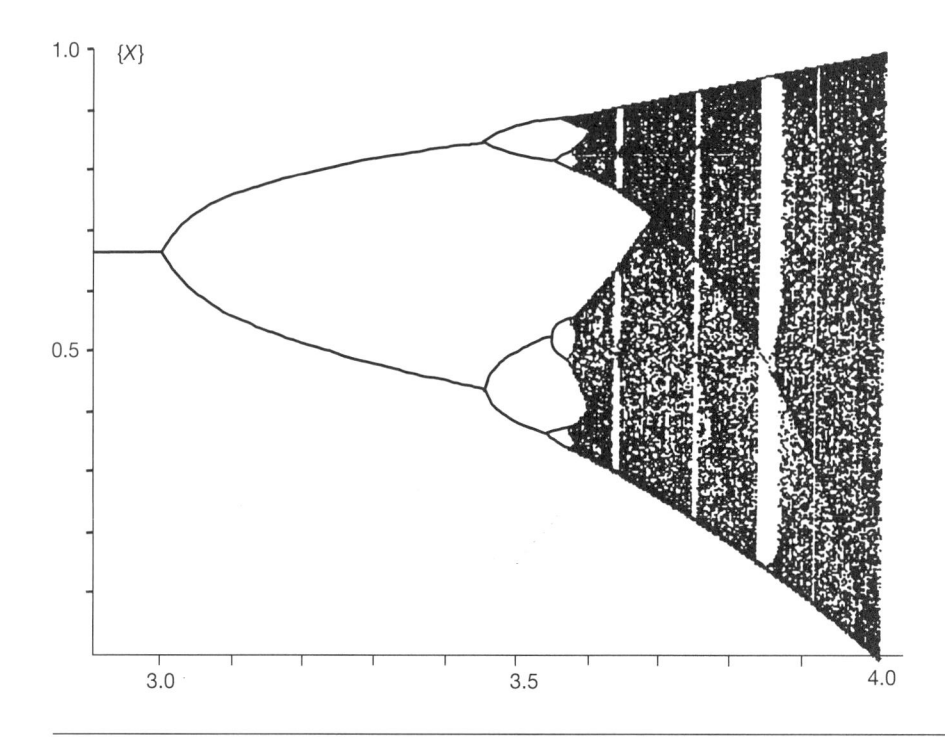

Fig. 3.8.
The asymptotic (long-time) dynamics of the logistic map, {X}, over the entire range of allowed values of c.

Different Types of Population Dynamics

▮ Think 3.5C

In constructing Fig. 3.8 and subsequent figures involving different values of c, some thought should be given to the question **"which values of c?"**. Remember that computers (as well as us!) cannot manipulate real numbers but only some limited amount of information. As you proceed, try to refine this concept as it influences our study of complex forms of dynamics. ▮

There are clearly a number **mysterious dynamic regions** of c values in Fig. 3.8, which beckon to be explored. To begin, there are a number of new features to be found in the region just above c_∞. An example of this is shown in **Fig. 3.9.**

Among the striking features in this figure are the **"bands" of x values in which the dynamics is constrained,** and how the **bifurcations** now involve the changes in these band structures. Then, of course, one sees mysterious regions where there are (apparently) **no bands,** which can also be seen in **Fig. 3.8.** There are many things to be explored!

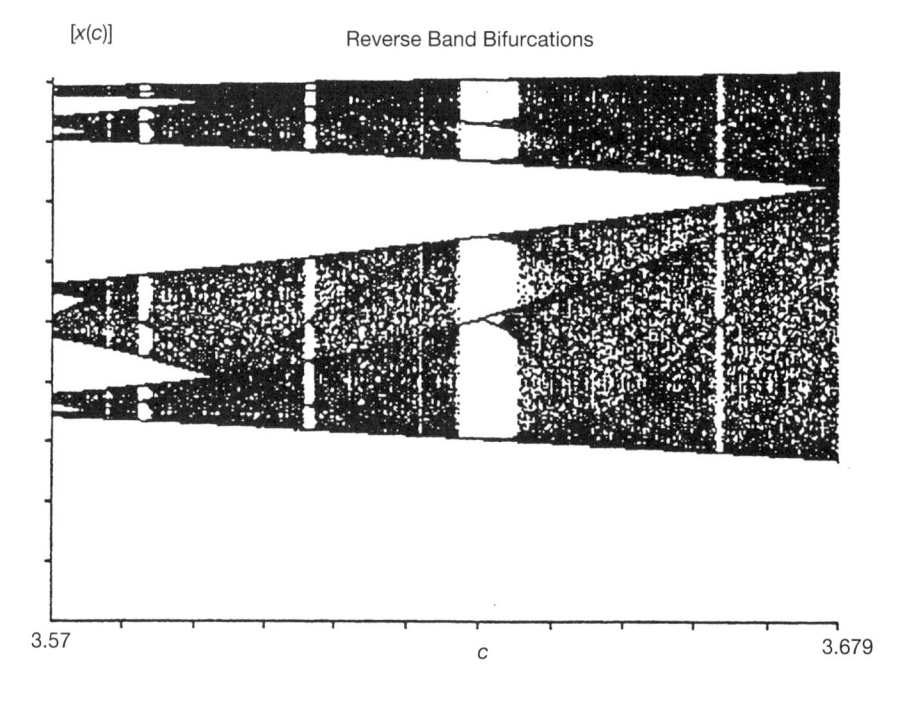

Fig. 3.9.
The "reverse band bifurcations." What is "reverse" about these bifurcations?

Explore 3.5D

To understand the significance of these bands, slow down this dynamics (e.g., by inserting "For $i = 1$ to N: next" at the beginning of the dynamic loop and selecting a large value of N). In this way, you can see where the individual steps are moving. Is the motion **"something like" periodic?** How does it differ? Is this dynamics anything like the temperature values in different seasons of the year? What is the difference? Note what happens near $c = 3.679$. Now check out the value of the (unstable) **period-one fixed point** at this value of c. How does this x^* relate to the result in **Fig. 3.9?** What about the other values of this $x^*(c)$? Where do they lie in **Fig. 3.8?**

Explore 3.5E

If you use the slower dynamics of Explore 3.5D, but now speed it up so that you can observe a large number of points being generated, you will be able to understand the development of the **"veil-like" structures,** which you can see in **Figs. 3.8 and 3.9.** If you consider the likelihood or **probability** that a **particular x value** will occur, then what is happening near the **edge of a veil?**

Next, you might consider what is going on in those regions in both Fig. 3.8 and Fig. 3.9 where there is **periodic dynamics.** These regions appear as relatively **"clear" regions** in the x direction. For this reason, such periodic regions are called **"windows."** The largest window (i.e., extends over the largest range of c values) is the so-called **period-3 window in Fig. 3.8.**

Explore 3.5F

By generalizing the search in Fig. 3.9, list in the order of increasing c values **the periods of the windows** that you find. Do you see any order in these periods? Possibly not, but **there *is* an ordering** (see Sharkovsky's theorem, discussed in Jackson [1], p. 166).

Explore 3.5G

You might look in more detail **within the period-three window** and see how that dynamics compares with **the entire Fig. 3.8.** Do pieces of the period-three dynamics look rather like a form of **"dynamic déjà vu"?** How many pieces are there? What happens if you blow up this region of c values? Is there another "period-three-like" window in there somewhere? And **what if ... ? A microcosmos?**

▪ Think 3.5D

You may also note what is happening with the **veil-like structures** in Figs. 3.9 and 3.10, particularly **in the regions near the windows.** Referring back to **Explore 3.5E,** you may note that this says something about the probability distribution of dynamic points as c approaches a window (from below). What is this information? This will prepare you to **"hear" the consequences of this dynamics in Section 3.6.** ▪

After all, **seeing is only one of our senses,** so as important as the above figures are to our understanding of dynamics, **listening to the dynamics may also be quite revealing.** So let us begin to **explore this approach to understanding.**

Reference

1. E. A. Jackson, *Prospectives of Nonlinear Dynamics,* Vol. 1, Cambridge University Press, Cambridge, 1991.

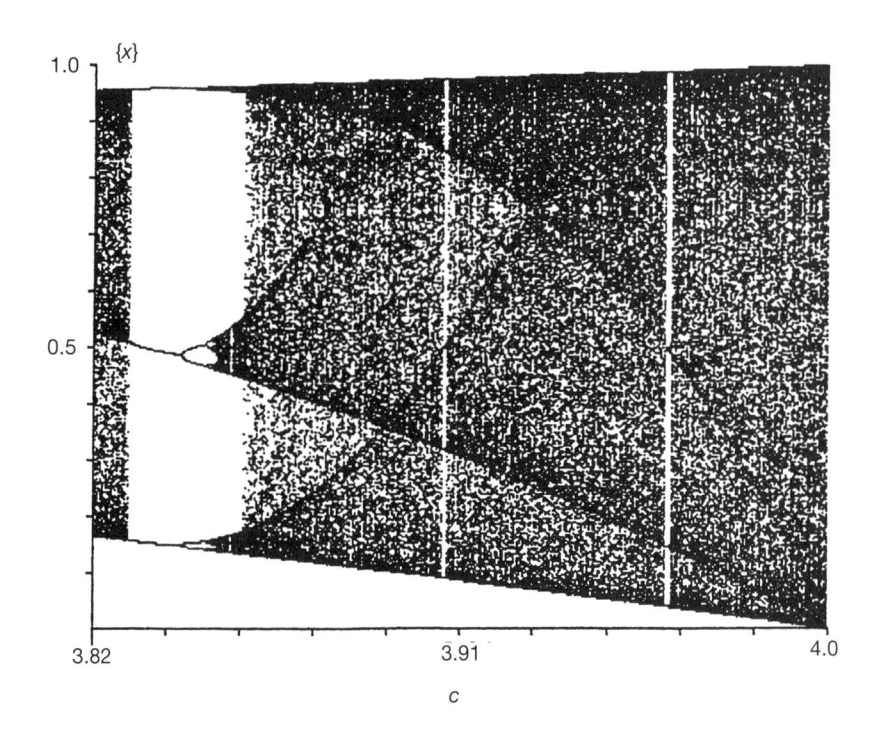

Fig. 3.10.
This figure shows more details of the dynamics in the range $3.82 < c < 4.0$. This includes the period-three window as well as several other visible windows, and more "veil-like" structures.

3.6 THE SOUNDS AS WELL AS SIGHTS OF LOGISTIC DYNAMICS—USING TWO SENSES TO "UNDERSTAND" WHAT IS GOING ON

Here you will be able to explore the possible effects of the combined sight (including colors) and sound representations of the various logistic-map dynamics on your imaginations and "understandings." You will get out of such explorations only what you put into them by way of **your receptiveness** to such new insights.

After you have seen what is offered by way of examples on the floppy disk, you may hopefully be inspired to try some of your own explorations. If you are not so inspired, then you have a long way to go in developing your imagination! You can clearly make new color and tonal associations, as well as more selective and detailed examinations of the control parameter regions. And, of course, there is no law that says that you cannot change the dynamic equations!

Computer **Program 3-6** contains this **entire introduction** to the development of tones. This is the longest program in this book. it is a composite of many programs, simply patched together, so that they are **largely "stand alone" programs,** which can be modified locally. (Experienced programmers are invited to visit **Explore 3.6D.**)

To begin, start by reading the discussion in **Appendix C6, "Let There be Tones!" Program 3-6** begins with an opportunity for you to listen to a little Bach interlude (you will need to modify the program), which is based, of course, on the use of **standard musical notes** (a second example will close this program).

However, for the uses in **dynamics,** it is more useful to use more **continuous tones,** designated by arbitrary **frequencies.** This will be done for the logistic map, by associating the numerical value of $x(t)$ with a frequency ($freq$), such as

$$freq = 500 + 2000 \, x(t) \qquad (3.6.1)$$

You can obviously change these numbers to something that is more appealing to you. One then introduces the command

$$\texttt{sound freq, dur} \qquad (3.6.2)$$

for the **tone (freq)** and the **duration** of the note **(dur),** which is best understood simply by trials. The details are explained further in Appendix C6 and by looking into Program 3-6, where the **sound commands are identified.**

Different Types of Population Dynamics

▨ **Think 3.6A**

As you **use Program 3-6** you will find a number of opportunities to modify colors, initial conditions, or the values of the control parameter c in the logistic map

$$x(t + 1) = c \cdot x(t)[1 - x(t)]$$

For example, the **colors of $x(t)$** are often based on their magnitude, which has no particular significance; you may well want to make this **more meaningful.** ▪

NOTICE
An **important feature** to note is that you can turn the tones
OFF or ON at any time by keying <F> or <N>.

This is helpful after you have heard some presentation several times. To see how this is done, note that the notation SN = 0 corresponds to a tone, and SN = 1 is no tone (!). The **program runs much faster** when there is **no tone,** if you want to get through some segment fast (but also see the suggested by-pass in the program, for which you will have to do some programming).

The parameter **LM equals the number (length) of the mapping iterations.** Also the frequency in (3.6.2) may also be designated as

$$\textit{freq} = [(\textit{HF} - \textit{LF})x(t)] + \textit{LF} \qquad (3.6.3)$$

where *HF* and *LF* are the **high and low frequencies** to be used, since $0 < x < 1$. Thus, in (3.6.1), $HF = 2500$ cycles/sec.

Sections of Program 3-6

What follows is a short description of the **"hear-and-see"** of each portion of Program 3-6. First of all, **refer to Fig. 3.11** and you will see that seven values of c have been selected. These represent **four distinct types of dynamics,** described below in **sections I–IV.**

- **I.** The first type of dynamics exhibits the **four short-duration** approaches of $x(t)$ to the **period 2^n attractors, where $n = 0, 1, 2, 3$**. Here $x(t)$ is plotted versus t, and $HF = 1200$ and $LF = 200$ were arbitrarily selected for the frequency in (3.6.3). You should change these to your preference, once you find them in the program.

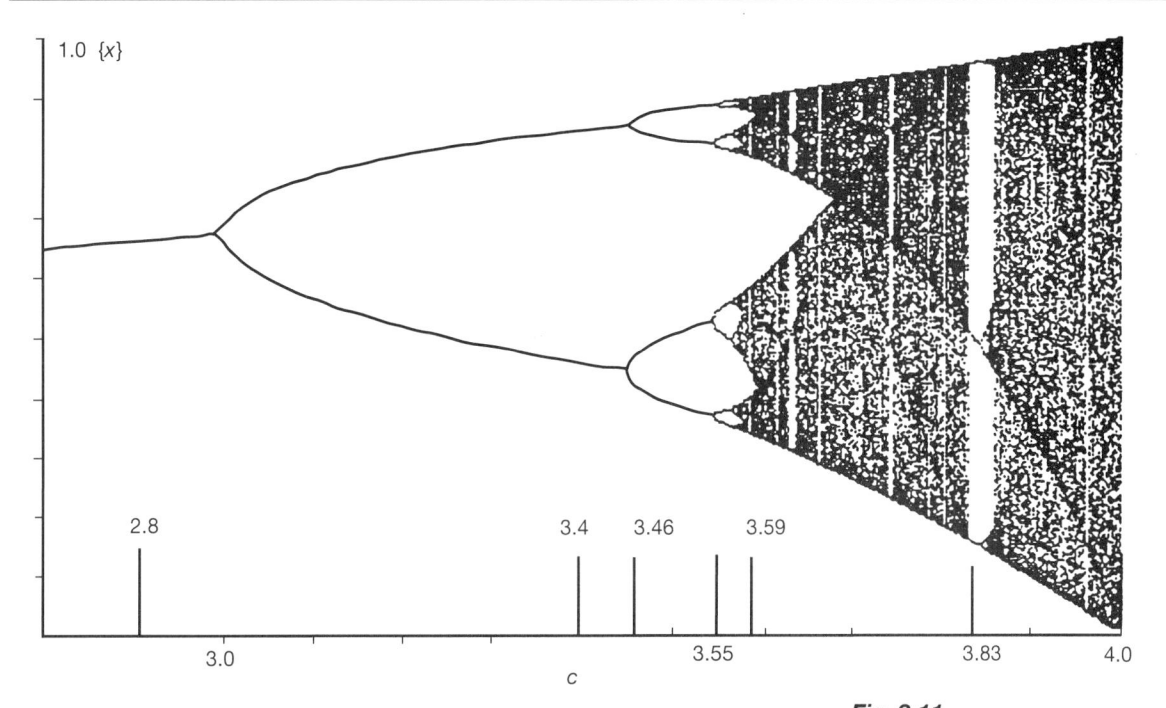

3.6 The Sounds as Well as Sights of Logistic Dynamics

Fig. 3.11.
The seven values of the control parameters that are used to generate the dynamics that **approaches the asymptotic behavior** {*x*}, illustrated above these *c* values.

Explore 3.6A
You should note that the **initial conditions** are taken near unstable periodic orbits. This selection makes it clearer **how the dynamics actually approaches the asymptotic states.** (a) Study how this is changed if the **initial conditions** are made **closer** to these **unstable periodic orbits.** (b) Also, in the case $n = 3$, see what happens if the initial condition is taken close to the **unstable period-two orbit, Eq. (3.5.8).**

II. The second type of dynamics is **the semiperiodic dynamics,** in which $x(t)$ jumps periodically between **bands of *x*-values. However, the values that it takes within these bands are chaotic.** In the selected example, $c = 3.59$, there are **only two** such bands.

Different Types of Population Dynamics

Explore 3.6B
Change the value of c to a region where there are **four bands, $3.57 < ?? < c < ?? < 3.59$**. Is the **dynamic ordering** between these bands similar to the **four-season temperatures in our weather?**

III. **The third type of dynamics** is the very novel and interesting dynamics **called intermittency.** As the control parameter approaches the **bifurcation into any window (i.e, from lower values of c),** the dynamics **sounds periodic for a while,** but then **goes "chaotic" for a time,** only to return at an **unpredictable time** to its old "periodic" behavior, then . . . ? This type of dynamics produces patterns illustrated in **Fig. 3.12.**

Fig. 3.12.
An example of **intermittency dynamics,** when c is **slightly smaller than $1 + \sqrt{8}$, just below (–0.00049) the period-three window.**

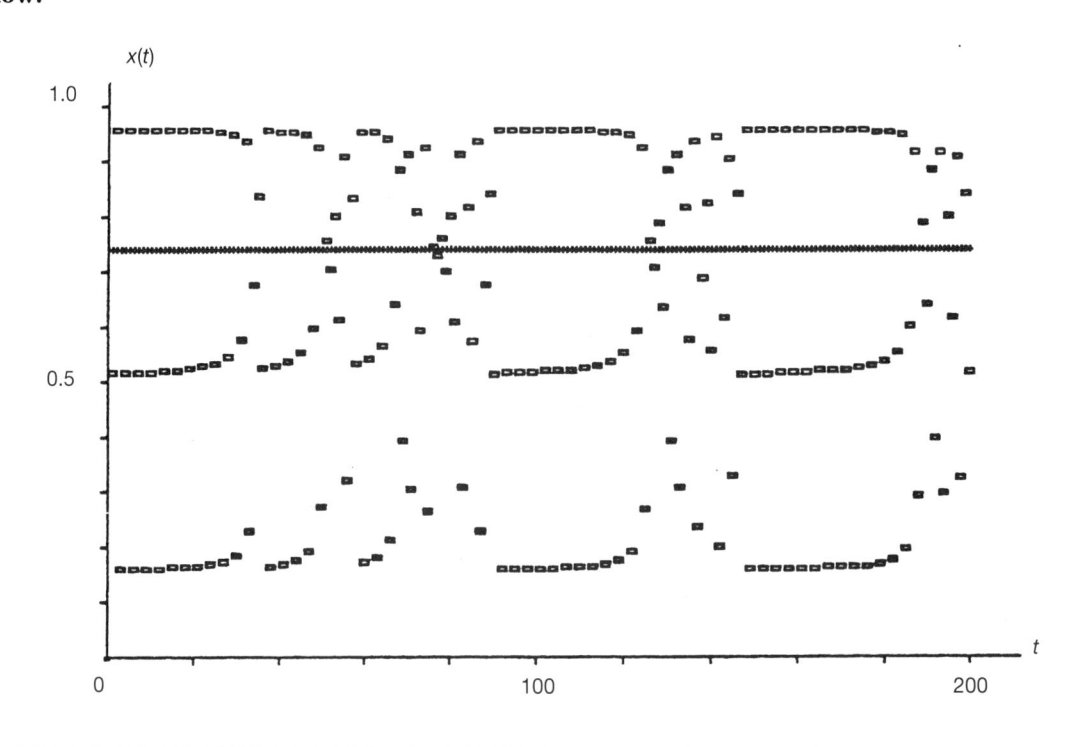

96

3.6 The Sounds as Well as Sights of Logistic Dynamics

In this type of dynamics, one's capability of **hearing the difference between tones** can sometimes be **more sensitive than** the visual perception of the **graphical difference** in values of $x(t)$. Moreover, in addition to a graphic characterization, this certainly gives an **added "holistic perceptual dimension"** of the **manner** in which there is a **breakup of a tonal periodicity.**

Think 3.6B

How does **intermittency** relate to the **veil-like structures in Fig. 3.11**? To **be specific,** consider the **distribution** of the values of $x(t)$ in this intermittency region, and **compare** that to the **figures in Section 3.5.**

Explore 3.6C

Here one can make some **careful** explorations of what happens as **c is brought closer and closer to the bifurcation point** of a window. If c_{bp} is the **bifurcation point** for some window, you can (a) first see **qualitatively** how the **"periodicity" pattern changes** as $(c - c_{bp}) > 0$ becomes smaller. (b) One can then advance to the problem of determining how the **distribution of the durations of the periodic segments** depend on $c - c_{bp}$. This is a more challenging exploration, for you will need to make a **histogram** of the durations of "periodic" lifetimes (defined by some accuracy criterion), and see how this distribution changes as $c - c_{bp}$ approaches zero.

Explore 3.6D

By looking in the program, you can explore the **intermittency** near the **period-six window.** You will need to adjust the values of c very carefully in this case. Obviously, one can also carry out explorations similar to those in **Explore 3.6C**.

IV. The **final type** of dynamic behavior illustrated **in Fig. 3.11** concerns **deterministic chaotic motion.** Note again that in **Fig. 3.11,** as in **Figs. 3.8, 3.9, 3.10, and 3.11,** there are **veil-like structures,** which implies an **uneven frequency** of occurrence of x-values, in all of the **chaotic regions.** (If you **did not Think 3.6B, think it now!**) Therefore, **all chaos is not identical.** Indeed, the number of veils for different values of c is not easy to determine (or even to define!), as can be appreciated from

Figs. 3.9 and 3.10. It appears **to the eye** that there are **no "veils" when c = 4** (i.e., no regions between $0 < x < 1$ that have **local peaks** in the frequency distribution of x-values), and this is **in fact the case.** This does not mean that the frequency distribution is uniform over all x, for **it is largest near x = 0 and x = 1,** but in several senses the **most widespread chaos occurs at c = 4.**

This dynamics can be heard in this program, **and its uneven character** over **short periods of time** can easily be detected. However, over longer periods of time the distribution becomes "smoother" and more evenly distributed. To be more specific requires a little analysis.

The Last Fig. 3.11 Dynamics

V. The **last dynamic issue** that is depicted in Program 3-6 concerns **two of the basic characteristics of chaotic motion:**

Va. The first concerns **the sensitivity of the dynamics** to the initial conditions. **The initial exponential separation of nearby initial states is one** of the very important signatures of **deterministic chaos.** It is what makes it **impossible to "predict" events into the "far" future, given the fact** that all initial states are only known to a **"limited" empirical accuracy.** There would seem to be three factors involved in such a discussion:

E = the magnitude of the empirical accuracy
P = the magnitude of the accuracy of the prediction
F = how far into the future one is trying to predict

However, since there is no scientific significance to predicting with any more accuracy than the empirical accuracy, the only significant scientific condition is to set $P = E$. Therefore, the question of interest is what is the relationship between F and E as a function of the **control parameters, C, of the system?** That is, **what is the function $H(E, C)$ such that**

$$F = H(E, C) \qquad (3.6.4)$$

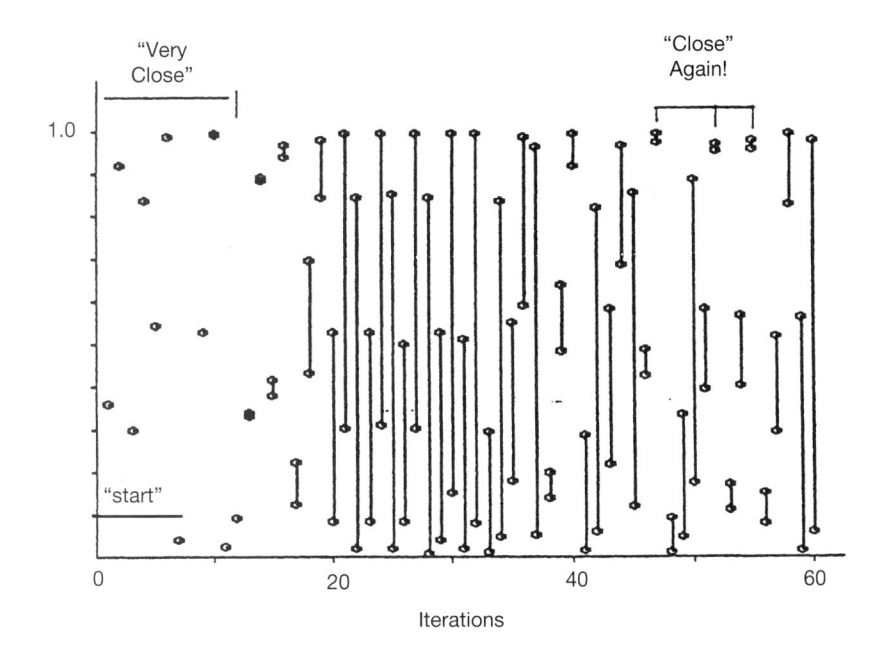

Fig. 3.13.
Sensitivity to initial conditions, and the near recurrence of chaotic dynamics, due to its bounded motion.

■ **Think 3.6C**

Consider the simple **linear dynamics,** Eq. (1.3.1), which has the solution Eq. (1.3.2). In this case, it is not very difficult to **determine the function** $H(E, C)$ in Eq. (3.6.4) by using two initial states, say $A(0)$ and $A^*(0)$, where $|A(0) - A^*(0)| = E$. In particular, determine the ratio of E/L in terms of C. [Some facts: if $x = \exp(y)$, then the "natural" $\log(x) = y$, and $\log(x) - \log(y) = \log(x/y)$.] ■

Vb. The **second essential feature** about chaotic motion is that it is **bounded (in magnitude).** It will be noted that the solution of the **linear dynamics,** while it has exponential sensitivity to its initial conditions, is **not a bounded dynamics,** hence it **is not chaotic.** What is illustrated in the program is how **two nearby initial states can rapidly separate** and remain **empirically** uncorrelated to each other for some time, but will **occasionally return near each other.** The essential point is that there is **nothing predictive about this behavior.** This **nonpredictive returning feature** is a necessary consequence of their **uncorrelated,** yet **bounded, dynamics.**

Explore 3.6D

For "concerned" computer programmers. Computer **Program 3-6** is the longest program in this book. It is a composite of many programs, patched together to illustrate dynamic concepts, but without concerns about redundancy, changing notation, and the like (i.e., it is a totally unprofessional job of programming!). If this bothers some of the **more proficient computer programmers,** they might want to take up the challenge to **clean up and generalize this whole production.** In doing so, see if you can **extend the dynamic insights** that are offered by tonal and color representations. I would be interested in the results of any such effort (**if** it remains user friendly).

3.7 MORE EMPIRICALLY REALISTIC MODELS OF POPULATION DYNAMICS: DIFFERENT "UNIVERSAL CLASSES"

The logistic map, although it captures some of the essential features of the population dynamics of many species, is not realistic in its treatment of a **cutoff at a finite population,** $N = 1 \rightarrow N = 0$. The other models, noted in (3.1.4), all extend to arbitrarily large values of N, decaying at different rates. These models have been widely used in ecology to represent the dynamics of a number of species.

The class II model in (3.1.4) is

$$N(t + 1) = R \cdot N(t) \exp[-A \cdot N(t)] \quad (3.7.1)$$

which has been widely used in the biological literature [1–3]. It replaces the logistic "finite-resources," or mortality factor of $1 - N$ with an exponentially decreasing factor, which decreases rapidly for large values of $N > 0$. This model is apparently plausible in those cases where **the mortality of a high-density population is regulated by epidemics.**

It should be noted that **the constant A,** although necessary to **quantitatively account** for the features of the empirical data, **does not influence the overall sequence of qualitative behaviors** of the dynamics of such species as R is varied. This can be seen by multiplying (3.7.1) by A, and defining a new variable $M(t) = A \cdot N(t)$, in which case

$$M(t + 1) = R \cdot M(t) \exp[-M(t)] \quad (3.7.2)$$

In other words, all species that satisfy (3.7.1), **when scaled to the variable $M(t) = A \cdot N(t)$,** satisfy the same dynamic equation, (3.7.2), and hence have **the same sequence of qualitative dynamic behaviors as the parameter R is varied.** This means that there is a **universal behavior** of the sequence of bifurcations for all of these species, governed by only **one parameter, R,** which is a noteworthy fact. Such universal behavior is sometimes referred to as a **"universal class"** of physical systems.

Explore 3.7A

Use **Program 3-7A** to explore the bifurcation sequence of (3.7.1), where A has arbitrarily been assigned a value. You can change the value to $A = 1$, which is equivalent to the form (3.7.2), but you may find challenges in scaling the graphics. You can modify the program to obtain the bifurcation diagram of the asymptotic dynamics, similar to Fig. 3.8.

The **class (III) model in (3.1.4)** is

$$N(t+1) = R \cdot N(t)/[1 + a \cdot N(t)]^b \qquad (3.7.3)$$

This **model** has be found to describe the population dynamics of **twenty-eight seasonally breeding insects, in which the generations do not overlap** [2]. It will be noted that (3.7.3) contains **three parameters,** but that the universality class of their bifurcation properties is governed by only two of these parameters (again by setting $M = a \cdot N$). Since **two parameters** are required to describe the qualitative behavior of this class of insects, they **do not belong to the same universality class as those satisfying (3.7.2).** To fully understand the qualitative behavior of this class of insects requires a study of the dynamics of (3.7.3) as a function of the **two parameters (R, b),** which is rather involved. Put another way, the bifurcation characteristics of this system can only be represented in a **two-dimensional bifurcation space,** with coordinates (R, b).

Explore 3.7B

Program 3-7B deals with dynamics of the form (3.7.3). The published forms of the boundaries between regions with different types of dynamics (monotonic damping, damped oscillations, stable limit cycles, chaos) is shown in **Fig. 3.14** [1,2]. This figure

Different Types of Population Dynamics

Fig. 3.14.
Regions of different dynamic behavior of **(3.7.3)**, compared with life table data of field populations (solid circles) and laboratory populations (open circles) **[2,3]**. The broken line separates the region of period-two cycles from higher-order cycles (analogous to C_2 in Think 3.5B, except now depending on the two parameters, R and b).

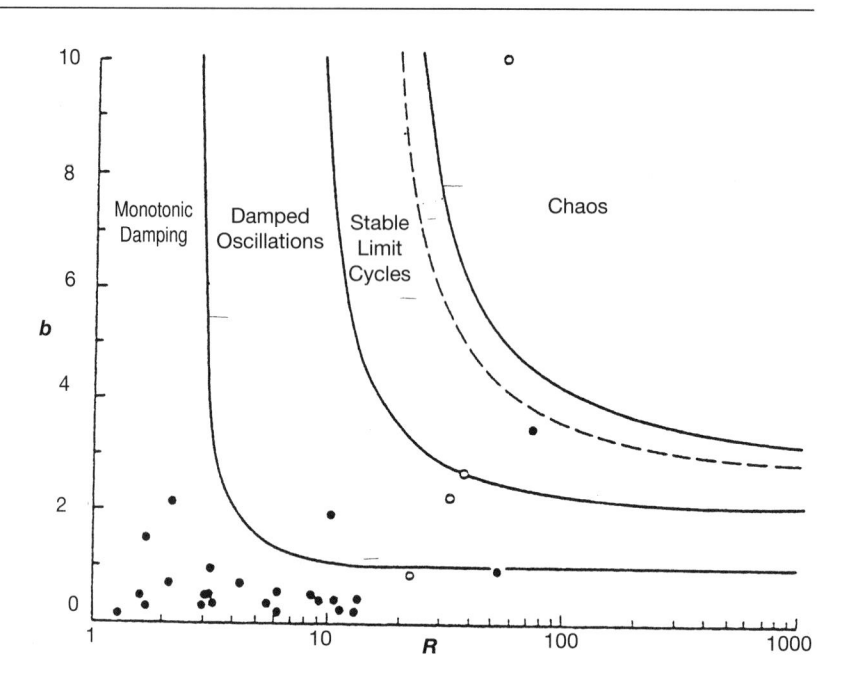

might suggest that these **qualitative behaviors** are governed by the **single parameter $Q = R^{1/b}$. Take this as a conjecture** and see if it appears reasonable or not. (You may be able to show that it is not correct, but can you prove that it is correct?) **This Is a real explore!**

Think 3.7
Begin by showing that the fixed points of (3.7.3) for all R, **and using the scaled variable $M = a \cdot N$, depend only on $Q = R^{1/b}$**.

Explore 3.7C
Return to **Program 3-7B** and modify it to be able to examine the full bifurcation spectrum of **(3.7.2)** as a function of the parameter R, and compare it with **Fig. 3.8**.

For some literature concerning the role of stable equilibria, periodic oscillations, chaos, density dependency, and spatial dynamics in ecology, see references **[4–8]**, and for general ecological issues see **[9–10]**.

References

1. R. M. May (Ed.). *Theoretical Ecology: Principles and Applications*, 2nd ed. Blackwell, Oxford, 1977.
2. R. M. May. Models for Single Populations. In *Theoretical Ecology: Principles and Applications*, 2nd ed., pp. 5–29. Blackwell, Oxford, 1977.
3. M. P. Hassell, J. H. Lawton, and R. M. May. Patterns of Dynamical Behavior in Single-Species Populations. *Theoretical Ecology: Principles and Applications*, 2nd ed., pp. 471–486. Blackwell, Oxford, 1977.
4. S. A. Levin (Ed.). *Frontiers in Population Ecology*, Part IV in *Frontiers in Mathematical Biology* Lecture Notes in Biomathematics, Vol. 100. Springer-Verlag, Berlin, 1994.
5. R. M. May. Spatial Chaos and Its Role in Ecology and Evolution. *Theoretical Ecology: Principles and Applications*, 2nd ed., pp. 326–344. Blackwell, Oxford, 1977.
6. H. Degn, A. V. Holder, and L. Folsen (Eds.). *Chaos in Biological Systems*. Plenum, New York, 1987.
7. R. May. Necessity and Chance: Deterministic Chaos in Ecology and Evolution. *Bull. Amer. Math. Soc. 32,* 291–308 (1995).
8. L. Glass and M. C. Mackey. *From Clocks to Chaos: The Rhythms of Life*. Princeton University Press, Princeton, NJ, 1988.
9. T. F. H. Allen and T. W. Hoekstra. *Toward a Unified Ecology*. Cambridge University Press, New York, 1992.
10. R. E. Ulanowicz. *Ecology, The Ascendent Perspective*. Columbia University Press, New York, 1997.

4
Dynamic Models Based on Differential Equations

4.1 FROM DIFFERENCE EQUATIONS TO DIFFERENTIAL EQUATIONS, AND BACK AGAIN!

As seen in Chapters 2 and 3, some dynamics are naturally expressed in terms of recursive relationships that involve discrete-time steps. As outlined in **Section 2.1,** Issac Newton's masterpiece of 1687, *Philosophiae Naturalis Principia Mathematica* (commonly referred to as ***Principia***), treats the planetary motions along curved paths (orbits) in an approximate geometric fashion, involving equal discrete-time steps. The classic example is shown in Fig. 1.4, which is reproduced here as Fig. 4.1. As is described in more detail in **Section 2.1,** the planetary orbit is approximated by straight-line segments between A, B, C, D, E, and F. At these points, an impulsive force acts on the planet, directed toward the sun (S) (called a **central force**), causing a change in the direction and speed of its motion. Using this construction, Newton could prove that **Kepler's relationship K1** is satisfied for **any** such central force. For more details and references, see **Section 2.1.**

Although this type of discrete-time analysis could sometimes be achieved rather easily, it often led to involved and confusing lines of reasoning [1, 2, 3]. Later, by considering **infinitesimal time**

Dynamic Models Based on Differential Equations

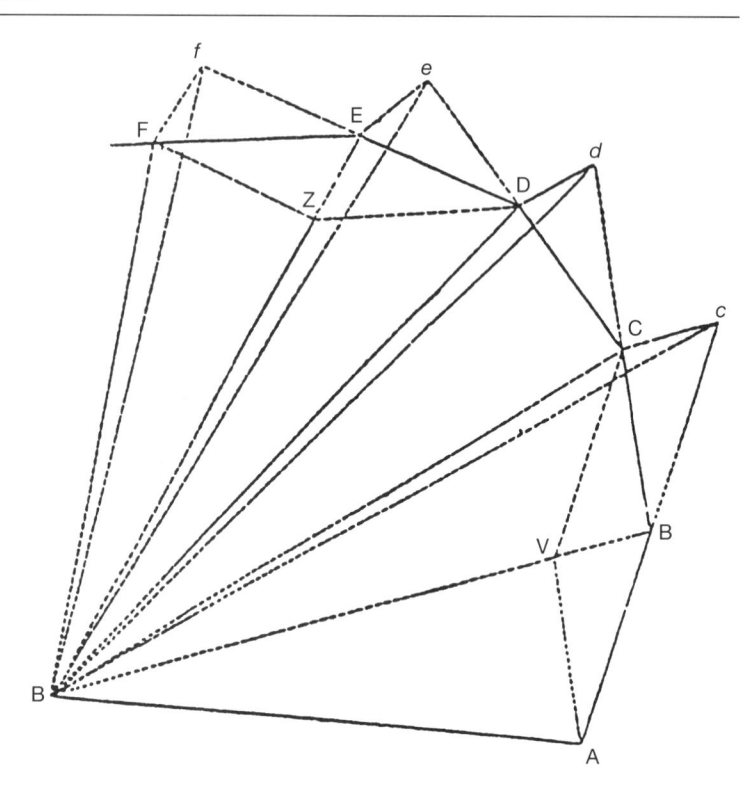

Fig. 4.1.
An early figure in Newton's *Principia*, used in the proof that **Kepler's relationship, K1,** is a consequence of **any** central force.

steps, Issac Newton first described the laws of mechanical dynamics in terms of "fluxons" that developed, thanks also to the mathematical contributions of Wilhelm Gottfried Leibniz (1646–1716) [4], into today's **ordinary differential equations of calculus.**

It turns out that the mathematics of calculus is much easier to use for the purpose of analysis, and therefore it is used for all "continuous" forms of dynamics. In this section, we will look into the **connection between differential equations and the discrete-time difference equations** used in the previous chapters. In making this connection, we will not only be interested in the mathematical connection but also in the approximations that each representation makes of **empirical observations in science,** which is the primary concern of science.

Dynamic models that are based on **discrete-time observations,** at **intervals of time equal to dt,** have the general form

$$x(t + dt) = F[x(t)] \qquad (4.1.1)$$

where $t = 0$, dt, $2\,dt$, $3\,dt$, ..., etc. Here we are **generalizing the time step** to be in units of an **arbitrary interval dt,** which will be

4.1 From Difference Equations to Differential Equations, and Back Again!

useful shortly. These **recursive relationships** are also referred to as a **map of $x(t)$ to $x(t + dt)$,** such as the "logistic map." Such maps are sometimes characterized by

$$M: x(t) \to x(t + dt)$$

It is sometimes convenient to write (4.1.1) in a form that expresses the value of $x(t + dt)$ in terms of its previous value, $x(t)$, plus its change. This leads to the **difference equation** representation of the dynamics

$$x(t + dt) = x(t) + G[x(t)]dt \qquad (4.1.2)$$

where $G[x(t)]dt = x(t + dt) - x(t)$ simply represents the **difference** that occurs in the value of $x(t)$ in the next time step, dt. It can be seen that the functions in (4.1.1) and (4.1.2) are related by $F[x(t)] = x(t) + G[x(t)]dt.$

To proceed from this representation of dynamics in the form of difference equations to the idea of differential equations, we write (4.1.2) in the form

$$\frac{x(t + dt) - x(t)}{dt} = G[x(t)] \qquad (4.1.3)$$

The idea of a differential equation involves what happens to $x(t)$ when dt **is very small.** It is here where the **arbitrary magnitude of dt** comes into play. **If we simply set $dt = 0$,** the numerator on the left side of (4.1.3) vanishes, $[x(t) - x(t)] = 0$, but the denominator also vanishes, so **the ratio is not a defined quantity.** However, if we let dt "smoothly" approach the value of zero, the ratio on the left side of (4.1.3) usually has a meaning. This smooth operation is called a **limiting process** and is denoted as **limit $dt \to 0$.**

An Illustration of limit $dt \to 0$

To see how this ratio idea works, consider the case when the function $x(t)$ is

$$x(t) = a \cdot t^2 \qquad (\text{I.1})$$

and construct the ratio on the left side of (4.1.3). We then obtain, by expanding out the quadratic term

$$\frac{a(t + dt)^2 - a \cdot t^2}{dt} = \frac{a \cdot t^2 + 2a \cdot t\, dt + a(dt)^2 - a \cdot t^2}{dt} \qquad (\text{I.2})$$

We see that the two $a \cdot t^2$ terms on the right **cancel** each other, and all the other terms in the numerator contain multiplicative factors involving dt. Thus, the numerator does indeed go to zero as dt goes to zero, as noted above. But now we see that the denominator cancels out some (not all) of these factors of dt, leaving for **the ratio (I.2)**

$$2a \cdot t + a\, dt$$

and the last factor will go to zero if we take the limit $dt \to 0$. We have therefore established the following **result (of calculus!)**

$$\text{limit } dt \to 0 \; \frac{a(t + dt)^2 - a \cdot t^2}{dt} = 2a \cdot t \tag{I.3}$$

If we use the **modern notation** for the left side, introduced by Leibniz, **(I.3) is written as**

$$\frac{d(a \cdot t^2)}{dt} = 2a \cdot t \tag{I.4}$$

The left side of (I.4) is called the **first derivative** of the **particular function, $x(t) = a \cdot t^2$.** A number of **other examples** of the **derivatives,** $dx(t)/dt$, of commonly occurring functions, $x(t)$, can be found in the mathematics **Appendix Ml.** For example, it is not difficult to show by the same process (sure, if you know how!) that

$$\frac{d(a \cdot t^n)}{dt} = n \cdot a \cdot t^{(n-1)} \tag{I.5}$$

for any positive integer $n = 1, 2, 3, \ldots$, which is a very useful fact. **Refer to Appendix Ml** if this is not clear after a few minutes thought.

Now let us return to the general case of any function $x(t)$ in (4.1.3), and take the limit $dt \to 0$

$$\text{limit } dt \to 0 \; \frac{x(t + dt) - x(t)}{dt} = \frac{dx(t)}{dt} \tag{4.1.4}$$

As before, $dx(t)/dt$ in (4.1.4) is now the **first derivative of $x(t)$ of any "differentiable" function $x(t)$.** The function $x(t)$ is called **differentiable if** the limiting value of **the ratio in (4.1.4) is finite,** so that the derivative "exists" (is finite). In this case, in the **limit $dt \to 0$,** we can **write (4.1.3)** in the form

$$\frac{dx(t)}{dt} = G[x(t)] \tag{4.1.5}$$

Equation (4.1.5) is called a first-order differential equa-

tion and is used as a dynamic model for "continuous" motions in time, involving one **variable,** $x(t)$ ("one-dimensional" motions). Generalization to more than one variable is easy, as will be discussed in the next section.

So now we have **two types** of recursive relationships to describe dynamics, **the difference equation**

$$x(t + dt) = x(t) + G[x(t)]dt \qquad (4.1.2)$$

and **the differential equation**

$$\frac{dx(t)}{dt} = G[x(t)] \qquad (4.1.5)$$

If one takes a small value of dt in equation **(4.1.2)**, this is frequently called the **Euler approximation** of the **differential equation (4.1.5)**. On the other hand, (4.1.5) is based on the limiting process **(4.1.4)**, in which $dt \to 0$. However, from the point of **view of the empirical sciences,** all observations are separated by finite values of dt. Therefore, any differential equation **(4.1.5)** is an **"overly refined" representation of any empirical knowledge.** As the famous physicist Richard Feynman expressed it in his Messenger Lectures at Cornell University in 1964, referring to the differential equations of such "laws" as those of Newton [5]:

> **It always bothers me that, according to the laws as we understand them today, it takes a computing machine an infinite number of logical operations to figure out what goes on in no matter how tiny a region of space, and no matter how tiny a region of time. How can all that be going on in that space? Why should it take an infinite amount of logic to figure out what one tiny piece of space/time is going to do?**

Leaving this for you to reflect upon, we should at least recognize that the two types of descriptions of dynamic phenomena, the difference equations, (4.1.2), and differential equations, (4.1.5), are related in very different manners (Fig. 4.2).

Despite such concerns as Feynman's, the **differential equation models** of natural phenomena are widely used, because they

The continuous approximation of recursive difference equations

$$x(t + dt) = x(t) + G[x(t)]dt$$

$$\frac{dx(t)}{dt} = G[x(t)]$$

The Euler approximation of ordinary differential equations

Fig. 4.2.
An extrapolation of scientific information, and an approximation of mathematical differential equations.

are amenable to **mathematical analyses,** which makes them of great practical value. (See **Explore 4.1B** for an example.) Nonetheless, one should keep in mind this dissimilarity with empirical knowledge. Indeed, any mathematical concept that depends on an **infinite** number of operations (as in **lim $dt \to 0$**), or an **infinite** precision (e.g., the distinction between a **rational and an irrational number**) has **no empirical basis.** As science tries to understand increasingly complicated dynamic phenomena, such considerations need to be kept in mind. Some examples of this will be encountered in **Chapter 6.**

As we have seen in Chapter 3, the use of **difference equations** can be of **fundamental importance** in representing some physical dynamics, without having any connection to differential equations (as in Fig. 4.2). In this context, difference equations are just as fundamental a representation of nature as are the classic uses of differential equations. In fact, since the **nonempirical** extrapolation, **lim $dt \to 0$,** is not used, an argument can generally be made that difference equations more accurately represent our scientific knowledge.

Think 4.1A

This issue can give rise to long discussions, and even debates, which are healthy exercises in raising your sensitivity to such issues as those which troubled Feynman (above quotation). **Open a conversation** with some colleagues, and see where it leads all of you!

One important fact is that a difference equation $x(t + \Delta t) = x(t) + H[x(t)]$ (note the **finite Δt!**) can be directly used in a **computer algorithm.** As noted above, $x(t + dt) = x(t) + G[x(t)]dt$ can **sometimes** be used as an Euler approximation for the differential equation $dx(t)/dt = G[x(t)]$. The **disadvantage** of this approach is that, in order to obtain a reasonably **good approximation** to the solution of the differential equation, dt must sometimes be taken quite small (e.g., possibly $dt < 10^{-4}$), which then requires many iterations to obtain a sizable value of t (in other words, it will **"run slow"** on a computer). However, there are times when a **rough approximation** of the solution is all that is needed, and $dt = 0.01$ can be used, in which case the **Euler approximation** may be **very useful** (this is particularly true if the dynamics involves **damping,** or any **attractor**).

In order to obtain **more accurate approximate solutions of differential equations,** there are a number of much **more accu-**

4.1 From Difference Equations to Differential Equations, and Back Again!

rate algorithmic methods for a computer. One computer algorithm that is commonly used is known as the **fourth-order Runge–Kutta method.** It will be used extensively in later calculations, beginning in **Section 4.2.** This method is discussed in the **Appendix C4,** and will be contained in many computer programs (see the **descriptions** of the these programs in the **Appendix C8**).

Explore 4.1A

To compare the relationship between a mathematical solution and the "Euler solution" of a simple linear equation, consider $dx/dt = x$ with its solution, $x(t) = x(0) \exp(t)$. Now compare this solution of (4.1.5) with the result obtained from its Euler approximation, (4.1.2), using different values for dt (say, starting with $dt = 0.01$). You can use the **"working" Program 4-1A,** which needs some completion, selecting the same initial conditions for both, and compare the results over **some range of the time** (you do the selecting, and exploring!). Under what conditions do the results differ by **less than 1%?** Can you insert this test in the program?

In the remainder of this section we will use the Euler approximation in order to explore how two historically important descriptions of population dynamics relate to each other. In **section 3.1** it was pointed out that the reproduction of populations of some species, $p(t)$, can be approximated by two forms of dynamics:

I. In the case in which the birth of each generation occurs at rather regular intervals of time, dt, one can model the dynamics in the discrete form (4.1.1). One example of this is like the **logistic-map dynamics,** for population $p(t)$

$$p(t + dt) = D \cdot p(t)[p_{max} - p(t)]$$

where p_{max} represents the **maximum population** that the environment can sustain. If we make the substitution $x(t) = p(t)/p_{max}$, and $c = D \cdot p_{max}$, we obtain the **"nearly standard"** logistic map

$$x(t + dt) = c \cdot x(t)[1 - x(t)] \qquad [0 < x(t) < 1] \qquad (4.1.6)$$

The difference from the standard logistic map is that we now have an **arbitrary time step, dt,** which we can vary, as above.

II. In the second situation, the populations **"continually"** reproduce. This is the situation with **human populations,** in which there is a birth at any minute (second?) of the year; indeed, there are **multiple births at random times.** In this case the average population at any time, $p(t)$, the population can be more closely represent in the form of a differential equation. When this is expressed in terms of the **average fractional population $x(t) = p(t)/p_{max}$,** we obtain a differential equation of the form (4.1.5), $dx(t)/dt = G[x(t)]$.

A famous model of human population dynamics was introduced in an 1845 memoir by the Belgian mathematician **Pierre-François Verhulst,** which remained unnoticed, only to be reinvented eighty years later [6]. Verhulst's suggestion was that the population growth would be limited by an influence that is proportional to $p(t)^2$. This led to a differential equation that is now known as the **Verhulst model**

$$\frac{dp(t)}{dt} = R^* \cdot p(t)[p_{max} - p(t)]$$

or in terms of the variable $y(t) = p(t)/p_{max}$, and $R = R^* \cdot p_{max}$, the normalized **Verhulst equation** is

$$\frac{dy(t)}{dt} = R \cdot y(t)[1 - y(t)] \qquad (4.1.7)$$

Equation (4.1.7) was used extensively in the early portion of the twentieth century to predict the future populations of various countries [6]. At the turn of the nineteenth century, **Robert Malthus (1766–1834)** predicted that the increase in food production could not keep up with the increase of populations, which would have disastrous results, causing famines, pestilence, and wars, as the **"mighty law of self-preservation"** would assert itself with a vengeance. (This concept was later to profoundly influence **Darwin** [7].) This gave rise to drastic social economic proposals, such as the abolition of the Poor Laws [8]. Modern concerns about the limited resources to sustain population growth are increasingly relevant [9, 10].

Think 4.1B

The Verhulst equation predicts no such disaster, because the population is limited. But how? **What does p_{max} depend upon?** How about fresh water, energy, jobs, etc.? For a list of nineteen factors, see reference [10], but you should be able to list at least five more factors. Moreover, if some of these are taken into account, how would they modify the Verhulst **model** through the behavior of p_{max}?

Think 4.1C

Show that **(4.1.6)** is the Euler approximation of **(4.1.7)** provided x and c have a specific relationship to y, R, and dt. [Hint: begin by introducing an unknown constant, A, and define $x = y/A$. Then equate the coefficient of x to c, and finally pick A to give the proper coefficient to x^2. You should find that $c = (1 + R\,dt)$. What is A in the relationship $x = y/A$?]

The stability of the difference equation implies the stability of the differential equation, but not vice versa [11].

Explore 4.1B

Based on the understanding of **Think 4.1C,** you can use **Program 4-lB** to explore the relationship between the dynamics of the Verhulst differential equation and the logistic map (difference equation), as dt is slowly increased [thereby increasing $c = (1 + R\,dt)$]. In this program, the **exact solution of Verhulst's equation** is used for comparison. This is one of those rare cases when a nonlinear equation can be solved analytically. The solution of (4.1.7) is

$$y(t) = \frac{y(0)}{y(0) + [1 - y(0)]\exp(-R \cdot t)} \qquad (4.1.8)$$

which is then used in Program 4-lB. The **dimensionless time scale Rt** is also used for the x-axis, which amounts to setting **R = 1**.

Figures 4.3 and 4.4 illustrate the difference between this exact solution and the Euler representation (4.1.6), where the correspondence requires that $c = (1 + dt)$. Thus, as dt is increased, the value of c increases, leading to the logistic-like bifurcations as c becomes greater than 3. Even though (4.1.6) is a map, the values of $x(t + dt)$ are connected by lines to $x(t)$ for purposes of comparison.

Summary of Concepts

- the mathematical concept of a first derivative of a function $x(t)$, denoted by $dx/d\tau$
- examples involving the functions $x(t) = a \cdot t^n$
- an ordinary differential equation, $dx/dt = G[x(t)]$

Fig. 4.3.
A comparison of the solution of the Verhulst differential equation, **(4.1.7)**, $R = 1$, and the logistic-map difference equation, **(4.1.6)**, when $c = 2.7$, or **$dt = 1.7$**, based on the Euler approximation in **Think 4.1C**.

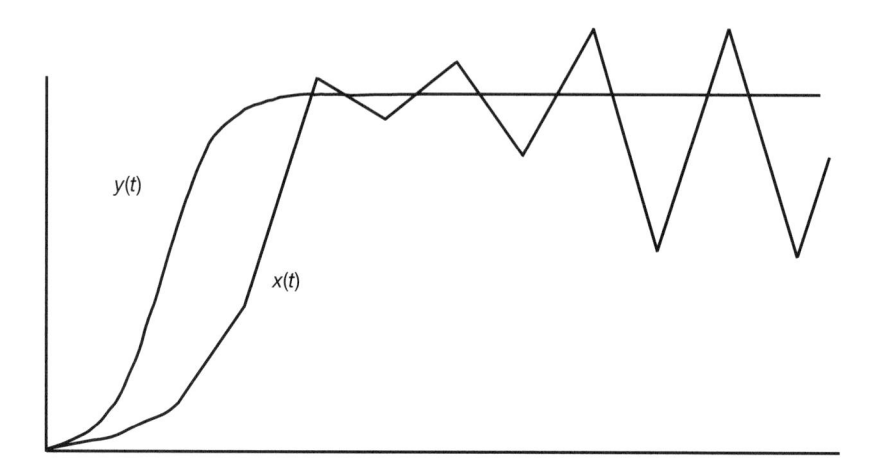

Fig. 4.4.
Similar to Fig. 4.3, except for **$dt = 2.6$ ($c = 3.6$)**.

- relationships between differential equations and difference equations, and their relationships to sciences' empirical information

References and Notes

1. D. L. Goodstein and J. R. Goodstein. *Feynman's Lost Lecture*. W. W. Norton, New York, 1996.
2. J. B. Brackenridge. *The Key to Newton's Dynamics*. University of California Press, 1995.
3. S. K. Stein. Exactly How Did Newton Deal with His Planets? *Math. Intelligencer*, *18*(2), 6–11 (1996).

4. See Chapter VI in M. Kline, *Mathematics: The Loss of Certainty*, Oxford University Press, Oxford, 1980.
5. R. Feynman. *The Character of Physical Law*, p. 57. M.I.T Press, Cambridge, MA, 1985.
6. S. E. Kingland. *Modeling Nature: Episodes in the History of Population Ecology*, 2nd ed. University of Chicago Press, 1995.
7. For a discussion of this point, see S. J. Gould, *Ever Since Darwin: Reflections in Natural History* W. W. Norton, New York, 1992, pp. 21–22.
8. For a informative brief survey, and additional references, see D. Winch, *Malthus*, Oxford University Press, Oxford, 1987.
9. J. E. Cohen. *How Many People Can the Earth Support?* W. W. Norton, New York, 1996.
10. L. R. Brown, G. Gardner, and B. Halweil. *Beyond Malthus: Nineteen Dimensions of the Population Challenge*. W. W. Norton, New York, 1999.
11. R. M. May. On Relationships among Various Types of Population Models. *Amer. Naturalist, 107,* 46–57, Jan–Feb, 1973.

4.2 HARMONIC OSCILLATORS: THEIR PROPERTIES, PHASE SPACE REPRESENTATION, AND COMPUTATIONAL ALGORITHMS

Periodic motions (those that repeat after regular intervals of time) are among the most important types of dynamics in nature. You only have to think of day and night cycles, the beat of your heart, and our dependence on clocks. Hidden within all of us are numerous biological clocks, which keep the **heart** going **(Section 5.5)** and influence a variety of hormonal activities (see the discussion of **circadian rhythms in Section 7.6**). Of course, none of these phenomena are strictly periodic, but this will be ignored in the present models of periodic motion (recall the discussion about models and reality in **Section 2.3**). To describe periodic dynamics requires at least two dynamic variables (such as a position and velocity), in contrast to the one dynamic variable considered in the last section.

So in this section we will consider a few aspects of **general systems** of differential equations that involve **two dynamic variables,** say $x(t)$ and $y(t),$ since more dynamic variables are treated in a similar manner. The recursive models of the behavior of two dynamic variables are typically based on two differential equations of the form

$$\frac{dx(t)}{dt} = F[x(t), y(t), t]$$
$$\frac{dy(t)}{dt} = G[x(t), y(t), t]$$
(4.2.1)

The system of equations (4.2.1) is called a **second-order system** or a **two-dimensional system.** The form of the functions $F(x, y, t)$ and $G(x, y, t)$ naturally depend on the particular physical system that is being modeled by (4.2.1). F and G are functions that generally depend on both of the variables, and sometime explicitly on the time, t. If F and G do not depend explicitly on t, the system is called an **autonomous system,** otherwise it is called a non autonomous system. Typical examples of a **nonautonomous system** are when **time-dependent outside forces** or actions of some type influence the system. These types of systems will be considered in **Chapter 7.** For the present, we will only consider the **autonomous systems**

$$\frac{dx(t)}{dt} = F[x(t), y(t)]$$
$$\frac{dy(t)}{dt} = G[x(t), y(t)]$$
(4.2.2)

A historically important example from classical mechanics is **Newton's equations of motion for a particle that moves along the x direction**

$$\frac{dx}{dt} = v$$
$$\frac{m \cdot dv}{dt} = F(x, v)$$
(4.2.3)

Here $x(t)$ **is the position** of the particle, $v(t)$ **is its velocity, m its mass,** and $F(x, v)$ **is the force** acting on it in the x direction, which may depend both on its position and its velocity (e.g., if there is friction).

A simple, but very important example of this dynamics, is the so-called (damped) **harmonic oscillator.** In this system, there is a **force** that is **proportional to** the particle's **displacement,** $x(t)$, **from the point $x = 0$** (where it can remain at rest—an **equilibrium point**), and a frictional force that acts in the **opposite direction** to its velocity. Then

$$F(x, v) = -\kappa x - \mu v \qquad (\kappa, \mu > 0) \qquad (4.2.4)$$

where κ is the "force constant" (**e.g., a "spring constant"**), and μ

4.2 Harmonic Oscillators: Their Properties, Phase Space Representation, and Computational Algorithms

is the **"coefficient of friction,"** which has a larger value for rougher surfaces on which the mass m is sliding (see Fig. 4.5). Substituting (4.2.4) into (4.2.3) yields

$$\frac{dx}{dt} = v$$
$$\frac{m \cdot dv}{dt} = -\kappa x - \mu v$$
(4.2.5)

We will return to the more general case of **nonlinear equations (4.2.2)** in the next section. For the present, let us explore the properties of the harmonic oscillator, and how it can be **represented graphically** and **simulated on a computer.**

Let us first consider the case of an **undamped harmonic oscillator**, $\mu = 0$ (no friction), so (4.2.4) becomes

$$\frac{dx}{dt} = v$$
$$\frac{m \cdot dv}{dt} = -\kappa x$$
(4.2.6)

To make things simple, consider the case where the mass is initially at rest [so $v(t = 0) = 0$], and is released from the point $x(t = 0) = A$. In this case, the **amplitude of the oscillations** in space will be A. The corresponding solution of (4.2.6) is

$$v(t) = -(A\omega)\sin(\omega t) \quad \text{and} \quad x = A\cos(\omega t) \quad (4.2.7)$$

Note that these functions indeed **satisfy the initial conditions,** $v(0) = 0, x(0) = A$.

Think 4.2A

Consider another solution of (4.2.6), in which $x(t)$ has the same amplitude A but a **different initial condition,** namely

$$v(t) = -(A\omega)\sin(\omega t + \varphi), \quad \text{and} \quad x(t) = A\cos(\omega t + \varphi)$$

where φ is an arbitrary constant. What are the possible values of φ

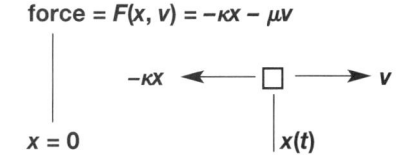

force = $F(x, v) = -\kappa x - \mu v$

Fig. 4.5.
Two forces (vectors) act on the mass, m, at $x(t)$. Note that since $x > 0$, the restoring force points left, as shown. Figure out how the friction force, $-\mu v$, contributes to the total force if v is in the direction shown. What if v is in the opposite direction?

if $x(0) = 0$? What are the two corresponding initial conditions that yield the amplitude A for $x(t)$, and **where are they located on the orbit in Fig. 4.6?**

Think 4.2B
Use the **property P1 in Appendix M1** and the results **(M3.7)** in **Appendix M3** to show that (4.2.7) is in fact a particular solution of (4.2.6), provided that the so-called **angular frequency,** ω, is given by

$$\omega = \sqrt{\frac{\kappa}{m}} \qquad (4.2.8)$$

To do this, identify both a and x in (M3.7), and how P1 is used.

Equation (4.2.8) shows that the frequency is larger for a **stiffer spring (larger κ)** and smaller mass. Moreover, **the period, P,** of this dynamics (the time interval between repeated dynamics) is given by $P = 2\pi/\omega$ (seconds), and is **independent of the amplitude, A**. This is a very important feature of a harmonic oscillator. For example, because a **pendulum behaves like a harmonic oscillator** when it swings back and forth over small angles, **it keeps regular time,** even if it slowly decreases its amplitude. Galileo is said to have realized from his observations of the swinging chandelier in the Cathedral of Pisa that it behaved like a simple pendulum. To what degree he also appreciated the above regularity in its period is unclear [1]. In any case, we will examine **general (nonlinear) pendulum motion in Section 4.5.**

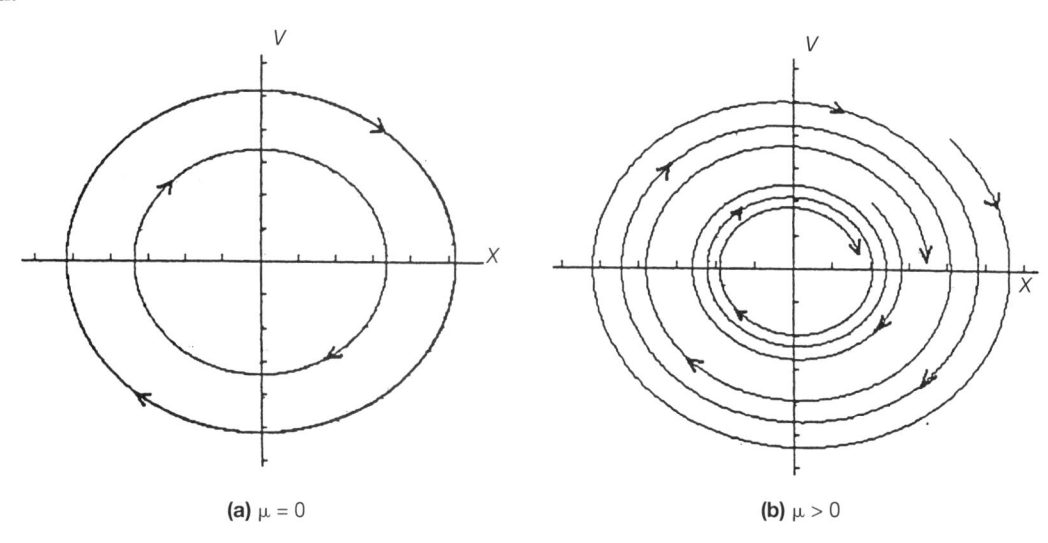

Fig. 4.6.
Dynamics represented in the **phase space,** with **coordinates x and v.** Two orbits are shown, both for the undamped (4.5a) and damped (4.5b) harmonic oscillators. In both cases, there are two initial conditions, $v(0) = 0$ and $x(0) = A$ or $2A$. Note that in **(a)** the orbits are **finite,** but **infinite in (b),** so only **a portion of each orbit is represented.**

(a) $\mu = 0$

(b) $\mu > 0$

Phase Space

To graphically represent dynamics, we now introduce the concept **of a phase space.** This concept was employed by **Poincaré** in the 1890s to uncover new **(topological)** modes of "understanding" the contents of ordinary differential equations, as we will see. The present phase space is a two-dimensional plane, with a coordinate system consisting of **two orthogonal axes,** representing **values of x and v.** Therefore, the **state of the system** at any time, which is specified by the values of $x(t)$, $v(t)$, **is represented by a point in this space. As the time changes,** so does the location of this point, and **it traces out a curve** in this space. To indicate the **direction** that the point moves in with increasing time, an **arrow** is attached to the curve, and this **oriented curve is called an orbit.**

To emphasize these features, consider the following characterization:

Dynamic System	**Represented in a Phase Space**
state of system at time t, $[x(t), v(t)]$	$\Rightarrow$ by a **point** in phase space
the history of the dynamics **for all t**	$\Rightarrow$ by a **unique** "orbit" = a curve containing **all** points, **oriented** for **increasing t**

In the case of the **general harmonic oscillator dynamics,** (4.2.5), this **phase space representation** is illustrated in **Fig. 4.6a (if $\pi = 0$),** and in **Fig. 4.6b (if $\pi > 0$)**. In both cases **two orbits** are shown, corresponding to the **two initial conditions $[v(0) = 0, x(0) = A$ or $2A]$**. Let us consider the undamped and damped cases separately.

Case $\mu = 0$: It can be seen from **(4.2.7)** that these **orbits** are simple **circles** passing through the initial-condition point. Moreover, they are **oriented in a clockwise direction,** provided that in this **phase space** we use x for the horizontal axis, and v/ω for the vertical axis.

Think 4.2C

Draw a straight line along the x axis, crossing these two circles at their initial states, $x = A$ and $x = 2A$. Now assume that this line is really a **flexible rubber band,** which pivots freely at the origin. Using the property implied by (4.2.8), discussed in the text, decide how this band will change its shape if it remains attached to the future points of these two systems.

■ Think 4.2D

What would be the shape of these orbits if one used the **phase space** (x, v), **and** $\omega = 2$? [**Hint:** consider the location of the points (x, v) at the time $t = 0$, and also at $t = \pi/\omega$, then connect these by a smooth closed curve.]

***Case* $\mu > 0$:** Because the equations (4.2.5) are linear (contain only the first power of the variables), it is again possible to obtain the solution. As might be expected, it is more complicated, but for the initial conditions $v(0) = 0, x(0) = A$ it reduces to

$$x(t) = A \exp\left(\frac{-\mu t}{2}\right) \cos[(\sqrt{(\omega^2 - (\mu/2)^2)}\,t] \qquad (4.2.9)$$

Note that this reduces to (4.2.7) if $\mu = 0$. Also note the fact that $\mu > 0$, rather than being negative, is important.

■ Think 4.2E

If you doubt (4.2.9), or are simply curious, you can check it out by using the definition $v(t) = dx/dt$, the properties P1 and P2 in Appendix M1, and the differential relationships (M3.4) and (M3.7) established in Appendix M3.

As can be seen from (4.2.9), $x(t) \to 0$ as t becomes large, and the same is true for $v(t) = dx/dt$. On the other hand, this shows that the motion also continues to oscillate due to the trigonometric term. Both of these properties are, of course, pretty obvious, given the forces acting on the mass m. This combined dynamics is then referred to as a **damped harmonic oscillator,** which is illustrated in **Fig. 4.6b.**

It will be noted that both initial states (and any other initial state for that matter) tends to go to the origin of the phase space, $(x = 0, v = 0)$. This point is an **equilibrium,** or a so-called **fixed point,** of this system, because (4.2.6) shows that $dx/dt = 0$ and $dv/dt = 0$ at that point. This also a fixed point if $\mu = 0$, but in this case the orbits simply circle this fixed point. In the case, $\mu > 0$ all states are "attracted" to the origin as time increases, hence this **fixed point** is referred to as **an attractor.** The set of **all** the points that are attracted to an attractor is referred to as its **basin of attraction** (analogous to what a river basin is to a river). In the present case, this is pretty simple, for the basin of attraction

4.2 Harmonic Oscillators: Their Properties, Phase Space Representation, and Computational Algorithms

is **the entire phase space.** More interesting cases will show up shortly.

General Features of Phase Space Representation

In the phase space representation of the dynamics there is **no indication of the value of the time,** such as contained in the solutions (4.2.7). Rather, the **phase space representation** of the solutions **is for all values of the time** (forward or backward in time, for any selected initial condition). In addition, **all possible solutions** of (4.2.7), corresponding to any initial condition, can be represented in the phase space. In the case of no damping, $\mu = 0$ in (4.2.6), this yields all **possible circular orbits (oriented clockwise).** We cannot draw such an infinite set of circles, so what is done it to **represent a few, implying that everywhere else the orbits have the same character.** Moreover, in the case $\mu > 0$, the **orbits are infinite in length,** so now we cannot even represent **any orbit** in its entirety, but only for a finite duration of time. Once again, what is represented implies that the rest of the orbit, and all orbits, have the same character. In all cases, the **family of all solutions** of equations of the form (4.2.2) [not of (4.2.1)], when **represented as oriented orbits in phase space,** is referred to picturesquely as a **phase portrait.** This limitation in actually drawing all orbits does not restrict in any sense the precision of this concept, which can be formalized quite precisely mathematically. In this sense, the phase space representation is **holistic in character.** The mathematical way of saying this is to call the phase space representation a **"topological representation,"** which roughly means that this representation gives general **qualitative, rather than quantitative, information.** We will return to and develop this concept in what follows. It is important to recognize that this **holistic representation** of dynamics, which was developed by **Poincaré** toward the **end of the 19th century,** is an entirely **new holistic characterization** of dynamics, quite distinct from the pioneering concept introduced by **Kepler** in the **early 17th century** (recall **Kepler's three relationships,** K1, K2, and K3 in **Section 1.2**).

Poincaré's new **topological form of holistic understanding** proved to be invaluable in comprehending complicated forms of dynamics **(see, e.g., Sections 7.4, and 8.2).**

Finally, expanding upon the brief discussion in **Section 2.1,** an essential fact about the solutions of (4.2.2) is the following:

Existence and Uniqueness of Solutions: If the functions $F(x, y)$, and $G(x, y)$ can be differentiated for all x and y, and do not increase "too rapidly" for large values of $(x^2 + y^2)$, then there is a **unique solution** of (4.2.2) for any **initial condition [$x(0)$, $y(0)$],** which satisfies this initial condition, and **exists** (remains finite) **for all $t > 0$ [2].**

No Intersecting Orbits: An important consequence of this in terms of the phase orbits is that **no two orbits can intersect.** For if they did, we could take the **intersection point as an initial condition,** and **two solutions would be coming out of that point.** But then there would not be a unique solution, **in violation of the above theorem.**

Computational Algorithms

Because the equations of motion (4.2.5) are linear, it is possible to obtain analytic solutions, such as (4.2.7) and (4.2.9). However, with nonlinear equations, we need to rely on computers to uncover the features of their solutions. It is useful to look into this issue in the case where the solutions are already known, so as to be able to assess the accuracy of different computational algorithms. We will compare two such algorithms.

The Euler Approximation

Considering the simplicity of the Euler approximation, we first explore the issue of what happens if one uses the **Euler approximation to simulate the dynamics of second-order differential systems,** in contrast to the one-dimensional Verhulst equation in **Section 4.1.**

The Euler approximation of (4.2.6) is

$$x(t + dt) = x(t) + v(t)dt$$
$$v(t + dt) = v(t) - (\kappa/m)x(t)dt$$
(4.2.10)

Explore 4.2A

Program 4-2A can be used to study this approximation. We know from the study in Section 4.1 that the Euler approximation deviates from the solution of the ordinary differential equation as dt is increased. However, in this case, the Euler approximation **retains one essential feature of the Verhulst solutions,** namely that the motion **remains bounded** for all time. Now use Program

4.2 Harmonic Oscillators: Their Properties, Phase Space Representation, and Computational Algorithms

4-2A to see what occurs in the case of the Euler approximation of the harmonic oscillator, (4.2.10), if it is **undamped (μ = 0).** From this experience, it is clear that a better approximation must generally be used for simulating the **two-dimensional dynamics,** represented by two first-order differential equations.

Runge–Kutta iteration method

Now a widely used method will be introduced, which is very adequate for most dynamic systems of **any dimension.** The method is known as the **fourth-order Runge–Kutta method,** and the general program can be found in **Appendix C4.** The **"fourth-order"** refers to the fact that the method is accurate to order $(dt)^5$, so that it is very accurate for small values of dt. Usually the value **dt = 0.01** is quite adequate for most simulations.

(A Major) Explore 4.2B

The application of the Runge-Kutta method to a "damped" (?) harmonic oscillator is given in **Program 4-2B.** To simplify matters, (4.2.6) has been transformed to the form

$$dx/d\tau = y \qquad dy/d\tau = -x - My \qquad \text{(E1)}$$

where **$y = v/\omega$, $\tau = \omega t$, and $M = \mu/(m\omega)$,** so τ and M are dimensionless parameters, and x and y both have the dimensions of distance. This program shows **the dynamics of (E1) in the (x, y) phase space.** It therefore contains the **graphic subroutine** for these coordinates, and numerical pixels on the axes. The variables in the above differential equations, (E1), are represented as $(x, y) = [X(1), X(2)]$ in the program. The variables $[Y(1), Y(2)]$ are temporarily used as their corresponding **dummy variables,** which are **changed during each of the four iterations** in the fourth order Runge–Kutta program (as noted in this program). Thus, in **Program 4-2B,** the equations **(E1)** are written in the form

$$K(1, J) = Dt*Y(2),$$
$$K(2, J) = Dt \cdot [-Y(1) - M \cdot Y(2)] \qquad \text{(E2)}$$

At the end of these four iterations the values of identified $[Y(1), Y(2)]$ yield the new values of $[X(1), X(2)]$. It is not difficult to see how this is done in the program, if you are interested, but these details are not necessary to explore further. **What you do need to recognize** is the **relationship** between equations such as **(E1)** and the form **(E2),** which is reasonably obvious. This relationship is generally true for **any system of differential equations.**

■ Think 4.2F

In **Program 4-2B,** the parameter *M* can be arbitrarily **increased or *decreased*** from the keyboard. This raises the question of whether this is a **physical model** or an **exploratory model (review Section 2.3).** Thus, **Program 4-2B** is a rich program that **requires some careful study** if one wants to modify or write future programs to **develop your own ideas [3].** A general **text** of the **Runge–Kutta program** can be found in **Appendix C4,** and on the floppy disk, in **Program C4,** there is illustration of an **unexpected form of nonlinear dynamics,** in contrast to the linear dynamics in **Fig. 4.6b.**

Concepts

- harmonic oscillators have **amplitude-independent** frequencies
- phase space representation
- orbits are **oriented** curves, each representing **one** solution for **all** time
- **phase portraits** indicate the character of **all** orbits
- a **holistic** (topological) characterization of a system
- fixed points
- **attractor**
- **existence** and **uniqueness** of solutions
- **no intersections** of orbits in phase space
- fourth-order Runge–Kutta iteration method

References and Notes

1. See, e.g., the discussion in R. J. Seeger, *Galileo Galilei, His Life and Works*, Pergamon, Oxford, 1966, p. 170.
2. For a simple introduction to these ideas, and more detailed references, see E. A. Jackson, *Perspectives of Nonlinear Dynamics,* Vol. 1, Cambridge University Press, Cambridge, 1991, pp. 30, 34.
3. For references to more complete information on Basic programming, consult the reference in Appendix C2.

4.3 PROPERTIES OF NONLINEAR OSCILLATORS

In the last section, we considered the **harmonic oscillator** dynamics

$$\frac{dx}{dt} = v$$
$$\frac{m \cdot dv}{dt} = -\mu v - \kappa x \quad (\kappa > 0)$$
(4.2.5)

which is a **linear-dynamics** example of Newton's more general equations of motion (4.2.3)

$$\frac{dx}{dt} = v$$
$$\frac{m \cdot dv}{dt} = F(x, v)$$
(4.2.3)

The dynamics of (4.2.5) is only **periodic** if it is undamped ($\mu = 0$) **(recall Fig. 4.6)**. Moreover, in this case the **orbits** in the **phase space** are both **closed curves** (hence **periodic dynamics**) and **symmetric** about the **fixed (equilibrium) point** at $x = 0$ (and necessarily, $v = 0$) (Fig. 4.6). This symmetry is due to the fact that the force $F(x) = -\kappa x$, an **odd power of x**. Thus, if x is positive or negative, the restoring forces (toward $x = 0$) are equal in magnitude, but opposite in direction (− or +, respectively).

Now we will consider systems in which the **restoring force is a nonlinear function of x**, and is also **an odd function of x**, namely

$$F(x) = -\kappa x - \alpha x^3$$
(4.3.1)

where α now indicates the strength of the nonlinear restoring force. Note that x^3 has the same sign as x, regardless if x is positive or negative. When the **frictional damping force**, $-\mu v$, is included, the equations of motion, **(4.2.3)**, become

$$\frac{dx}{dt} = v$$
$$\frac{m \cdot dv}{dt} = -\mu v - \kappa x - \alpha x^3 \quad (\kappa > 0, \mu \geq 0)$$
(4.3.2)

This type of nonlinear dynamics is referred to as a (damped, if $\mu > 0$) **Duffing oscillator**. This oscillator can describe **two types of nonlinearity**, depending on the **sign of the constant α**:

1. **$\alpha > 0$.** In this case, as can be seen from (4.3.1), as x varies, the nonlinear force always acts in the same direction as the linear force. In other words, in this case, the magnitude of **the total force is larger than the harmonic force.** Therefore, this type of nonlinearity is referred to as a **hard nonlinearity.**

2. **$\alpha < 0$.** In this case, the nonlinear force always acts in the opposite direction to the linear force, so **the total force is less than the harmonic force;** this type of nonlinearity is referred to as a **soft nonlinearity.**

These two types of forces are illustrated in Fig. 4.7.

Think 4.3A

Recall that the fixed (equilibrium) points in the phase space are those points where ***dx/dt* = 0 and *dv/dt* = 0.** Referring to (4.3.1) and (4.3.2), we see that this requires that ***v* = 0 and *F(x)* = 0.** Show that in the case of the hard nonlinearity, **$\alpha > 0$,** there is only one fixed point, whereas there are **three fixed points** if **$\alpha < 0$.** In both cases, determine their locations in terms of κ and α. Only nonlinear equations can have more than one fixed point in the phase space [1].

Explore 4.3A

Program 4-3A (where **$\kappa = 1$, and $\alpha = A$**) illustrates in the phase space the influence of a hard nonlinearity, **$A > 0$,** on the periodic motion. See how the larger nonlinearity influences the orbits [which all have the initial conditions $x(0) = 0$, $v(0) = 2$]. **Some** of

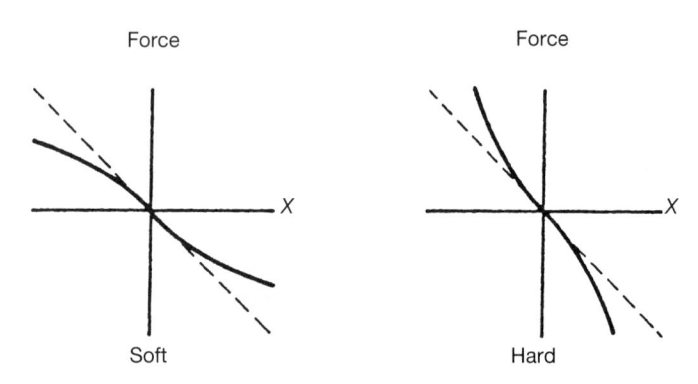

Fig. 4.7.
An illustration of the deviation of a **hard and soft nonlinear force** from a harmonic force (the dashed line).

these orbits are illustrated in **Fig. 4.8.** Others are illustrated in Fig. 4.9.

Explore 4.3B
Modify Program 4-3A to extend this study to the case of soft nonlinearities $A < 0$. To do this **insert a prime, '**, at the beginning of the lines labeled #####, which removes these actions. However, from **Think 4.3A,** we know that there are now **three fixed points,** so the "global" dynamics in the phase space is now very different. This change will become very evident as the value of A is decreased.

In the cases of no damping, $\mu = 0$, it is possible to understand the general dynamics of the system by introducing the concept of the **total energy** of the system, E. E is the sum of the **kinetic energy,** $(m/2)v^2$, and the **potential energy,** $V(x) = (\kappa/2)x^2 + (\alpha/4)x^4$, which is defined in terms of the force **(4.3.1)** by the relationship $dV(x)/dx = -F(x)$. Therefore

$$E(x, v) = \left(\frac{m}{2}\right)v^2 + \left(\frac{\kappa}{2}\right)x^2 + \left(\frac{\alpha}{4}\right)x^4$$

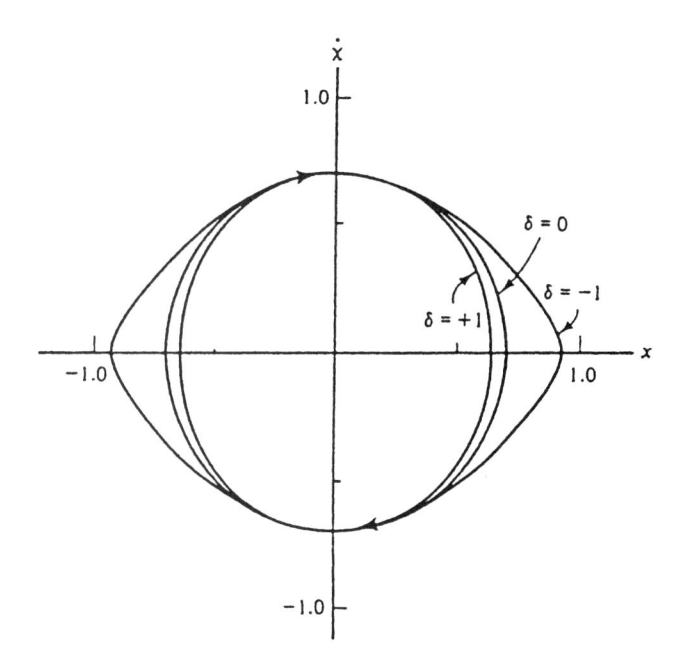

Fig. 4.8.
Orbits of the **hard and soft nonlinear oscillator** near the origin. Identify them from Explore 4.3A and 4.3B (where $\delta \equiv \alpha$).

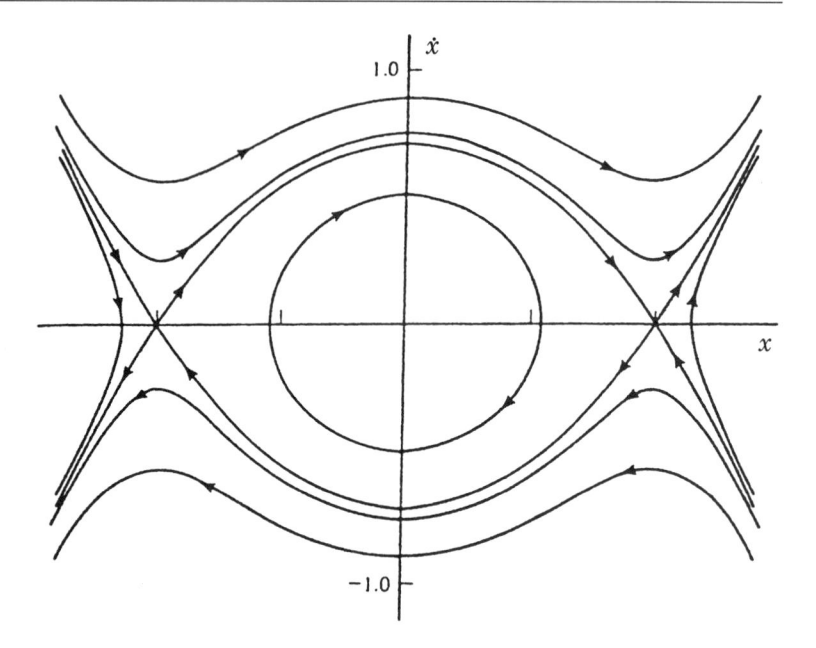

Dynamic Models Based on Differential Equations

Fig. 4.9.
The phase portrait of the soft nonlinear "oscillator," whose curves are determined in **Program 4-3B.**

To simplify matters, we choose the unit of mass $m = 1$, so that

$$E(x, v) = \tfrac{1}{2}(v^2 + \kappa x^2 + \tfrac{1}{2}\alpha x^4) \qquad (4.3.3)$$

The importance of this concept is that **it does not change with time,**

$$\frac{dE}{dt} = \frac{v \cdot dv}{dt} + (\kappa x + \alpha x^3)\frac{dx}{dt} = 0$$

Think 4.3B

Show that $dE/dt = 0$ by using the equations (4.3.2), when $m = 1$ and $\mu = 0$. A function of (x, v) that does not change with time, such as $E(x, v)$, is called a **constant of the motion.** Such a **constant of the motion is very important,** because of its **generality.** It remains constant for **any solution of (4.3.2)** (when $\mu = 0$), even if the solution is not known, which is generally the case. Put another (geometric) way, the function $E(x, v)$ **is constant along any orbit in the phase space,** with **each orbit** being associated with **some value of $E(x, v)$** (not necessarily unique).

Explore 4.3C

To see how $E(x, v)$ can be used, **study Program 4-3B,** where for simplicity we take $\alpha = -2$ and $\kappa = 2$, and set $2E = K$. Then

$$(v^2 + 2x^2 - x^4) = K \qquad (4.3.4)$$

gives an algebraic expression for **all orbits everywhere in the phase space. Program 4-3B** illustrates a number of **curves** in the phase space. These curves are obtained from the **constant of the motion (4.3.4)**, using differing values of K. This constant of the motion **does not indicate the direction of motion** along these curves. So first of all, make a sketch of some of these curves and add an arrow indicating the **orientation** of the dynamics along these curves, using the **first relationship** in (4.3.2) (so $dx = v \cdot dt$ indicates the direction when $dt > 0$). When you figure this out you will be doing better than many professors who carelessly mess this up! The resulting family of orbits (oriented curves) is referred to as a **phase portrait** of this system of equations.

Note that **some orbits** (green in the program) tend toward or away from one or more of the **fixed points. These fixed points** are called **limit points** of these orbits as $t \to +\infty$ or $t \to -\infty$, respectively (∞ means "infinity," a number larger than any finite number). Using the fixed point information, determine the value of K, (4.3.6), on each of these orbits. Now if you go into the program and remove the prime at the beginning of the line #####, the program will print the value of K for each orbit.

Topological (Holistic) Concepts

(I) Topologically Equivalent Orbits

Now we will discuss some very important concepts that are illustrated in the phase portrait, using **Program 4-3B,** which is now displayed (without the colors) in **Fig. 4.9. First** of all, note that the two (green) orbits each approach the **same** two limit points (fixed points) as $t \to +\infty$ **and** as $t \to -\infty$ (but in reverse temporal order). They **enclose a region** in which all the solutions are **periodic. Outside** of this region, **none** of the solutions **are periodic.** Therefore, **this pair of orbits** separates two regions of phase space with **"different types" of dynamics.** For this reason, these two orbits, which are the **boundary** between these different type of dynamics, are referred to as a **separatrix.**

Notice that this is entirely a **qualitative (holistic) feature of the phase space orbits** (numbers do not matter, but rather the collection of all orbits in different regions of the phase space). It is an example of a **topological (holistic) understanding of dynamics,** which was **introduced by Poincaré.**

To make this more precise, we need to define what it means for two orbits to be **"topologically different."** There is a nice easy way to picture this idea. It begins by drawing one orbit on a **rubber sheet.** Now, if you can **stretch this rubber sheet** (no tearing) so that is can be **laid on top of the another orbit (with the same orientation),** then the two orbits are said to be **"topologically equivalent."** If you **cannot** do this, they are not topologically equivalent. (This is an example of why topology is sometimes referred to as **rubber-sheet geometry.**) In the above case, there is, of course, no way to stretch an infinite (nonperiodic) orbit to match up with a periodic orbit, so these are indeed topologically different types of orbits. Notice also that two topologically equivalent orbits can represent dynamics with quite **different quantitative features** (e.g., "speeds," periods, etc.). It is the overall qualitative feature of the orbits that are being compared by this concept.

(II) Bifurcation Points

In Section 3.3, there is a discussion of bifurcation points of the **logistic map.** These are the value of $c = c^*$ such that the dynamics is **qualitatively different if $c > c^*$ or $c < c^*$.** Here the idea is much the same, except we now are dealing with a phase space representation of continuous orbits. The **rubber-sheet geometry** idea can now be extended to the **entire phase space,** rather than simply comparing two orbits, as in (I) above. The idea now is to distinguish **two dynamic systems** as being "topologically" equivalent or not. To do this we look at the **entire phase portrait** of these two systems. **Two phase portraits** are then said to be **topologically equivalent** if they can be stretched in any way such that **all their orbits coincide (including orientations).** Now the concept of a **bifurcation point** is the value of a parameter, such as $\alpha = \alpha^*$ in (4.3.2), for which the **phase portraits are not topologically equivalent** when $\alpha > \alpha^*$ and $\alpha < \alpha^*$. In other words, as in the case of the map, the dynamics on either side of a bifurcation point is qualitatively different.

■ Think 4.3C

Can two phase portraits be topologically equivalent if they have a different number of fixed points? Using the information in Think 4.3A and Figs. 4.8 and 4.9, determine a bifurcation value α^* for (4.3.2).

Nonlinear versus Linear Oscillators

Finally, we need to consider an important way in which the nonlinear oscillator differs from that of a harmonic oscillator in its temporal behavior. It will be recalled that in the case of the **harmonic oscillator,** the **frequency** of the oscillation is **not influenced by the amplitude** of the oscillation. In the case of **nonlinear oscillations,** this is no longer true; the **period (and frequency)** of the oscillations now **depends on the amplitude A.**

Although it takes some mathematical manipulations to estimate what this dependence actually is, it is not difficult to **physically understand** why this should be the case. Since $\alpha > 0$, **the nonlinear restoring force,** (4.3.1), **is larger** than the linear restoring force, **for any value of x.** Hence, it makes sense that it will move **back and forth faster** than the linear oscillator (i.e., have a **shorter period** than the harmonic oscillator, for the same amplitude). Moreover, this **period will decrease more for larger values of the amplitude** (this follows from the fact that the nonlinear effect is greater, and that even though the distance moved is greater, we know from the harmonic oscillator that this doesn't affect the period in that case). A **simple analysis** is given in the mathematical Appendix **M4.** This predicts that, for **small values of α,** the **angular frequency** is given approximately by

$$\omega = \sqrt{\kappa/m} + \frac{3\alpha A^2}{8\sqrt{\kappa m}} \qquad (4.3.5)$$

where A is the amplitude of the oscillation. Then, since the **frequency** is $f = \omega/2\pi$, **the period of the oscillation** is approximately

$$P = \frac{2\pi}{\omega} \qquad (4.3.6)$$

Explore 4.3D

Use **Program 4-3C** to see how the **computed period** compares with the **theoretical period,** (4.3.6) and 4.3.5), as a function of α **and the amplitude.** Begin with a **small value of α and increase the amplitude** to see when the **theoretical period deviates** significantly from the **computed period.** Repeat this with **larger values of α.**

▪ **Think 4.3D**

Notice that **the computed period in Program 4-3C does not remain constant.** Why? See how it is determined. Is this an error, or a possible characterization of **empirical reality?** What "reality"? ▪

For a more general exploration of the dynamics of the Duffing equation, including damping, and problems associated with unbounded solutions, you can use the following Explore as **a starting point upon which to build.**

Explore 4.3E

The general character of the Duffing dynamics can be explored in **Program 4-3D.** The nonlinear parameter α is represented by A in the program. Both α and the coefficient of friction, μ, **can be increased or decreased by keying <I>/<D> or <H>/<L> (for "higher/lower"), respectively.** See what happens if you try to make these negative. These start out at $A = 0$ and $\mu = 0$. **First increase A** and see the nonlinear modification of the circular linear orbit. Do you understand why this happens? Now **increase μ** and see how the orbit changes as the magnitude of the **amplitude** of the oscillation decreases. Note that you can clear the screen at any time, or insert a new initial condition, which is selected randomly. Modify the program for a specific initial condition.

Concepts

- hard and soft nonlinear oscillators
- the energy constant of the motions ($\mu = 0$)
- "rubber-sheet" geometry
- topological equivalent orbits
- separatrix orbits
- topologically equivalent phase portraits
- bifurcation points
- amplitude-dependent frequencies

Note

1. In connection with **Think 4.3A,** consider a system of linear ODE. **If there are two distinct fixed points,** say x_1 and x_2, then $(ax_1 + bx_2)$ would also be a fixed point, for **any values of a and b.** What would this

imply about other fixed points? (For ODEs with more variables, all these quantities are vectors, with vector products, and the consequence would be the same.)

4.4 DYNAMICS IN A "DOUBLE-WELL" POTENTIAL

The equations of motion (4.3.2),

$$\frac{dx}{dt} = v$$
$$\frac{m \cdot dv}{dt} = -\mu v - \kappa x - \alpha x^3 \qquad (\kappa > 0, \mu \geq 0)$$
(4.3.2)

were studied in the last section for the cases $\alpha > 0$ and $\alpha < 0$ (respectively, soft and hard nonlinearities), but always when **$\kappa > 0$**. This last condition means that the **force near the origin** is always **directed toward the origin,** which makes **this motion periodic** about the origin, regardless of the sign of α. This was illustrated in **Fig. 4.8.**

In this section we consider a different system, in which the **force is away from the origin for small x** (so **$\kappa < 0$**), but toward the origin for sufficiently large values of x (so **$\alpha > 0$**). Therefore, the present equations of motion are

$$\frac{dx}{dt} = v$$
$$\frac{m \cdot dv}{dt} = -\mu v - \kappa x - \alpha x^3 \qquad (\kappa < 0, \alpha > 0, \mu \geq 0)$$
(4.4.1)

In this case, the force, $F = -\kappa x - \alpha x^3$, vanishes at three points, so there are **three fixed points**

$$x = 0, \qquad x = +\sqrt{(-\kappa/\alpha)}, \qquad x = -\sqrt{(-\kappa/\alpha)} \qquad (4.4.2)$$

These are at the **same locations** as found in the case of the soft nonlinear oscillator in **Think 4.3A.** However, the dynamics will be seen to be entirely different.

To understand the physical content of this dynamics, consider the **potential energy, $V(x)$,** that is associated with the force $F = -\kappa x - \alpha x^3$. As was discussed in connection with the total energy, (4.4.3), the potential energy is related to F by the definition $dV(x)/dx = -F$, which yields for the potential energy [1]

$$V(x) = \tfrac{1}{2}\kappa x^2 + \tfrac{1}{4}\alpha x^4 \qquad (\kappa < 0, \alpha > 0) \qquad (4.4.3)$$

Dynamic Models Based on Differential Equations

If $V(x)$ is plotted as a function of x, it looks something like **Fig. 4.10,** depending on the magnitudes of $\kappa < 0$ and $\alpha > 0$.

The dynamics of this system is similar to a mass that slides on such a surface, under the influence of a downward gravitational force. **If there is no friction,** one would expect it to be able to oscillate in either of these wells, provided it did not have too much speed (kinetic energy). However, if it has **enough energy** it will slide over the center hill, and then **oscillate between the outer "walls."**

Explore 4.4A

Run **Program 4-4A,** which illustrates all of the **topologically distinct types of orbits** in this system [2]. You should identify some of the dynamics **near the fixed points** with those shown in **Fig. 4.11.** You will also see **strange orbits** that tend to the **same fixed point** as $t \to +\infty$ and $t \to -\infty$, not shown in this figure [3]. A **complete "zoology"** of all the possible linear dynamics near fixed points can be found in **Appendix M5,** together with the associated mathematical stability analysis, establishing when each type occurs.

Think 4.4A

Recall from the last section that if $\mu = 0$ the total energy

$$E(x, v) = \frac{mv^2}{2} + \kappa x^2/2 + \tfrac{1}{4}\alpha x^4$$

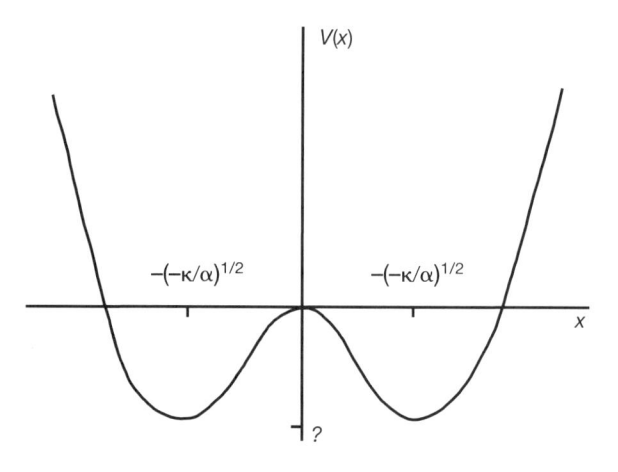

Fig. 4.10.
The shape of $V(x)$ inspired the name **"double-well potential."**

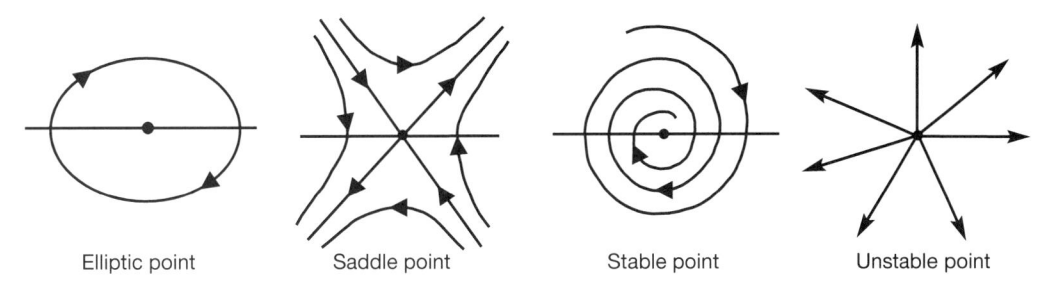

4.4 Dynamics in a "Double-Well" Potential

Elliptic point   **Saddle point**   **Stable point**   **Unstable point**

Fig. 4.11.
Some types of linear dynamics near **different types of fixed points (with their names)** in a two-dimensional phase space.

does not change with time for any solution of (4.4.1) (it is a "constant of the motion"). What is the **total energy** of the **yellow orbits** in Program 4-4A? What is the **range** of energies for the **red orbits?** Finally, what is the range of energies for the remaining periodic orbits? **Add this information to the phase portrait in Fig. 4.12.**

The phase portrait in Fig. 4.12 illustrates examples of both the **elliptic point** and **saddle point** types of fixed points. Now we will see how these are modified when **damping is included.**

Explore 4.4B

Again use the **program 4-4A,** but now **modify it** on the line indicated by ######, so that $\mu = 0.1$ (i.e., remove the apostrophe before MU = 0.1). Notice that now there are **two fixed point attractors,** which are **stable nodes (Fig. 4.11),** and still **one sad-**

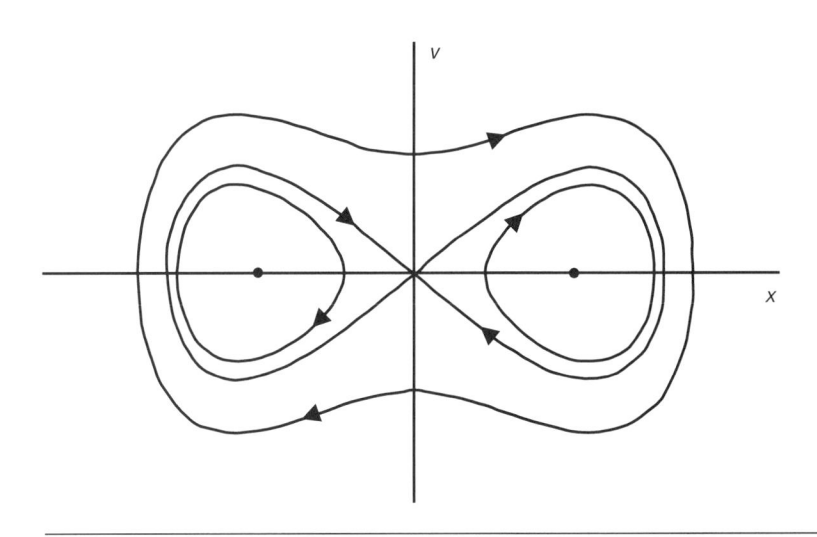

Fig. 4.12.
The phase portrait of the "double-well" system, when $\mu = 0$.

dle point. The stable nodes are "attractor" fixed points. Note that one of the previous "external orbits" goes to one fixed point, and the other to the second fixed point. The **set of all points** that tend to one of the attractors is called its **basin of attraction** (somewhat similar to a river basin), so this system has **two basins of attraction. Between any two such basins,** there must be a **boundary orbit** that belongs to neither basin.

Think 4.4B

From **Explore 4.4B,** make a sketch of what you think the **phase portrait** of this system looks like, paying particular attention to where all of the orbits that contact the saddle point go in this phase portrait. **Answer the question** in the **modified** Program 4-4A. Attempt to **sketch** the portion of **these two basins of attraction** that are in the region of the origin.

Explore 4.4C

Use **Program 4-4B** to study this damped dynamics a little **more systematically.** This program allows you to see what happens to **initial states** on the axis $x = 0$, for various initial values $v(0) > 0$. As these values of $v(0)$ are increased, you will see that the orbits change from one attractor to the other, then back to the first, etc. Perhaps you will want to revise your sketch in **Think 4.4B.** You can **reduce the step size of** $v(0)$ to discover where the **two incoming orbits** of the **saddle point cross the axis** $x = 0$. What do these orbits have to do with the **basins of attraction?**

A "Backward" Device!

One way to get **a more complete phase portrait** of such **basins of attraction** is to use a device that **does not** follow the orbits forward in time. Instead, if we take $dt < 0$, the dynamics will **run backwards,** and if we take the **initial conditions near** the fixed points, $x = \pm\sqrt{(-\kappa/\alpha)}$, they will move backwards within their respective basins of attraction. So the idea is to take initial conditions on a circle about each fixed points $x = \pm\sqrt{(-\kappa/\alpha)}$, **and keep track of where each solution goes** (within a limited region, of course, since they tend to infinity). If we **give them different colors,** then the two basins of attraction can be clearly seen. **This is a very general device for exploring basins of attraction.**

In the present case, the equations (4.4.6) have one other prop-

4.4 Dynamics in a "Double-Well" Potential

erty that can be used to speed up this process. The equations (4.4.6) have a **symmetry property;** namely, **the equations are not changed** if one makes the transformation $(x, v) \Rightarrow (-x, -v)$. This means that if we start one solution at $[x(0), v(0)]$ and another at $[x^*(0) = -x(0), v^*(0) = -v(0)]$, then **for all values of t** they will satisfy $[x(t), v(t)] = [-x^*(t), -v^*(t)]$. This is even true if we run the dynamics backward in time. What this means is that we only need to compute one of these solutions, and simply use the last identity to display the other. These ideas are now all collected in the next Explore.

Explore 4.4D

Use **Program 4-4C** to see how the two basins of attraction can be readily generated and displayed in this case. Here note that $dt < 0$; the integration of each initial condition is terminated by **limiting the magnitude of $x(t)$,** then changing the initial conditions to go **around a small circle** about an elliptic fixed point. Finally, there at **two graphic commands,** which differ only by the above identity.

A rough characterization of these interwoven basins of attraction is illustrated in **Fig. 4.13 [4]**.

Think 4.4C
Referring back to the discussion of bifurcation points in the last section, determine **all of the bifurcation points** of the parameters κ, α, and μ in (4.4.1).

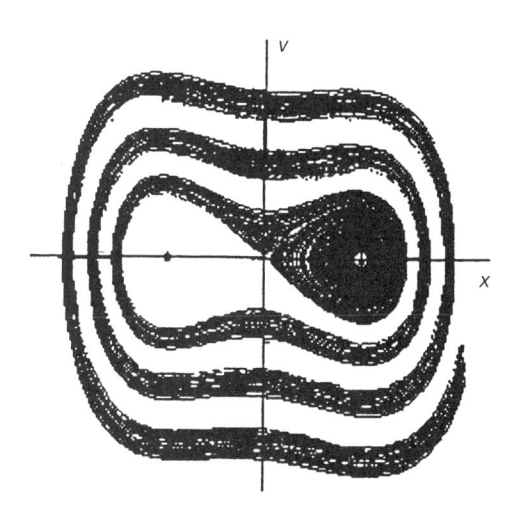

Fig. 4.13.
A phase portrait of the two interwoven basins of attraction in a damped "double-well" system ($\mu > 0$).

Notes

1. Note that $V(x)$ is only defined up to an additive constant, since that constant does not change F. Here the constant has been taken to be zero.
2. This program contains several new programming devices. One involves setting up a **sequence of examples,** by directing the program to a selected sequence of initial conditions. This involves setting up a "count," $Z = Z + 1$, then following this, inserting a line: **ON Z GOTO A, B, C.** The result of this is that if $Z = 1$ the program goes to A, if $Z = 2$, it goes to B, and so on. The second device is to use a **SLEEP #,** which **delays the program for a time** dependent on the number #. An alternative method is to insert a loop: **FOR I = 1 TO #: NEXT I.** Because computers are fast, the number # must be large. Both of these are best understood by experimenting.
3. Some **mathematical** jargon: such an orbit is called a **"homoclinic" orbit.** If the orbit goes to **two different fixed points** when $t \to +\infty$ and $t \to -\infty$ the orbit is called a **"heteroclinic" orbit** (there are two of these in Fig. 4.9), which may impress somebody. It is more impressive if you recognize that a computer cannot carry out this limiting process, since it only runs a finite time, with a finite accuracy.
4. To produce this figure, 50 backward orbits were run with different colors than shown in Explore 4.4D. The orbits that ultimately ended up as white obviously do not show up in this figure, accounting for some of its roughness.

4.5 A PENDULUM THAT CAN *REALLY* SWING (SOMETIMES)!

In this section, we will consider the dynamics of a pendulum, but not the one found in a **grandfather clock.** This pendulum consists of a rod of **length L and negligible mass,** with **a large mass, M,** at one end. The rod is pivoted at the other end in such a way that the rod **can freely rotate all the way around the pivot point** (**this pendulum** can really swing! Later see Explore 4.5C). This pendulum is illustrated in **Fig. 4.14.**

There are **two variables** that describe the state of such a pendulum. One is the **angle θ** between the rod and the downward vertical axis through the pivot. The other variable is the rate at which this angle is changing in time, the so-called **angular velocity, $d\theta/dt = \Omega$.**

Only the component of the gravitational force that is **perpendicular to the rod, $mg \sin(\theta)$,** causes the angular velocity, Ω, to change (the gravitational force along the rod is countered by an op-

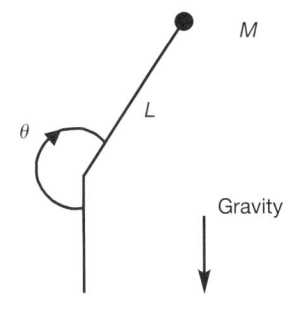

4.5 A Pendulum that Can Really Swing (Sometimes)!

Fig. 4.14.
A pendulum that can swing in the vertical plane, all around a pivot point, P, with a gravitational force, mg, acting downward.

posite force generated by the "rigid rod"). Therefore, if there is **no frictional damping,** the equations of motion are of the form

$$\frac{d\theta}{dt} = \Omega$$
$$\frac{d\Omega}{dt} = -A\sin(\theta)$$
(4.5.1)

The constant A turns out to be $A = (g/L)$, which is **independent of the mass M** [1].

If the initial displacement of the pendulum is small, **θ is small.** Then since (see the definition in Equation **M3.7** in Appendix **M3**)

$$\sin(\theta) \cong \theta - \tfrac{1}{6}\theta^3 + \text{higher powers } \theta$$

the terms θ^3 and beyond can be neglected. In this case (4.5.1) becomes an equation for **harmonic oscillator dynamics,** for which the **period is independent of the amplitude** of θ [**recall (4.2.8) and (4.3.5)**]. This feature of a pendulum may have been discovered by Galileo [2]. In any case, this is a very important feature, because as the pendulum looses energy and decreases its amplitude of oscillation, its period changes very little, which obviously makes it very useful in keeping time (for a grandfather clock or Big Ben).

Think 4.5A

What is the period of this clock? When the temperature increases, the length, L, of a metal rod in a pendulum also increases. How does this influence the period? Effects like this had to be compensated for in the construction of Big Ben in London [3].

It is not difficult to show that the dynamics (4.5.1) has the **constant of the motion** (the energy)

Dynamic Models Based on Differential Equations

$$E = \frac{\Omega^2}{2} - A\cos(\theta) \qquad (4.5.2)$$

Think 4.5B
Using the relationship $d\cos(\theta)/dt = -\sin(\theta)d\theta/dt$ [see (M.3.8) in Appendix M3], show that $dE/dt = 0$, by using (4.5.1).

Now we will see how this **constant of the motion** gives a **(holistic)** form of **understanding** about **all solutions** of (4.5.1). Consider some particular dynamic case with a **given value of E**. At the bottom of the swing, say where $\theta = 0$, the potential energy, $-A\cos(\theta)$, is at its lowest value, $-A$. In order for E to be constant, Ω^2 must be at its maximum value (Ω being either positive or negative, depending on the direction of its swinging). As θ increases in magnitude, Ω^2 must decrease, and if $\Omega^2 = 0$ for some value of θ, the rotation will stop and then reverse its direction. In other words, periodic motion will occur. But if $\Omega^2 = 0$, we see that $E = -A\cos(\theta)$ at that point, and this **can always be satisfied for some θ if $E < A$**. However, if $E > A$, then $\Omega^2 = 2[E - A\cos(\theta)] \neq 0$ for any θ. Thus, we come to an important **holistic understanding** of the dynamics of (4.5.1):

> **The dynamics of (4.5.1) is periodic if $E < A$,**
> **and is not periodic if $E > A$.** (4.5.3)

Think 4.5C
Show that $E = A$ corresponds to fixed points. What physical situations do these represent?

Now we generalize the dynamics (4.5.1) to include the **influence of friction.** Just as in the case of the damped oscillator, the influence of friction will be taken proportional to the "velocity," which here is the angular velocity Ω. In this case, the equations of motion will be taken to be

$$\frac{d\theta}{dt} = \Omega$$
$$\frac{d\Omega}{dt} = -\mu\Omega - A\sin(\theta) \qquad (4.5.3)$$

where μ is the friction (due to the pivot or air drag).

In the present case, where the pendulum **can swing around the pivot,** we generally need to keep track of the angle θ for all its possible values. Thus, the **phase space** for the system, involving the variables θ and Ω, ranges over the values $-\infty < \theta < \infty$ and $-\infty < \Omega < \infty$. In this case, (4.5.3) has an infinite number of **fixed points** in this phase space, at the points

$$\theta = n\pi \quad (n = 0, \pm 1, \pm 2, \ldots) \quad \text{and } \Omega = 0 \quad (4.5.4)$$

All these fixed points either correspond to the pendulum hanging downward at rest, or precariously balanced in the upright position at rest.

Think 4.5D

Distinguish which of the fixed points (4.5.4) belong to which of these physical configurations. Are these fixed points different for (4.5.1)? Which of these fixed points are elliptic points (if $\mu = 0$), and a stable node (if $\mu > 0$)? (recall **Fig. 4.11**). Can you figure out what is the type of the other fixed points? You will have help soon if this is obscure.

Following the same analysis as in Think 4.5B, it is not difficult to show that when the equations of motion are (4.5.4), the "energy," $E = \Omega^2/2 - A\cos(\theta)$, decreases at the rate

$$\frac{dE}{dt} = -\mu\Omega^2$$

and, except for singular cases, not only does Ω go to zero, but also $\cos(\theta) \to 1$; in other words, this is the minimum of the potential energy (the "bottom of the potential well"). Thus, while the phase space has an infinite range of θ, any solution always tends to rest in a finite range of θ, but this can be quite variable depending on the initial energy. To deal with this **in computer graphics,** the following type of **"wrap-around" phase space** is employed.

Explore 4.5A

Program 4-5A shows examples of the **dynamics,** (4.5.3), of the **physical pendulum,** together with the **dynamics** in the "**6π wrap-around**" phase space. All **initial states** start at $\theta = 0$, but at **random values of Ω,** with values of μ to be selected. After seeing this dynamics, modify the initial conditions so as to be able to answer the last question in **Think 4.5D**. Also, the number of "wrap-arounds" can be modified at the line ######.

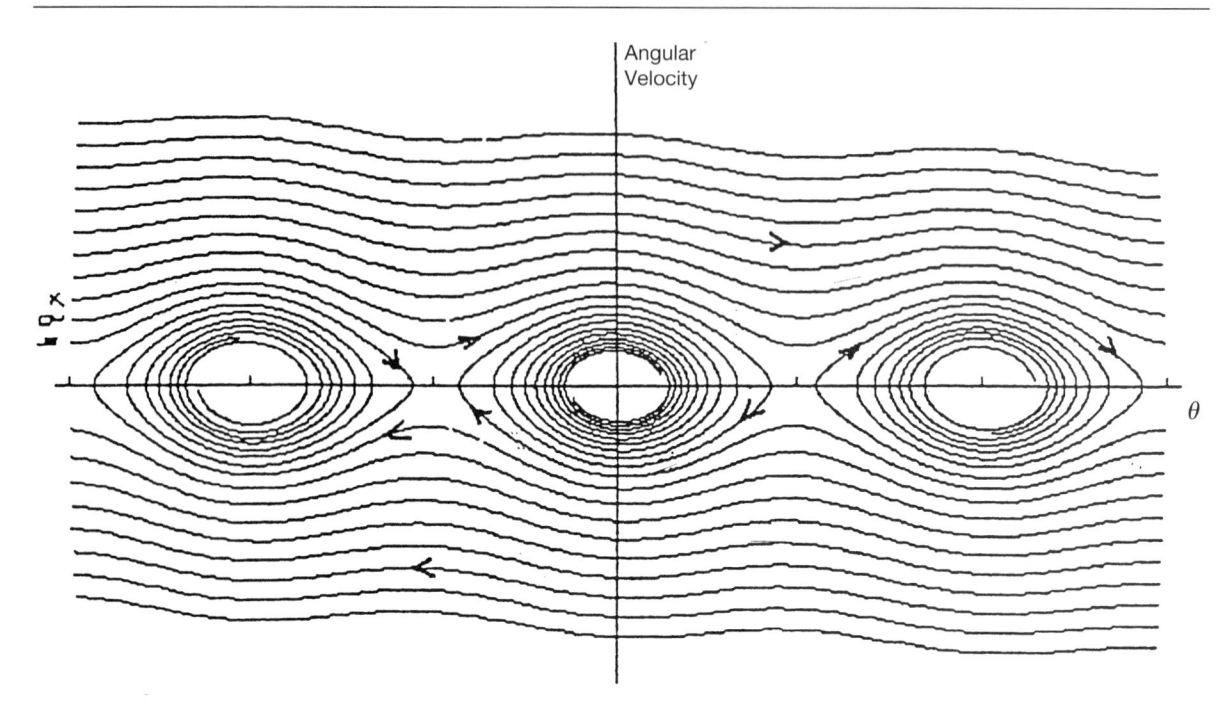

Dynamic Models Based on Differential Equations

Fig. 4.15.
A phase portrait in the "6π wrap-around" phase space. The points at $(\theta = 3\pi, \Omega)$ and $(\theta = -3\pi, \Omega)$ are treated as being identical, for all values of Ω. **See Explore 4.5B**

Explore 4.5B
Program 4-5B shows a **sequence of dynamic solutions,** which together give a fairly good representation of the **phase portrait** in this wrap-around phase space. **Missing** from this phase portrait is a particular set of orbits, again related to **Think 4.5D** and **Explore 4.5A.**

▨ **Think 4.5E**
In Fig. 4.15 you will find several orbits marked on the left margin. Identify the **basins of attraction** that they belong to **in this phase space.** More challenging, what would be the final **change in θ,** in this phase space for the orbit at the top (with an arrow), before coming to a rest (starting at the left edge)? Note that in the

usual (θ, Ω) **phase space,** where both axes extend to infinity, there are an **infinite number** of basins of attraction.

Explore 4.5C

If a coil is attached at the pivot point of a pendulum, producing an **additional force of** $-\kappa\theta$ in (4.5.1), then the **number of basins** of attraction becomes **finite** in the usual (θ, Ω) phase space. To explore this effect, **activate this term in Program 4-5B.** You may also want to use a **"$2n\pi$ wrap-around" phase space,** where n is larger. Before doing that, determine for what **value of κ there** is only **one fixed point,** then decrease the value of κ slowly. This will determine the value of n that is needed (and how you will need to decrease the x-magnification, *MX*).

References and Notes

1. The component of the force, $mg \sin(\theta)$, is called the **torque,** and the usual mass is replaced by the "moment of inertia," mL, which indicates how difficult it is to rotate the system. Now the usual momentum, mv, is replaced by the new "mass" and the angular velocity, to yield the **"angular momentum,"** $mL \cdot \Omega$. Newton's equation now becomes **"the rate of change of the angular momentum equals the torque,"** or $mL \cdot d\Omega/dt = -mg \sin(\theta)$. Canceling out the mass yields **(4.5.1).** Note that $\Omega > 0$ is clockwise motion, and θ is defined in Fig. 4.14. From this, justify the last minus sign.
2. See the discussion, beginning on p. 170, in R. J. Seeger, *Galileo Gallilei: His Life and Works*, Pergamon, Oxford, 1966.
3. J. Darwin, *The Triumphs of Big Ben*, Hale, London, 1986.

5
Interacting Systems Producing (Near) Periodicity

This chapter will contain a number of examples of **interactions between different types of systems,** which gives rise to some form of periodic, or near-periodic dynamics. This type of response to interactions may be due to competitive feedback effects, or simply to a system's inherent "give-and-take" relationship with its environment, or perhaps an environmental influence that dictates a selective response from a system. We will look at examples of all of these in what follows [1].

5.1 PREDATOR-PREY SYSTEMS: THE LOTKA–VOLTERRA MODEL

As has been noted before, the mathematical modeling of the growth in human populations dates from the first half of the nineteenth century, with the work of the Belgian mathematician P.-F. Verhulst. Here the concept involved the **self-limitation of populations** due to the feedback effect of **finite resources.** Around 1920, the interest in population dynamics again grew, and extended into the study of the **limitations of populations** caused by the **interactions between predators and their prey.** This type of interaction between species is now frequently referred to as **predator–prey dynamics.** The names of Thompson, Lotka, Volterra, Nicholson, and Bailey were often associated with these studies in the 1920s and 1930s, each dealing with dynamic models that incorporated varying degrees of realistic treatments of biological systems. For an informative history of the development of these ideas see reference [1].

The simplest equations that attempted to describe this interaction between two species are now known as the **Lotka–Volterra equations.** Lotka expressed the relationships between the predator and prey in the following fashion [1,2]:

[The change in the number of **prey, N_1,** per unit time] = [The natural increase of prey per unit time] (minus) [Destruction of prey by predators per unit time]

[The change in the number of **predators, N_2,** per unit time] = [The increase in predators per unit time, due to ingestion of prey] (minus) [Death of predators per unit time]

If we reorder the second equation so that the linear term appears first on the right side, it yields the following mathematical form of the **Lotka–Volterra equations:**

$$\frac{dN_1}{dt} = R_1 N_1 - K_1 N_1 N_2 \qquad (5.1.1)$$

$$\frac{dN_2}{dt} = -R_2 N_2 + K_2 N_1 N_2 \qquad (5.1.2)$$

The positive constants (R_1, R_2) represent the rates of growth and death for the prey and predators, respectively, which accounts for the negative sign in front of R_2 (i.e., the predators would die off unless there are prey present, $N_1 \neq 0$). Similarly, the positive constants (K_1, K_2) represent the nonlinear interaction strength between the two species, which accounts for their destruction and growth, respectively (hence the negative sign in front of K_1).

If the two species do not interact, so $K_1 = K_2 = 0$, then since $K_1 = 0$, the number of prey would simply grow exponentially, because this model takes **no account of the limited resources (e.g., vegetation) for the prey.** On the other hand, since $K_2 = 0$, the number of predators would decay exponentially, yielding the two dynamics

5.1 Predator–Prey Systems: The Lotka–Volterra Model

$$N_1 = N_1^0 \exp(R_1 t); \qquad N_2 = N_2^0 \exp(-R_2 t) \qquad (5.1.3)$$

as illustrated in **Fig. 5.1.**

To explore the nonlinear dynamics of $N_1(t)$ and $N_2(t)$ as a function of time, when **$K_1 \neq 0$, $K_2 \neq 0$,** see **Program 5-1A.** This representation of the dynamics, shown in **Fig. 5.2,** might be referred to as an **empirical representation,** for it is similar empirically to

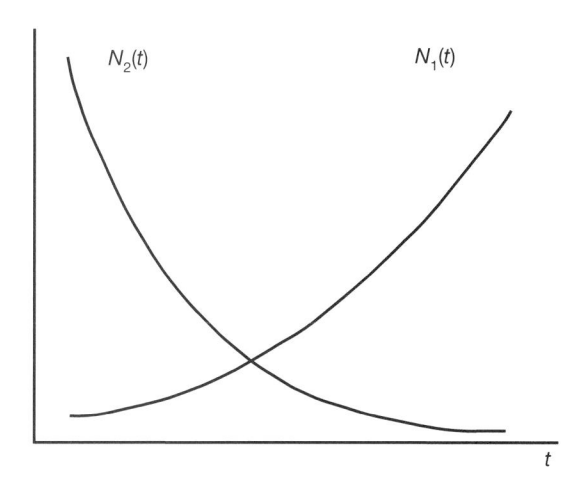

Fig. 5.1.
The behavior of the two predator and prey species, if there are **no interactions** between them, $K_1 = K_2 = 0$ in (5.1.1) and (5.1.2).

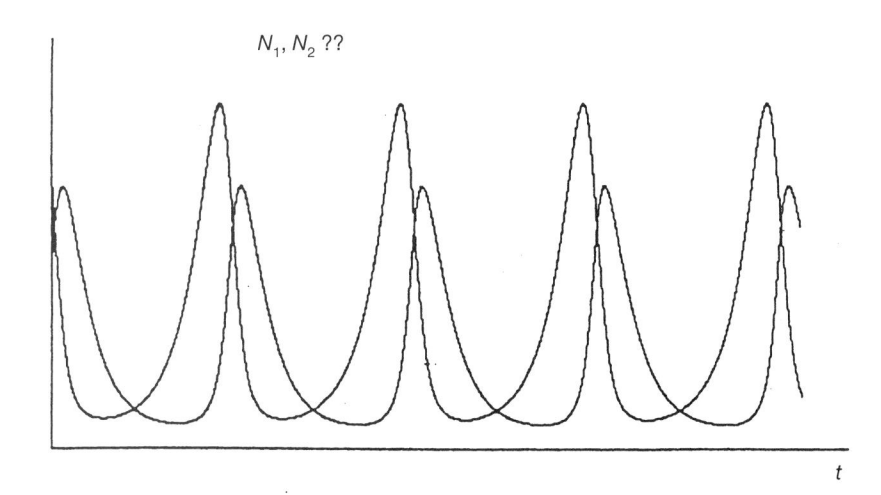

Fig. 5.2.
The temporal behavior of $N_1(t)$ and $N_2(t)$ for particular values of (R_1, R_2) and (K_1, K_2) given in the program. See **Explore 5.1A.**

Interacting Systems Producing (Near) Periodicity

recorded historical data, shown in **Fig. 5.3.** While this representation is sensitive to the values of all of the physical parameters (R_1, R_2, K_1, K_2), it will be seen below that scaling concepts can simplify the representation and thereby the **understanding** of such nonlinear systems.

An example of the dynamics, when $K_1 \neq 0, K_2 \neq 0,$ that is obtained from Program 5-1A is shown in **Fig. 5.2.**

Think 5.1A

Which curve in Fig. 5.2 represents the number of predators? Be sure that you understand this type of interaction. If you have lingering doubts, look into **Program 5-1A.**

Think 5.1B

If $K_2 \to 0,$ develop a line of reasoning that can determine if the long-time behavior of N_1 and N_2 is again given by Eq. (5.1.3).

Explore 5.1A

Now you can run **Program 5-1A** to see the dynamics illustrated in Fig. 5.2. Moreover, you can now use **keyboard interactions** in this program, which will allow you to **increase or decrease** the nonlinear **interaction parameters** (K_1, K_2). To accomplish this, you will need to **modify the program as indicated.** In this way, one can obtain a better appreciation of the details of the dynamics involved in **Think 5.1A** and **Think 5.1B**. This will make clear a

Fig. 5.3.
The number of trappings of Canadian lynx and snowshoe hare, according to the records of the Hudson Bay Company.
See Think 5.1C.

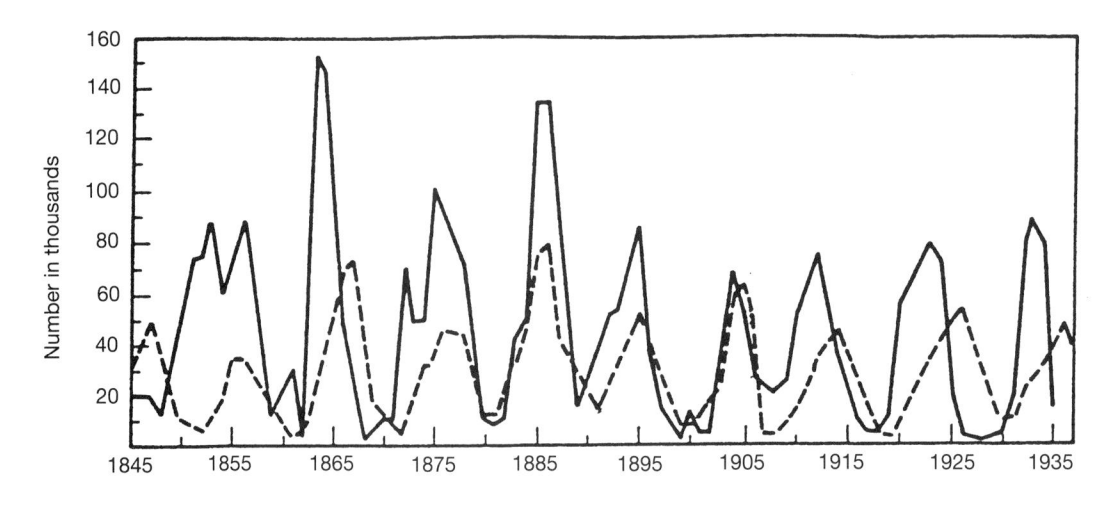

5.1 Predator–Prey Systems: The Lotka–Volterra Model

fundamental difference between the $K_1 \neq 0$, $K_2 \neq 0$ dynamics and $K_1 = K_2 = 0$ dynamics shown in **Fig. 5.1.**

The interaction between the **Canadian lynx** and its prey, the **snowshoe hare,** has long been of interest to population ecologists and others. This has been the case, in part, because of the availability of historical records related to their populations. An interesting historical example of a such a record is illustrated in **Fig. 5.3.** This is based on trading records from the Hudson Bay Company in Canada, over the period 1845–1935. It indicates the number of catches that were traded by trappers, for both the Canadian lynx and the snowshoe hare, over this extended period. It is clear that there is quite a difference between this record of trades and the Lotka–Volterra dynamics shown in **Fig. 5.2.**

Think 5.1C

Which curve represents the lynx and the hare in Fig. 5.3? **Are you sure?** Look at what happens around 1850, 1860, and 1870. Did hare eat lynx? Consider the source of this data, and see if you can think of some possible causes for this "strange result." On the other hand, think of the simplification used in the Lotka–Volterra model, and suggest another possible cause. For an early discussion of this affair, see Gilpin [3]. For references to a subsequent history of ideas, and a more recent approach and spatial generalization of this predator–prey system, see reference [4].

Think 5.1D

The **number of parameters** in equations (5.1.1) and (5.1.2) can be **reduced by "scaling"** the magnitude of variables to new variables, (M_1, M_2), through the relationships $N_1 = S_1 M_1$, $N_2 = S_2 M_2$. Show that if these are substituted into (5.1.1) and (5.1.2), one can obtain the equations

$$\frac{dM_1}{dt} = R_1 M_1 (1 - M_2) \qquad (5.1.4)$$

$$\frac{dM_2}{dt} = -R_2 M_2 (1 - M_1) \qquad (5.1.5)$$

provided that (S_1, S_2) are properly selected. Determine these constants in terms of (R_1, R_2, K_1, K_2). Note that this can only be accomplished if K_1 and K_2 are nonzero. What is the **fixed point** of (5.1.4), for **any** (K_1, K_2)? Verify that (N_1, N_2) is then the fixed point of (5.1.1) and (5.1.2), using the values for (S_1, S_2).

Think 5.1E

Reflect on the discussion of **topologically equivalent phase portraits,** given in **Section 4.3,** and, in particular, the idea of **rubber-sheet geometry.** Can the above scaling be viewed as stretching (or compressing) a rubber sheet? With these thoughts you may see that if neither K_1 or K_2 vanish the **phase portrait of (5.1.4) and (5.1.5), and** hence of **(5.1.1) and (5.1.2),** are **topologically equivalent** for all other values of (K_1, K_2). In other words, by drawing **the phase portrait of one on a rubber sheet, it can be stretched in such a fashion as to be identical with the other phase portrait (ignoring the coordinate system in this comparison).**

Explore 5.1B

Rather than display the Lotka–Volterra dynamics as a function of time, as in Fig. 5.2, it can be displayed in either the (N_1, N_2) or the (M_1, M_2) **phase space.** Use **Program 5-1B** to first see the display of the dynamics in the (N_1, N_2) phase space. You will see that it is qualitatively similar to what is shown in **Fig. 5.4,** which is in the (M_1, M_2) phase space (note the location of the fixed point). In this latter phase space the orbits are only changed by changing (R_1, R_2). **Modify Program 5-1B** so as to obtain the (M_1, M_2) phase portrait. This is simply done by introducing the equations (5.1.4), and (5.1.5), into the Runge–Kutta program [**note** that you can even **use the notation $N(1)$ and $N(2)$ for $M(1)$ and $M(2)$,** so this change need not be made everywhere in the program]. Now the orbits will only change when (R_1, R_2) are changed.

Think 5.1F

Complete Fig. 5.4 by indicating the direction the dynamics takes along the illustrated orbits (i.e., clockwise or counterclockwise?). Note that whereas the phase space representation is often the favorite from a mathematical and **scaling** point of view (see **Think 5.1B**), it does not give any temporal information, which is obviously of concern to ecologists.

Explore 5.1C

Notice that when you run Program 5-1B the **"velocity"** varies significantly over different portions of the orbits. It is this information that does not show up in a usual phase portrait. One way that this information can be introduced is by **making the modification** found at the line marked #######.

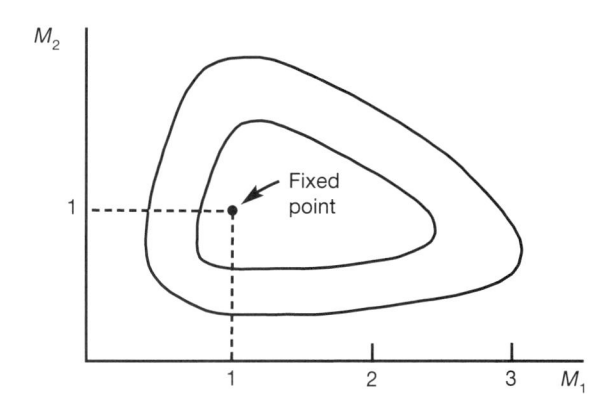

Fig. 5.4.
The incomplete "orbits" of the interacting predator and prey, represented in the (M_1, M_2) phase space. Complete this in **Think 5.1D.**

■ **Think 5.1F**
What other scaling can be used to reduce (5.1.4) and (5.1.5) to equations with only **one parameter?** There are two simple possibilities. Do you find that this sheds any light on physical features of the system? ■

General Aspects

The Lotka–Volterra dynamics illustrated in Fig. 5.4 is referred to as a **neutral form of stability** (as is the **undamped harmonic oscillator, Fig. 4.5a**). In both of these cases, if the system is perturbed from its form of oscillation, it will simply oscillate through the newly assigned point in phase space. It does not return to the original oscillation, nor move off and not oscillate through the newly assigned phase point. It is "neutral" in this sense. This type of dynamics is very "fragile," and indeed is quite simplistic, only representing real dynamics for quite a limited duration of time (e.g., until damping, environmental, or other "internal" factors influence the dynamics).

Explore 5.1D
You can learn some interesting results by studying the dynamics of the Euler approximation of (5.1.4) and (5.1.5). See if this gives a "good" representation of the phase portrait, Fig. 5.4, **for any small value of dt.** [Recall that the dynamics of the **Verhulst equation** and the dynamics of its Euler approximation (**logistic equation**) were very similar if (and only if) dt is sufficiently small

(Figs. 4.2 and 4.3 in Section 4.1)]. A **general study** of the connection between the dynamics of **a continuous model** and that of its **Euler approximation** was made some time ago in **[4]**. In particular, a detailed discussion is given concerning the **Lotka–Volterra system,** and the **predator–prey system,** to be discussed in the next section.

As noted previously, several environmental constraints are missing from the present Lotka–Volterra model. The importance of considering such constraints in dealing with conservation issues is illustrated in reference **[5]**. The additional importance of considering spatially inhomogeneous populations, as well as the vegetation dynamics, is studied in reference **[6]**, which contains many useful references. Even larger-scale climatic influences have been studied **[7]**. So one can see that the lynx–hare dynamic system has many enthusiasts!

For a **very broad study and modeling of the interaction between species, see [8].**

Concepts

- population control by predator–prey interactions between species
- spatially homogeneous modeling
- no limited resources for the prey
- new modes of understanding from scaling features
- topological equivalent phase portraits
- continuous and discrete dynamic representations

References and Notes

1. For a beautifully illustrated and broad coverage of ecological dynamics, see T. J. Case, *An Illustrated Guide to Theoretical Ecology*, Oxford University Press, New York, 2000.
2. S. E. Kingsland, Modeling Nature: *Episodes in the History of Population Ecology,* 2nd ed. University of Chicago Press, 1988.
3. M. E. Gilpin, Do Hares eat Lynx? *Amer. Nat. 107,* 727–30, 1973.
4. R. M. May, On Relationships among Various Types of Population Models. *Amer. Nat., 107,* (#953), 4–57, 1973.
5. The low population of snowshoe hare in Colorado may have led to the

death of many Canada lynx when they were introduced back into Colorado; see K. Kloor, Lynx and Biologist Try to Recover after Disastrous Start, *Science,* July 16, 1999, p. 320.

6. B. Blasius, A. Huppert, & L. Stone, Complex Dynamics and Phase Synchronization in Spatially Extended Ecological Systems. *Nature,* 399, 27 May, 1999, 354–359 (with extensive references).

7. N. C. Stenseth, et al., Common Dynamic Structure of Canada Lynx Populations within Three Climatic Regions. *Science,* 285, 13 Aug. 1999, p. 1071.

8. S. L. Pimm, *The Balance of Nature: Ecological Issues in the Conservation of Species and Communities.* University of Chicago Press, Chicago, 1991.

5.2 HOST–PARASITE INTERACTIONS: UNCOVERING MODEL LIMITATIONS

In this section we will consider a type of dynamics that limits the populations of two species by a type of interaction that appears to be quite simple in its structure, but leads to some interesting and **basic issues** concerning the **modeling of natural dynamics.**

The model is based upon the dynamics in a **large area, A,** over which an arthropod **host** are **uniformly distributed,** with a **density $H(t)$**. Each host lays K **"eggs"** (larvae or pupae), and then dies. The **parasites** considered here (parasitoids) are insects, which are likewise described by their **density $P(t)$** over this large area. They **independently search over a small area** for the hosts' eggs, and lay their eggs in or on the eggs of the hosts. These parasitized eggs are destroyed, each egg giving rise to N **parasites.** The unaffected host eggs go on to produce the next generation of hosts.

The parasites hunt for these eggs, and this has been modeled by introducing the concept of the **inefficiency** in hunting for eggs, $E[P(t)]$, as defined by

$E[P(t)]$ = **the fraction of eggs that are missed by the parasite population $P(t)$**   (5.2.1)

That is, the **larger is $P(t)$,** the **smaller is $E[P(t)]$** (the **fewer eggs** that are **missed,** which represents a smaller **inefficiency** in egg hunting).

A famous model, which is due to Nicholson and Bailey [1], assumes that this efficiency obeys a Poisson distribution

$$E(P) = \exp(-a \cdot P) \quad (5.2.2)$$

which is in keeping with the idea that **the larger the population of parasites the fewer eggs that are missed.** The **constant a** must have the **dimension of area,** because ($a \cdot P$) must be dimensionless. This area is related to the **search area** of each parasite, but is limited by the assumed independence of the search area of each parasite.

Since the hosts benefit if the parasites miss the eggs, their density in the **next generation** is given by

$$H(t + 1) = K \cdot H(t) \exp[-a \cdot P(t)] \quad (5.2.3a)$$

On the other hand, the parasites suffer from missing the eggs; the density of parasites in the next generation is given by

$$P(t + 1) = N \cdot K \cdot H(t)\{1 - \exp[-a \cdot P(t)]\} \quad (5.2.3b)$$

The equations (5.2.3) are very nonlinear, due to the **exponential function,** which contains all **powers of $P(t)$** [refer to the definition **(M3.1) in Appendix M3**].

Having obtained the equations (5.2.3), Nicholson and Bailey [1] recognized that the **character of the dynamics** of this system **does not depend on the constant a.** This can be seen by multiplying both equations in (5.2.3) by a, and introducing the **"scaled" variables**

$$H^* = a \cdot H \quad \text{and} \quad P^* = a \cdot P \quad (5.2.4)$$

The equations (5.2.3) then yield equations **independent of a,**

$$\begin{aligned} H^*(t + 1) &= K \cdot H^*(t) \exp[-P^*(t)] \\ P^*(t + 1) &= N \cdot K \cdot H^*(t)\{1 - \exp(-P^*(t))\} \end{aligned} \quad (5.2.5)$$

This "scaling" has also changed the **dimension of the new variables** to **the number of host and parasites** in the regions "associated" with the parasites' search region. This "association" is rather vague (see reference [1]), but of no dynamic consequence in this model. It only influences the magnitudes of the densities [$H(t)$, $P(t)$] through (5.2.4), which scales the graphical axes.

We will now **simplify the notation of (5.2.5) to**

$$H(t + 1) = K \cdot H(t) \exp[-P(t)] \quad (5.2.6a)$$

$$P(t + 1) = N \cdot K \cdot H(t)\{1 - \exp[-P(t)]\} \quad (5.2.6b)$$

keeping in mind, of course, that these are the functions **(H^*, P^*) in (5.2.4).** It is worth noting that from the **point of view of mathematics** one would come directly to the equations (5.2.6), and not have discussed the density variables. After all, we have not changed the character of the **mathematical dynamics** by this change of variables, so **why bother with the physical concepts involved in the equations (5.2.3)?** Well, let's proceed and see.

When dealing with nonlinear equations, it is always useful to search for the (nontrivial) fixed points:

▌ **Think 5.2A**

Show that the **fixed point of (5.2.6), (h, p)** is at

$$p = \ln(K) \quad \text{and} \quad h = \frac{\ln(K)}{N(K-1)} \qquad (5.2.7)$$

▪

▌ **Think 5.2B**

Consider the **dynamics near the fixed point,** by substituting

$$P(t) = p + \Delta P(t) \qquad H(t) = h + \Delta H(t)$$

where [$\Delta P(t)$, $\Delta H(t)$] are relatively **very small** quantities. Substitute these into (5.2.6) and expand all the factors in powers of $\Delta P(t)$ and $\Delta H(t)$. One then obtains the **linear equations for [$\Delta P(t)$, $\Delta H(t)$], by ignoring terms that are higher than the first power in these variables.** In particular, $\exp[-P(t)] \approx \exp(-p)[1 - \Delta P(t)]$, according to the exponential property (M3.1) in **Appendix M3**. Thus, for example **(5.2.6a)** yields, after **using (5.2.7)**

$$\Delta H(t+1) \approx K \exp(-p) [\Delta H(t) - h \cdot \Delta P(t)] = \Delta H(t) - \frac{\ln(K)}{N(K-1)} \Delta P(t)$$

Obtain the analogous equation for $\Delta P(t+1)$ from (5.2.6b). These equations can be used to determine if [$\Delta P(t)$, $\Delta H(t)$] → (0, 0) as t → ∞. If this happens, the fixed point is an **attractor,** and it is called a **stable fixed point.** If this does not happen it is called an **unstable fixed point.** The details of this linear analysis can be found in **Appendix M5.** Refer to **Fig. 4.10 in Section 4.4** for the possible different forms of linear dynamics near fixed points.

▪

As shown in Appendix M5, the **fixed point (5.2.7) is unstable if $K > 1$,** as will also become clear in **Explore 5.2A.** The question then is what is the **character of this dynamics for large times?** Nicholson and Bailey determined a number of basic features of the dynamics by a **mathematical analysis of (5.2.6).** This is something that today is immediately turned over to computer "solutions" of (5.2.6). However, as was discussed in **Section 2.2,** the mathematical and computational representation of dynamics are **logically distinct,** and that will be seen clearly in what follows. This, indeed, is the **primary lesson to be learned from the present study of this modeling of natural dynamics.**

Let us first look at some of the important **mathematical deductions** that were made by Nicholson and Bailey:

(**I**) The fixed point (5.2.7) is unstable.

(**II**) The small oscillations ultimately become large oscillations.

(**III**) For any values of $P(t)$ and $H(t)$ they both oscillate perpetually about this fixed point, with $P(t)$ lagging by about one quarter "period" behind $H(t)$.

(**IV**) Thus, $P(t)$ and $H(t)$ have no asymptotic limit point except for the fixed point (5.2.7). (This, of course, occurs in the limit $t \to -\infty$). They also gave an independent proof of this fact.

To study the detailed character of the **large oscillations,** they employed arithmetical means (no computers back then!), which made use of a "competition curve" developed by Nicholson [2]. An example of their findings by this method is shown in **Fig. 5.5**

As characterized by May [3], "As shown by Nicholson and Bailey, the resulting dynamics are ever-diverging oscillations, result-

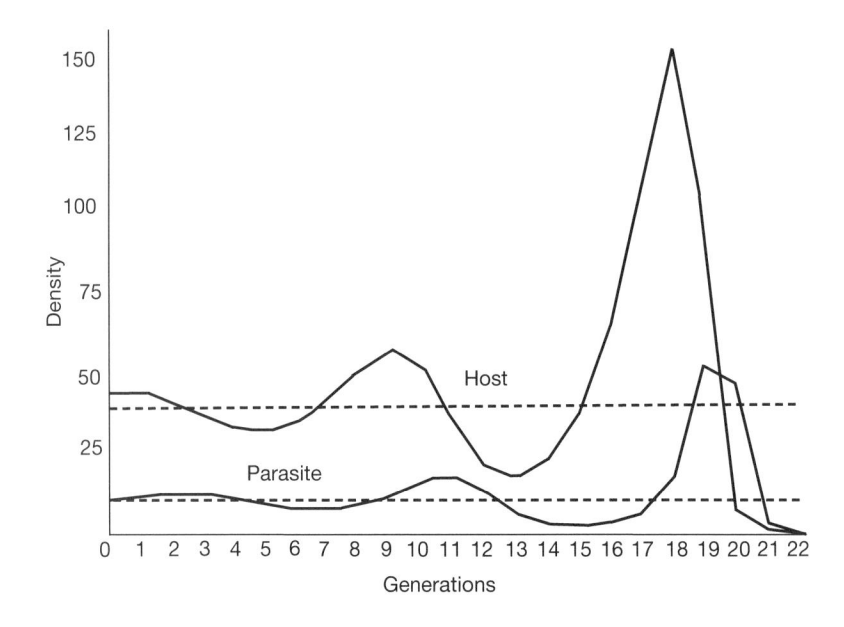

Fig. 5.5.
(Fig. 1 in reference [1].) A specific parasite ($N = 1$) and host ($K = 2$) interaction. The parasite curve is drawn to half the vertical scale of the host curve. The dotted lines are the steady (fixed point) densities. The density numbers have been scaled based on an "area of discovery of parasite 0.035."

ing in extinction first of the hosts, and consequently of the parasitoids." Despite the apparent agreement of this description with Fig. 5.5, and even more so if $K = 4$, Nicholson and Bailey did not come to this conclusion, and rightly so. Their conclusion (III) **excludes this possibility** on a **mathematical basis.** However, they did not rely on this, but rather turned to **the physical limitation of the assumptions used in the model.**

Think 5.2C

Before we see what they had to say about there result, in which their numerical results showing **that $H(t)$ and $P(t)$ approach near-zero values** in a few tens of generations, we will look see how the **computer characterizes** this dynamics. Before we return to Nicholson and Bailey's (and others since then) critique of their model, **see if you can think** of some **neglected physical aspects** that could come into play in such a situation.

Before seeing **what is the computational characterization of this physical system,** let us recall some basic issues discussed in **Section 2.2. Computational dynamics and mathematical dynamics are two logically distinct representations of nature.** Mathematics is, as Leibniz put it, **"the science of the infinite,"** involving an infinite amount of information. In contrast to this, **computational dynamics** always takes place on a **finite set of integers,** depending on the round off of the computer, and it runs for a finite time These properties of **computational dynamics** are **basically closer to our empirical knowledge,** with all of its **implicit uncertainties,** than is the dynamics described by mathematics, which ignores both our uncertainties and, moreover, various environmental influences, including noise. Indeed, the implicit uncertainties of computational dynamics **more closely represents what we know** than the precise mathematical representations. This idea is not wildly welcomed by a science that has been sure since the time of Newton that it has a deterministic understanding of nature. But, nonetheless, it is true. Put another way, if the computational dynamics **deviates significantly** from the mathematical dynamics (for example, in a topological sense with two nonequivalent forms of dynamics), then this may well be an indication that an **implicit assumption** is contained in the model, **which is incorrect.** This then requires some **new reflections** on the basis of the model being used. **Such is the case in the following Explore.**

Explore 5.2A

The **computer algorithm** based on the **physical reasoning** used for **(5.2.6)** is found in **Program 5-2A,** where the values $K = 2$, $N = 1$, and $a = 1$ have been selected (agreeing with those in **Fig. 5.5**). What you should explore is what happens for **different initial conditions.** Do you find results that are **(a)** "**self-consistent**" or **(b)** "**consistent**" with the above **mathematical results**? Returning to **Think 5.2C,** do you think that the mathematical or computational characterization of the dynamics **indicates the shortcoming** of the assumptions that were made in the physical modeling? **Why?** Possibly, you may gain some insight by exploring the character of the computational dynamics for other values of N and K, which is readily done in this program.

You may have come up with **several ideas** concerning physical oversimplifications in this model. One relates to assuming that the system can be described by **spatially homogeneous population variables.** This shortcoming was recognized by Nicholson and Bailey, and their intuition in this matter is worth quoting **[4]**. Notice how they reject the mathematical conclusions as not being physically meaningful.

A Second Simplification

The dynamics (5.2.6) does not take into account the important issue of **limited environmental resources for the host.** In these large oscillations, the population of the host also becomes very large at times, and the reproduction of the hosts may be influenced by such limited resources (as in the logistic and other dynamics of Chapter 3). This can be accounted for by changing the reproduction constant K to a function of H, such that **$K(H)$ decreases as H increases.** How rapidly it decreases will depend on the condition of the environment. So each environment will be modeled as having a certain **host capacity,** expressed by some constant H_c, related to the maximum number of hosts it can support. The parasites live off the hosts, not the environment directly, but whatever influences $H(t)$ naturally influences $P(t)$. So now explore the **very rich variety of dynamics** that can result from a simple model of this environmental influence.

Explore 5.2B

Now go into **Program 5-2B,** which is similar to Program 5-2A, except that **K has been replaced by K/[1 + (H/H$_c$)]**, thereby modeling a decreasing reproduction rate as *H* increases. Also **change the initial conditions** to be more general. You will find **very different** types of dynamics **from either the Lotka–Volterra or Nicholson–Bailey** dynamics, characterized by **different types of attractor dynamics,** depending on the value of H_c.

(1) Explore this dynamics as H_c is slowly increased. When H_c^* ≥ H_c you will find one type of attractor, and when H_c ≥ H_c^* you will discover an entirely new type of **two-dimensional attractor.** This **attractor** is called a **(discrete) limit cycle.** See what happens when you use **random initial conditions** in this case. The **bifurcation** from a **fixed point attractor to a limit cycle** is known as a **Hopf bifurcation,** and H_c^* is called the **bifurcation point. Determine** the approximate value of H_c^*.

(2) What else happens as H_c is varied? This is a **rich dynamic system,** so **ask your own questions, and explore!**

References and Notes

1. A, J. Nicholson and V. A. Bailey. The Balance of Animal Populations. Part I. *Zool. Soc. (London), Proc.* 3, 551–598 (1935).
2. A. J. Nicholson. *J. Animal Ecology, ii,* 1, Fig. 1 (1933).
3. R. M. May. Spatial Chaos and Its Role in Ecology and Evolution, pp. 327–344 in *Frontiers in Mathematical Biology,* S. A. Levin (Ed.). Springer-Verlag, Berlin, 1994.

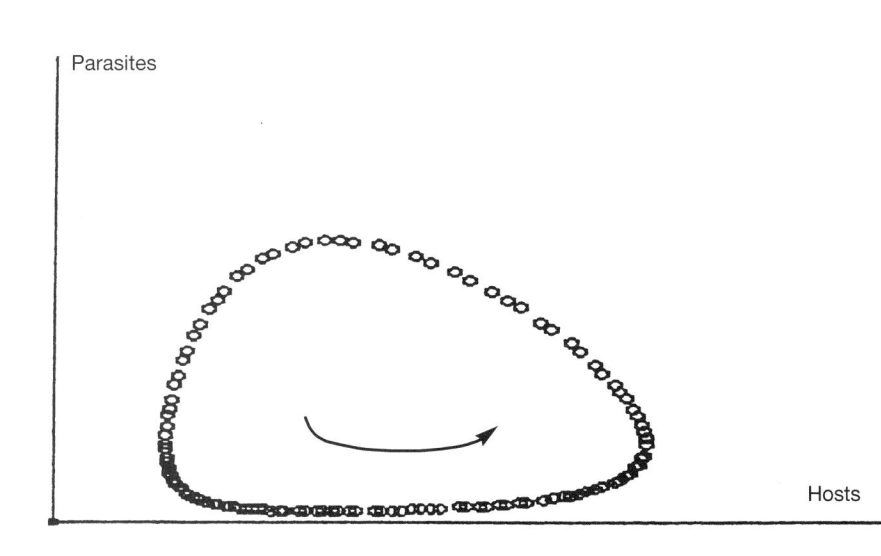

Fig. 5.6.
The occurrence of a **stable limit cycle** when $H_c > H_c^*$.

4. The physical understanding of the real implications of such large oscillations as shown in Fig. 5.5 is made clear in the following quotation from section 4 of reference [1].

> When considering large fluctuations, however, account must be taken of other factors besides those already used in this investigation.
>
> Each female is the centre of diffusion of her offspring, so when the density of a species becomes very low as a result of violent oscillation the zones of diffusion from the various females can seldom overlap and may frequently be separated from one another by great intervals. Thus the interacting animals exist in numerous disconnected small groups, within each of which interspecific oscillation follows its course independently of that in the other groups. Since the numbers of animals in these groups are very small compared with species-populations, the great reduction in numbers soon produced by increased oscillation frequently exterminates the groups, but meanwhile some hosts will have migrated into the surrounding previously unoccupied country and have established new groups there. If follows that:
>
> A probable ultimate effect of increasing oscillation is the break up of the species-population into numerous small widely separated groups which wax and wane and then disappear, to be replaced by new groups in previously unoccupied situations.

This conclusion is reinforced today [5], and expressed in reference [3]:

> So, what factors contribute to the persistence of host–parasitoid associations in the natural would? ... in field situations, the emerging consensus is that spatial patchiness, combined with a sufficient degree of variation in levels of parasitism among patches, may be the essential mechanism that enables the persistence of host and parasitoid population at roughly steady overall densities"

5. J. L. Maron and S. Harrison. Spatial Pattern Formation in an Insect Host–Parasitoid System. *Science, 278,* 28 Nov. 1997, p. 1619.

Implicit Assumptions and Forms of "Understanding" in Modeling

B. Cipra. Revealing Uncertainties in Computer Models. *Science, 287,* 11 Feb 2000, p. 960. ("Uncertainties" refer to unknown implicit assumptions.)

Can Biological Phenomena be Understood by Humans? *Nature, 403,* 27 Jan. 2000, p. 345. (The issue of our forms of **"understanding"** is becoming recognized as a central factor in the future of science, as it was in the past, as discussed in **Sections 1.2 and 2.1.** Some positive signs: *Nature, 403,* p. 335–338, 339–342, 2000; analyze natural networks; design and build biological networks to implement desired functions.)

Impacts of Foreseeable Science. *Supplement to Nature, 402,* 6761, 2, Dec. 1999.

References to other forms of biologically interactive dynamics that are actively studied follow.

General Biological Dynamic Models

For a beautifully illustrated and broad coverage of Ecological dynamics see T. J. Case, *An Illustrated Guide to Theoretical Ecology,* Oxford University Press, New York, 2000.

T. J. Case. Complex Systems. *Science 284,* pp. 80–109, 2 April 1999.

J. Murray. *Mathematical Biology.* Springer-Verlag, New York, 1989.

L. Edelstein-Keshet. *Mathematical Models in Biology.* Random House, New York, 1988.

S. A. Levin (Ed.). *Frontiers in Mathematical Biology* Lecture Notes in Biomathematics, Vol 100, Springer-Verlag, Berlin, 1994.

R. M. May. Spatial Chaos and Its Role in Ecology and Evolution. Ibid., p. 327–344.

J. Maynard Smith. *Mathematical Ideas in Biology.* Cambridge University Press, Cambridge, 1968.

Dynamics of Infectious Diseases and Methods of Controls

R. M. Anderson and R. M. May. *Infectious Diseases in Humans: Dynamics and Control.* Oxford University Press, Oxford, 1991.

R. May. Simple Rules with Complex Dynamics. *Science, 287,* 28 Jan. 2000, pp. 601–602.

D. J. D. Earn, P. Rohani, B. M. Bolker, and B. T. Grenfell. A Simple Model for Complex Dynamical Transitions in Epidemics. *Science, 287,* 28 Jan. 2000, pp. 667–670.

5.3 ENVIRONMENTAL FLUXES: AUTOCATALYTIC OSCILLATORS

All living systems depend on an exchange of energy and matter with their **environments** in order to sustain their dynamics, growth, and reproduction processes. The times during which this exchange takes place can vary greatly, since they have a variety of internal processes that may be altered for various reasons. Thus, we can eat over a wide range of times, depending on our desires, and also breathe at different rates in order to swim or run, or for no practical reason. It will be noted that the timing for this call on environmental resources is dictated by the system; it is an **"auto" process.**

In the **inanimate world,** there are, likewise, systems whose dynamics is sustained by energy, and possibly material exchanges with their environment. The **simplest** of these systems have **periodic dynamics,** and, therefore, exchange energy with the environment at regular intervals of time, dictated by a certain phase in their motion (in order to keep things periodic). In other words, **the system's state dictates** when it will draw energy from the environment, and also when it will return energy to the environment (again an **"auto" process**). Moreover, this **exchange of energy** is responsible for **sustaining the dynamics** (it acts as a **catalyst**), which would otherwise die out due to its **internal energy consumption** (a damping effect). This combined exchange process is referred to as an **autocatalytic oscillator.**

To understand this idea in more detail, let us begin with the usual damped harmonic oscillator, discussed in Section 4.2, containing the equations

$$\frac{dx}{dt} = v$$

$$m \cdot \frac{dv}{dt} = -\mu v - \kappa x$$

(4.2.6)

In frictional cases, when $\mu > 0$, the oscillations are damped and the system **loses energy,** which it gives to its environment in the form of heat. However, if it happens that somehow there is an interaction with the environment that can be modeled with $\mu < 0$, the oscillator would **gain energy** from the environment, and its amplitude of oscillation would increase. Here, by **"energy,"** we simply mean the frictionless energy ($\mu = 0$) given by $E(x, v) = \frac{1}{2}(mv^2 + \kappa k^2)$.

The **autocatalytic oscillator** is a **nonlinear oscillator** that both extracts energy from the environment $[\mu(x) < 0]$ and ejects an equal amount back to the environment $[\mu(x) > 0]$, because the dynamics is periodic. Note that the **sign of $\mu(x)$ depends on the value of x (the "auto" part).** Moreover, because of this, the modified equations (4.2.6) must be **nonlinear.**

5.3 Environmental Fluxes: Autocatalytic Oscillators

To simplify the notation, we will **scale several variables** in (4.2.6) and introduce

$$y = \sqrt{(m/\kappa)}\,v \quad \text{and} \quad \tau = \sqrt{(\kappa/m)}\,t$$

which transforms (4.2.6) into

$$\frac{dx}{d\tau} = y; \quad \frac{dy}{d\tau} = -M \cdot y - x$$

where $M = \mu/\sqrt{(m\kappa)}$. Finally, we will put this into a general form, by returning τ **to** t and replacing the constant M with a **function of x,** calling it $\Phi(x)$. Thus, the **general equation** we will consider is

$$\frac{dx}{dt} = y$$
$$\frac{dy}{dt} = -\Phi(x)y - x \quad (5.3.1)$$

To make this an autocatalytic oscillator requires that $\Phi(x) < 0$ **when x is small** (for then the system gains energy" from the environment), and since we want the dynamics to remain bounded, this requires that $\Phi(x) > 0$ **when x is large** (so that it will give up "energy" to the environment), as illustrated in **Fig. 5.7.** "Energy" in the context of (5.3.1) is simply the function "E" $= x^2 + y^2$, which one wants to be periodic in time.

Before considering a specific example of (5.3.1), it should be noted that this type of autocatalytic dynamics occurs in a wide variety of physical systems. A few physical examples are illustrated

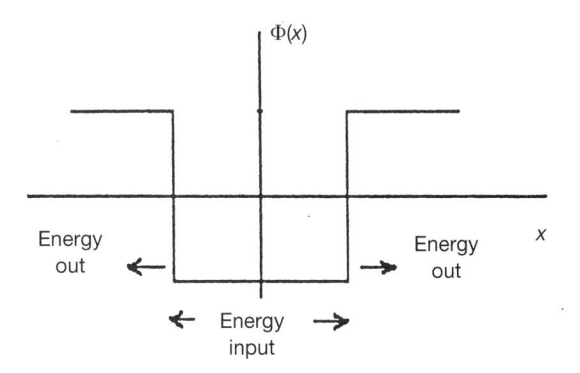

Fig. 5.7.
The function $\Phi(x)$, representing the **gain of energy** from a source when x **is small,** and the **loss of energy** when x **is large.**

in the **Fig. 5.8.** These examples involve such diverse systems as a **bowed violin;** an electrical circuit that **charges and discharges periodically; clocks; pulsating stars;** various **engines** (electric, steam, gasoline; an **impressive car** in this case); and the aesthetic and practical **Japanese shishiodoshi,** a bamboo teeter-totter that moves up and down as it slowly fills with water, becomes unbalanced, tips to rapidly empty, and falls back and strikes a rock, making a hollow sound (scaring birds from the garden?). Note that **all must have a source and a sink of energy,** and a number involve a **flux of matter (which?).**

▰ Think 5.3A

What are the physical sources and sinks of energy in each of the cases illustrated in Fig. 5.8? What regulates their periodicity? ▰

The **classic model** of an autocatalytic oscillator makes use of the function $\Phi(x) = \beta(x^2 - 1)$, where β is a positive constant, to agree with the above criteria. Then equation (5.3.1) becomes

Fig. 5.8.
Various examples of autocatalytic dynamic systems.

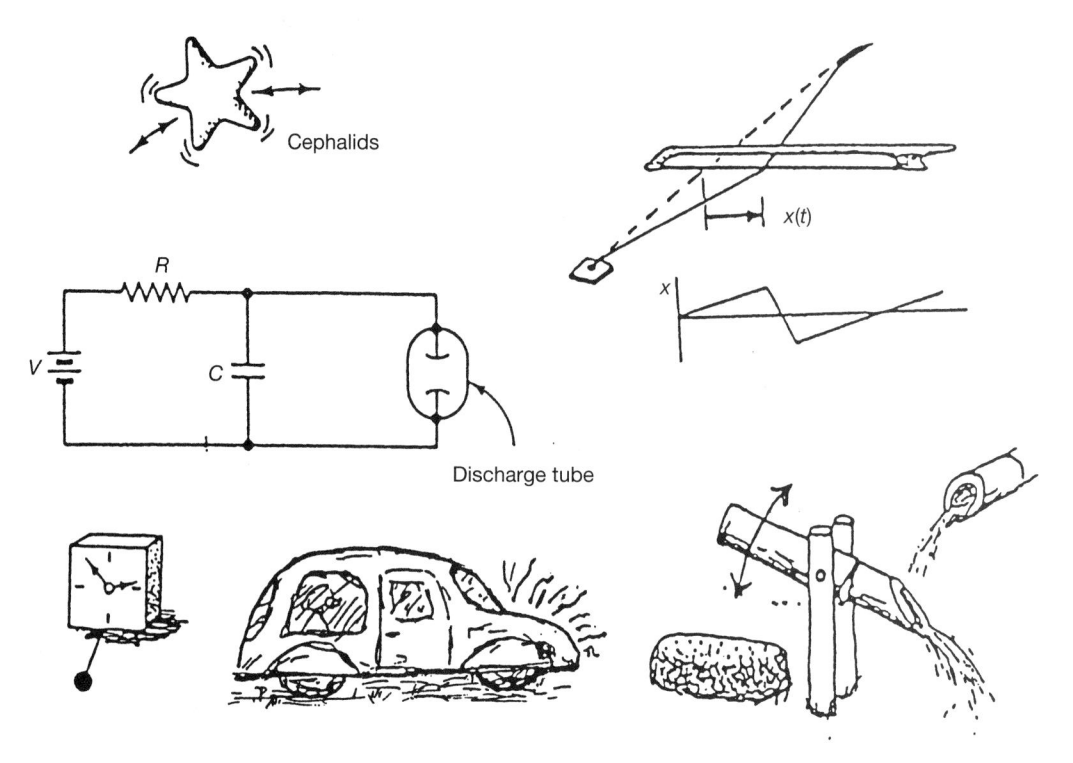

5.3 Environmental Fluxes: Autocatalytic Oscillators

$$\frac{dx}{dt} = y$$

$$\frac{dy}{dt} = \beta(1 - x^2)y - x \quad (\beta > 0)$$

(5.3.2)

Equation (5.3.2) is the famous **van der Pol equation,** named after the Dutch scientist who made extensive studies and applications of this type of dynamics, both to electrical circuits **(Fig. 5.8)** and in the modeling of the dynamics of the heart (see **Section 5.5** for a generalization to a heart model).

There is a very important feature of **the autocatalytic oscillator** (5.3.2). This has to do with the fact that not only does it oscillate, but it **is stable,** meaning that if you **change the variables** to some "nearby" values, **they will tend toward** the oscillating state. In other words, this oscillating state is a limiting state (for large times) of all "nearby" initial states. In the case of (5.3.2), this is, in fact true, for **all initial states,** regardless of their value; but more generally, it may only be true for some limited range of the values of (x, y) about the origin. For these reasons, the **periodic orbit** is known as a **stable limit cycle,** and the **range of attracted states** is called its **basin of attraction.** This is illustrated in **Fig. 5.9.**

Explore 5.3A

To investigate the dynamics of the **van der Pol oscillator,** begin with **Program 5-3A,** in which $K = \beta$. See how the **structure of the limit cycle** in this phase space is changed when the strength of the nonlinearity is changed, as determined by the **magnitude of K**. For each value of K, **try different initial conditions,** and see that they **all tend to one limit cycle,** which **depends on K.**

Think 5.3B

Referring to **Fig. 5.6,** and **Eq. (5.3.2),** for what values of x is **energy injected and extracted** from the oscillator?

Explore 5.3B

Using your insight from **Think 5.3B,** can you make a **simple modification** of **Program 5-3A** that will **increase the size of the limit cycle without changing K?** (Hint: change one number.) Try it out.

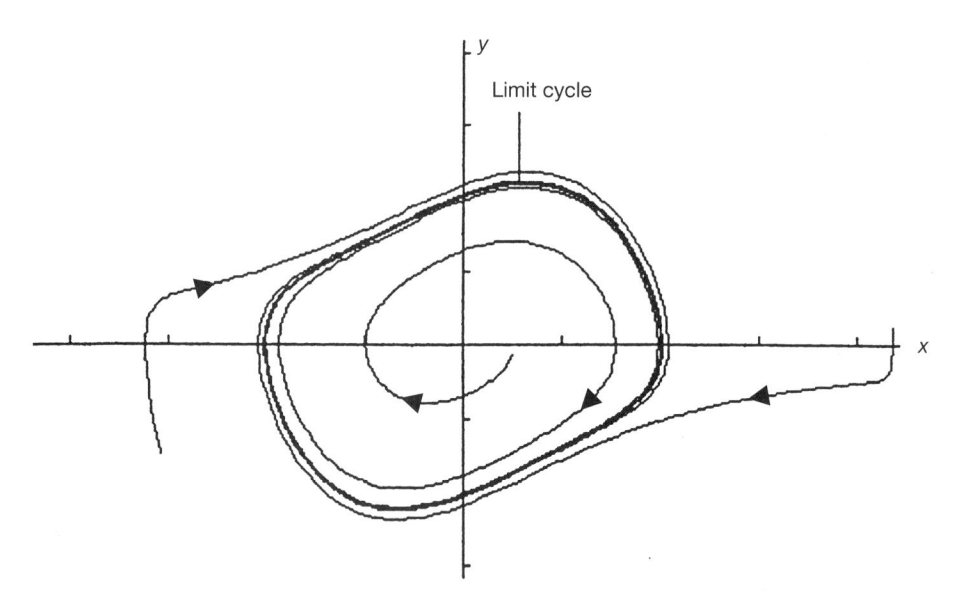

Fig. 5.9.
The **limit cycle** of an autocatalytic oscillator. **All states (except** the unstable fixed point at *x* = 0) are attracted to the limit cycle, forming the **basin of attraction of the limit cycle.**

For the purpose of representing the dynamics of neurons, it is useful to transform the van der Pol equations, (5.3.2), into another set of equations that do not have the distorted limit cycle when *K* is large, as found in Explore 5-3A. Two avoid this large distortion, another representation of the dynamics has been devised, known as the **Liénard phase space.** This phase space is defined by the **variables (*x*, *z*),** where *dz/dt* = –*x*, and, moreover, *z* = *dx/dt* + *F*(*x*). Here *F*(*x*) must be selected such that *dz/dt* equals –*x*. Now, according to (5.3.2), *dx/dt* = *y*, and therefore (from above) *dz/dt* = *dy/dt* + *dF*(*x*)/*dx dx/dt*. Using (5.3.2) this equals $\beta(1 - x^2)y - x + dF(x)/dx \, dx/dt$, or $\beta(1 - x^2)y - x + dF(x)/dx \cdot y$. Thus we see that *dz/dt* = –*x*, only if $\beta(1 - x^2)y + dF(x)/dx \cdot y = 0$, or in other words, if $F(x) = \beta(x^3/3 - x)$. Since *dx/dt* = *z* – *F*(*x*), we finally arrive at the **Liénard equations**

$$\frac{dx}{dt} = z + \beta\left(x - \frac{x^3}{3}\right)$$

$$\frac{dz}{dt} = -x$$

(5.3.3)

Explore 5.3C

Program 5-3B contains the limit cycle dynamics in both the **van der Pol phase space, (x, y),** and in the **Liénard phase space, (x, z) of Equation (5.3.3),** which are **now written** in the form

$$\frac{dX}{dt} = K\left(Y + X - \frac{X^3}{3}\right); \quad \frac{dY}{dt} = -\frac{X}{K} \quad (5.3.4)$$

This amounts to setting $\beta = K$ and $Y = z/K$, $x = X$ in (5.3.3). Now as K is increased, you will find that the **limit cycle is not badly distorted.** Also note that there are now clearly portions of the orbit that are **"fast" and "slow."** This type of oscillation is therefore referred to as a **relaxation oscillation.** This type of dynamics is characteristic of both **neural and heart dynamics,** as will be seen in the following sections, and hence this phase space is **particularly helpful** in understanding this type of dynamics.

This limit cycle is illustrated in **Fig. 5.10.** In addition, the curve on which $dX/dt = 0$, which is given by $Y + X - X^3/3 = 0$, has been included. This type of curve (one on which $dX/dt = 0$, or any variable) is called a **"nullcline."** It is a useful concept, because if the orbit is near this nullcline, then X changes **slowly,** whereas if it is not near this nullcline, then X changes rapidly.

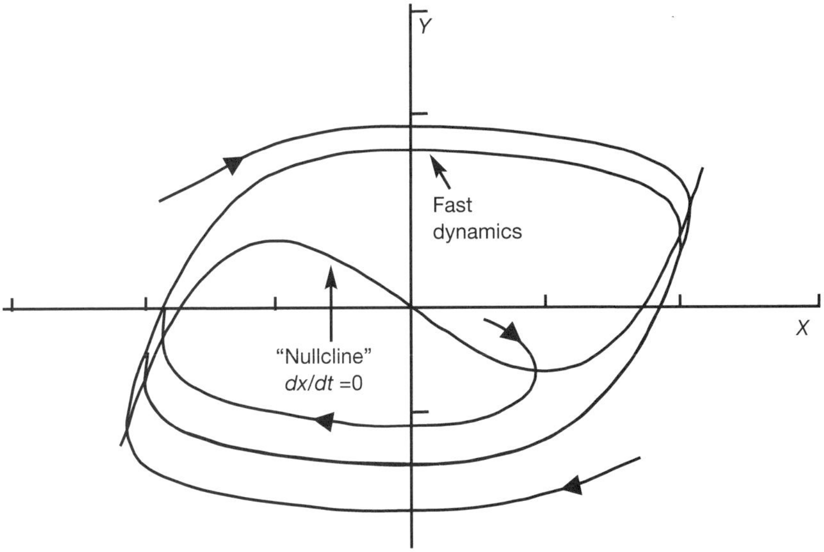

Fig. 5.10.
The limit cycle in the Liénard phase space, showing the the "nullcline" curve on which $dx/dt = 0$, so orbits must be "vertical" at these points. The regions of **"fast"** dynamics (producing **"relaxation"** oscillations) are also illustrated.

Think 5.3C

Locate the **stages of fast and slow motion** in the case of a **violin**, a **clock**, the **current in the discharge circuit**, parts of an **automobile engine**, and **the shishiodoshi (Fig. 5.8)**.

Now we proceed to **excitable neurons and then hearts**, which can be modeled with a generalization of the van der Pol model.

5.4 EXCITABLE NEURONS: THEY PASS ON INFORMATION, CONNECT US TO OUR ENVIRONMENT, LET US THINK, KEEP OUR HEARTS BEATING, ETC.!

The dynamics of higher forms of living systems rely upon special forms of biological cells called neurons. Neurons play an essential role in a vast variety of dynamic activities in all mammals, and in particular, us. Thus, all of our environmental sensory capabilities (touch, smell, taste, hearing, vision) rely upon special forms of neural networks at their sources of inputs. In addition, there are proprioceptors that provided us with information about ourselves (the position and movements of our limbs, muscular forces, our attitude and motion relative to the earth). This information is electrically transmitted to the brain, which consists of some trillion cells, 100 billion of which are neurons that are both supported and made operationally effective by many more glial cells. The neurons transmit electrical pulses along their length, and are linked into networks that somehow interpret this incoming information to give rise to our reactions to the world, involving our understanding, creativity, emotions, consciousness, and memory. We are just at the beginning of the process of trying to understand some of these wonders, but great progress has been made in locating the **various regions** of the brain that process different forms of sensory information, store memories, and generate the higher levels of intelligence [1–3]. However, the **dynamic processes** that are responsible for these accomplishments are **not understood in any detail**, and will likely remain a mystery forever. Hopefully, some more imaginative forms of "understanding" the above dynamic phenomena will be developed, since these massive details cannot really be important. There must be more **holistic modes of understanding**. So there is a **real challenge for the future.**

In this section, we will only be concerned with modeling the

5.4 Excitable Neurons

electrical dynamics of a single neuron when it is subjected to a "sensory" electrical stimulation. This is an important aspect of how a neuron passes on information; when this process fails, many dreaded afflictions result, so it is worthwhile acquiring some basic ideas of how a neuron collects and passes on information. The model we will use is a modification of the relaxation oscillator illustrated in **Fig. 5.10** of the last section, which will simulate the observed **electrical pulse forms of neuronal relaxation dynamics.** But first a little physiology of neurons. Their structures vary considerably in detail, and only the basic **generic features** are illustrated in **Fig. 5.11.**

The **neuron** is a physically extended cell that consists of a number of parts that perform different actions. At **one end** there are a large number of antenna-like **dendrites** (as many as 100,000), to which **chemical information is input** from other cells across synaptic gaps. This is then transmitted and collected in the **cell body (soma),** where all this incoming chemical information is electrically **"added up."** If this generates an electrical excitation that is **less than a threshold value, the neuron** does not respond but **remains in a quiescent state** (electrically inactive). However, if the excitation is **larger** than this **threshold,** molecular channels are opened up in the **adjoining portion** of the axon's membrane, allowing for ion exchanges with its environment. The neuron has become **"excited."** Systems that go into action **only**

Fig. 5.11.
A schematic illustration of the structure of a neuron and its connections to other neurons by chemical transmissions across synaptic gaps. The myelin sheaths increase the propagation speed by insulating the neuron.

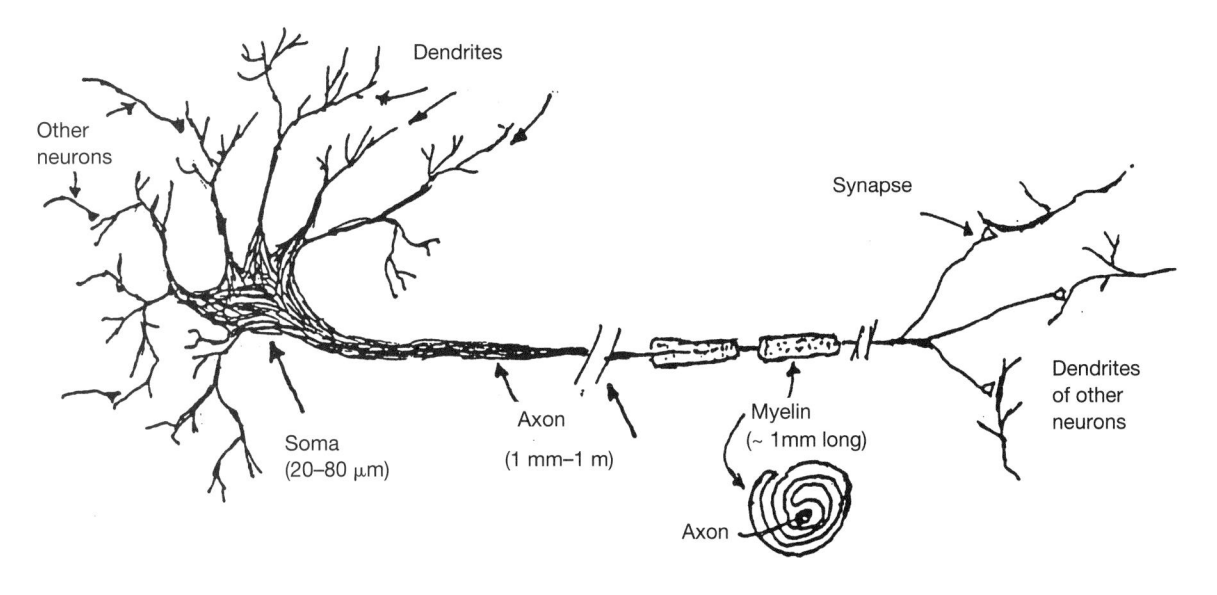

when some outside action **exceeds some threshold** are called an **excitable systems,** as are **neurons.**

This initial excitation near the soma initiates a process that produces an **electrical impulse (action potential)** that travels down the extended portion of the neuron, called the **axon,** at speeds as great as 100 meter/sec. This is assisted by myelin formed by Schwaan cells, all wrapped around the neuron, insulating it from the environment. The length of the axon varies greatly, depending on the type of neuron, ranging from 1 mm to over 1 meter in length (e.g., extending through the spinal cord).

At the **other end of the axon,** the neuron splits up into a **large number of synapses,** where action-potential information is sent to many other neurons. At the end of the synapses there are **vesicles,** which are electrically stimulated to **emit** a number of **specific neurotransmitters.** These molecules diffuse across a very narrow gap ("synaptic cleft") to the dendrites and body of the **adjoining neuron.** Different neurons can emit **"excitatory"** or **"inhibitory"** neurotransmitters, which then correspondingly influence the electrical potential in the body of the adjoining neuron, determining its ability to become excited, and so on through the neural networks.

This remarkable process of information propagation is influenced by a number of other factors. One important influence are the **billions of glial cells,** whose multiple roles continue to be discovered. Among them is their role in removing any "overflow" of neurotransmitters not absorbed by the neurons, and playing several important roles in the electrical activities of the brain, such as in the **myelin sheaths** around axons, which are essential for the effective propagation of the action potential. For more details, see references **[3, 4].**

The real-world complexity of neural networks, which generally involves the interaction of a great variety of different types of neurons, is more accurately captured in illustration of **Fig. 5.12**

It should be amply clear that we cannot model the detailed dynamic interactions of millions of different types of neurons in such networks. At best, we can only hope to model the activities of large composites of such neurons, and their interactions with other composites of neurons. This **average treatment** of large number of systems is, of course, the basis of most ecological, predator–prey, epidemic, and many other group models. It is a type of **statistical average** representation of such systems, which depends on the fluctuating influences of individual members being unimportant to the group behavior. Since there are undoubtedly **random firings of neurons** due to environmental noise (see **Explore 5.4A**), the

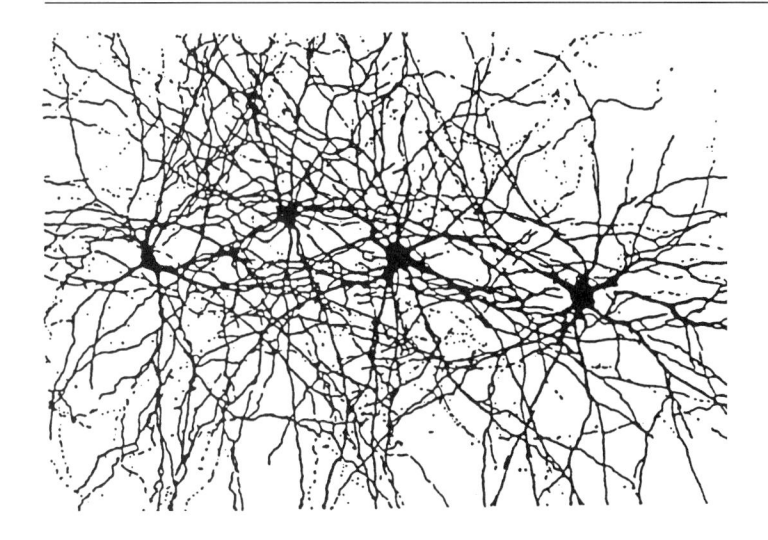

Fig. 5.12.
An example of a small portion of a neural network (approximately 1 by 1.5 mm).

large number of neurons in the brain may, in part, have been introduced by nature to ensure an average reliability to its information processing.

Whatever is the case, we now turn to examining the **dynamics of a single neuron.** The first detailed and accurate description of the dynamics of the axon was given by A. L. Hodgkin and A. F. Huxley in 1952, for which they won the Nobel Prize [5]. Several simpler models of this dynamics were subsequently proposed, which are intended to capture the essential features of the axon dynamics.

In 1961, Fitzhugh derived the following **Bonhoeffer–van der Pol (BvP) model** for the dynamics of a single neuron (only the dynamics involving the axonal part of a neuron is typically considered) [6].

$$\frac{dx}{dt} = c\left(x - \frac{x^3}{3} + y + z\right)$$
$$\frac{dy}{dt} = \frac{(a - x - by)}{c} \quad (a = 0.7, b = 0.8, c = 3)$$
(5.4.1)

This model is closely related to the so-called Fitzhugh–Nagumo model [7,8], which you may find more accessible than the other references.

In (5.4.1) x represents the negative **trans-membrane voltage** at a fixed location on the neuron (axon), and y is related to a particular chemical conductivity [6–8]. The **parameter z** physiologically represents the **external excitation,** corresponding to a membrane current in the neuron. Even this simplified model has a rather complicated phase portrait, depending on the value of z. The equations (5.4.1) always have only **one fixed point** (for the above parameter values), which is **stable only if $z \geq 0.3456$**. This stable state represents a **quiescent ("nonfiring") neuron.** Moreover, **if $z \geq 0.336$,** then **the basin of attraction** of this stable fixed point **is global** (i.e., all initial states converge to this fixed point). Between these two values of **z,** the phase portrait is complicated in that there are **two limit cycles** around the fixed point, the interior one being unstable and the exterior being stable (they **"annihilate" each other** at $z = -0.336$, which is a **pretty spectacular** type of **bifurcation!**). Fortunately, we do not have to worry about the details of the dynamics for the values of z in the small region between -0.336 and -0.3456, since the neuron never exists permanently in this region.

We will consider first the **quiescent neuron,** for which $z \geq 0.336$; therefore it has an infinite basin of attraction (**it always returns to a quiescent state,** no matter where it is initially perturbed). And perturb it we will! We are going to **excite** this quiet neuron, making it **"fire,"** which represents **sending the action potential wave down the axon.** But if we don't perturb it enough, it will not fire, which is characteristic of an **excitable system.** So we are now going to **generalize (5.4.1)** to allow for the ability to **excite the neuron,** and also to study the effects of **environmental noise** upon its dynamics.

Program 5-4 will be used for the following explorations (**Explore 5.4A–C**) of a neuron's dynamics, which is based on the following **generalization of Equation (5.4.1).**

$$\frac{dx}{dt} = c\left(x - \frac{x^3}{3} + y + z - EXC + NZ\right)$$

$$\frac{dy}{dt} = \frac{(a - x - by)}{c} \quad (a = 0.7, b = 0.8, c = 3)$$

(5.4.2)

Here, two forms of **interaction with the neuron** have been added:

> **EXC** = an **impulsive excitation,** whose strength, can be varied, and which is initiated by **<E>** at any time. As noted above, a large negative z makes the fixed point unstable, hence a minus sign is introduced in front of **EXC** (>0) in (5.4.2), in order to excite the neuron.

NZ = a continuous **environmental noise,** with a maximum amplitude **NMAX,** which can be turned on (`<N>`) or **off** (`<F>`). This noise relates to the fact that neurons are always subject to a variety of noisy signals from other cells.

In **Program 5-4,** the dynamics of (5.4.2) is displayed in the **phase space** (x, y), together with **two curves called nullclines.** The nullclines are **curves along which** $dx/dt = 0$ **and** $dy/dt = 0$ in (5.4.1), namely

$$x - \frac{x^3}{3} + y + z = 0 \quad (dx/dt = 0)$$
$$a - x - by = 0 \quad (dy/dt = 0)$$

(5.4.3)

The **fixed point** (quiescent neuron) is where these **two nullclines intersect.** These features are illustrated in **Fig. 5.13**

One such nullcline is illustrated in **Fig. 5.9** of the last section, where its relationship to the **fast and slow portions of a relaxation oscillation** was discussed in some detail. Moreover, this was illustrated dynamically in **Explore 5.3C.**

In the present case, there are **two nullclines,** and their **intersection point** (a fixed point of the dynamics) physically represents the **quiescent (stationary) state** of the neuron. Moreover, the location and **dynamic attractive properties** of this fixed point is influenced by the value of the **parameter** z, as has been partially indicated above. If the neuron is in a quiescent state, it can be excited out of that state by the **application of the excitation** *EXC* in (5.4.2). In this case, it **may** travel in a large orbit around the S-shaped nullcline in **Fig. 5.13,** similar to that in **Fig. 5.10.** Following this it returns to fixed point, until the next excitation, yielding the dynamics illustrated in **Fig. 5.14**

In all the following (Explore 5.4A–C) begin with **Program 5-4.**

Explore 5.4A

This exploration involves investigating the reaction of a **quiescent neuron** to the **excitation** *EXC,* which can be initiated **at any time.** The **stable fixed point** of the BvP equation **has an infinite basin of attraction if** $z \geq 0.336$. For this type of quiescent neuron, begin by **initially** setting $z = -0.25$, which of course can be changed later. For the present study, leave the noise off (*NZ* = 0), and using the suggested excitation strength (*EXC* = 4), see how this excitation causes the **neuron to "fire,"** that is, to go through

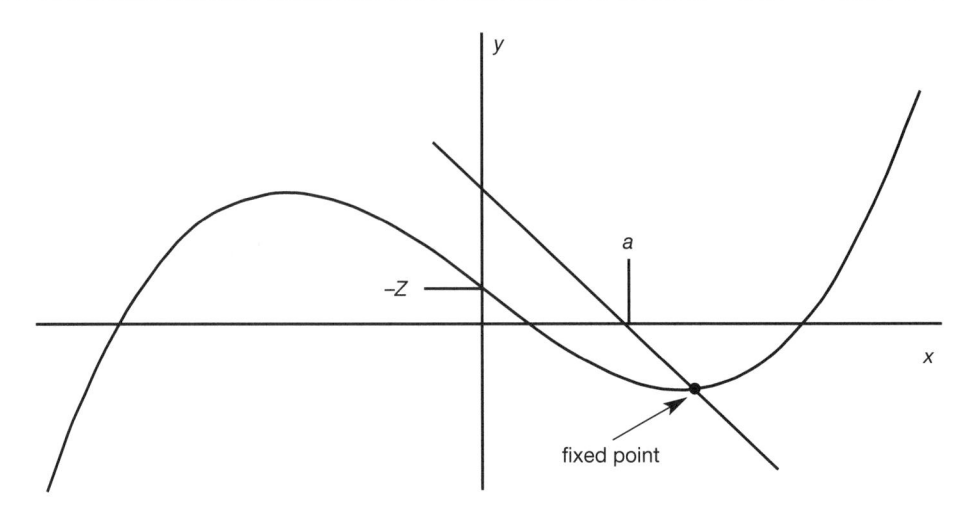

Fig. 5.13.
The structure of the two nullclines, (5.4.3), in the (x, y) phase space and the fixed point at their intersection.

Fig. 5.14.
Examples of the transmembrane potential of the axon, relative to the quiescent state, $-x(t) + x$ **(fixed point) versus t.** Arbitrarily timed excitations are indicated by E.

only one cycle of this dynamics, **returning to a quiescent state.** The phase space illustrates the **slow x-dynamics near** the **nullcline $dx/dt = 0$,** and the **rapid x-dynamics** when it is not near this nullcline. Notice also the representation of $x(t)$ **versus t** at the bottom of the graphics, similar to **Fig. 5.14.**

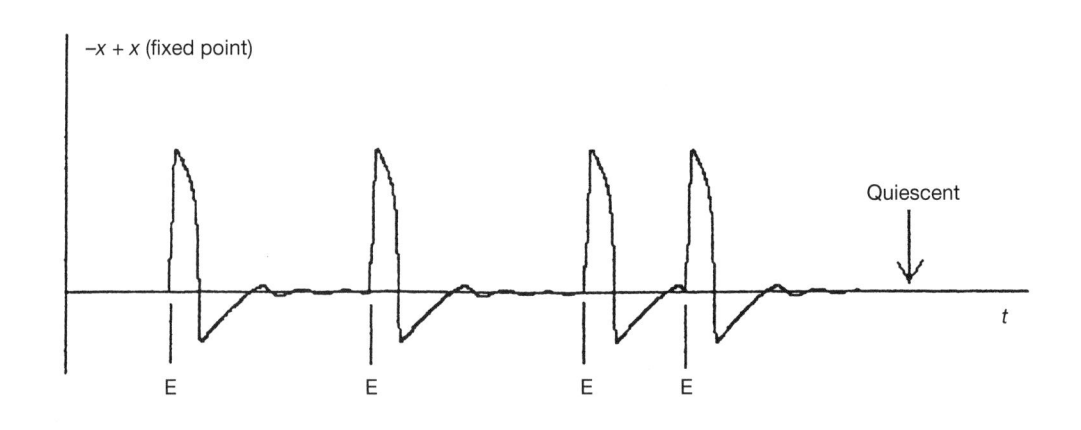

Next, see **the influence of several periodic excitations** of the neuron, with **varying degrees of rapidity.** Can you find situations in which **the neuron does not fire** at each excitation, as before? What are those conditions? Can you tell from the phase space why this is so?

Before proceeding, consider this:

▉ Think 5.4

Based on the **above numerical information,** try to predict the **minimum value of *EXC*** for *z* = –0.25 that can **excite the neuron to fire.** In **Explore 5.4B,** you will probably find that **you are wrong!** See if you can think of why this information is not sufficient to answer the question. ▉

Explore 5.4B

Now, for *z* = –0.25, slowly decrease the value of *EXC* by <D>. Determine the **minimum value of *EXC*** that is required to make the **neuron fire,** and compare it with the estimate in **Think 5.4A.** Which is larger, and why? [Hint: What do the above values *z* = **–0.3456 and *z* = –0.336** represent, and do these relate to the present question?

Explore 5.4C

Now that you have found that a rather large value of *EXC* is required to fire the neuron, **set *EXC* = 0,** but now **turn on the noise, *NZ*.** Note that the maximum value of the **noise is only *NMAX* = 0.3,** but watch what happens to the neuron **after some time.** Note that *NMAX* ≥ *NZ* ≥ –*NMAX*. Then **turn the noise off for a while,** and then **back on again.** Do you understand why it takes a while to fire, and at such a small value of *NMAX*? This illustrates the **disruptive effect of noise on individual neurons.** Why isn't this important for the information transmission of a **large group** of adjacent neurons?

Special Resource

There is now an impressive and very welcomed **free open-source software** called **PhysioNet** available on the web at **http://www.physionet.org.** This site offers **free access** to large collections of **recorded physiologic signals** and **related open-source software.** It is a public service of the **Research Resource for Complex Physiologic Signals,** funded by the National Center for Research Resources of the National Institute of Health.

References and Notes

1. For an introductory survey, see G. D. Fischback, Mind and Brain, *Scientific American,* Special Report, Sept., 1992. For **a few** diverse ideas **from 1999** concerning the brain's **dynamics** see: what maintains memory, *Science,* 15 Jan; Coffey study, *Neurology,* 13 July; use it or lose it, Science, 30 July, p. 661; spontaneous activity, *Science,* 23 July; cognition in a fruit fly (minibrain), *Nature,* 19 Aug; memory sequencing, *Science,* 3 Sept.
2. For a helpful discussion of neural mechanisms, sensory functions, motor functions, and higher functions (cerebral cortex), see: R. H. S. Carpenter, *Neurophysiology,* 3rd ed., Arnold, London; Oxford University Press, New York, 1996.
3. For those interested in more of the wonderful details of neural and other cellular processes, see: *Molecular Biology of The Cell*, 3rd ed., B. Alberts, D. Bray, J. Lewis, M. Raff, K. Roberts, and J. D. Watson, Eds., Garland, New York, 1994.
4. For a tour on a computer through many subjects in neurophysiology, but with no access to the programs used for modeling, see: R. J. S. Carpenter, *Neurophysiology,* 3rd ed., Arnold, London; Oxford University Press, New York, 1996; contains a PC-based NeuroLab disc.
4. A. L. Hodgkin and A. F. Huxley. A Quantitative Description of the Membrane Current and the Application to Conduction and Excitation in Nerve. *J. Physiology 117,* 500–544 (1952).
6. R. Fitzhugh. Impulses and Physiological States in Theoretic Models of Nerve Membranes. *Biophys. J. 1*:445–466 (1961).
7. J. Murray. *Mathematical Biology.* Springer-Verlag, Berlin, 1989.
8. L. Edelstein-Keshet. *Mathematical Models in Biology.* Random House, New York, 1988.

5.5 RELIABLE PACEMAKER, EXCITABLE HEART, AND ARRHYTHMIAS

The heart is clearly of central importance in our lives, and we take comfort in thinking that it pulses at a regular rate, which changes on demand depending on our activities. To accomplish this depends on the performance of a variety of dynamic factors, of which we will focus on two of the most basic elements. One concerns the dynamics of a **pacemaker,** and the other the response of systems of **excitable neurons,** discussed in the last section. The arrangement of the systems that are responsible for triggering the dynamics of the heart is shown in the schematic representation of the heart, **Fig. 5.15.**

5.5 Reliable Pacemaker, Excitable Heart, and Arrhythmias

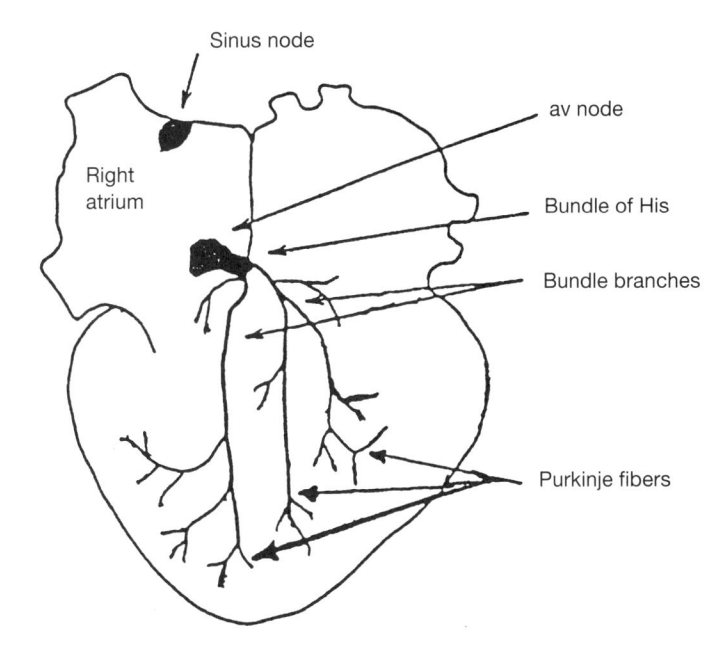

Fig. 5.15.
The pacemaker sinus node region, and excitable atrioventicular (av) node, each connected to various excitable neurons, which in turn cause the strong contraction of the muscular fibers in the lower ventricle walls.

The heart has four chambers. The right and left atria receive blood from the systemic and pulmonary blood vessels, respectively. They act as blood reservoirs for the right and left ventricles. These muscular chambers are the pumps that send the blood to the pulmonary and systemic systems, respectively, thus completing the sequential circulation of blood through both the lungs and systemic systems. The blood does not flow back into the left and right atria because of two one-way valves. A schematic diagram of the blood circulation is shown in **Fig. 5.16.** The details of the dynamic sequencing of the ventricle chamber is described in many references [1].

The cardiac rhythm is generated by a region in the right atrium, called the **sinus node,** which is a **pacemaker.** In humans, this pacemaker usually beats between 60 and 100 times a minute, varying according to the person's activities. The electrical impulse from the sinus node causes a contraction of its chamber, passing the blood to the right ventricle, and also exciting the atrioventricular node. This node then passes an electrical impulse through a collection of neurons called the Bundle of His, which branches out over the surfaces of the ventricles, and there causes the contraction of the muscle fibers in the walls of the ventricles. These are the power strokes that send the blood to the systemic and pulmonary blood vessels.

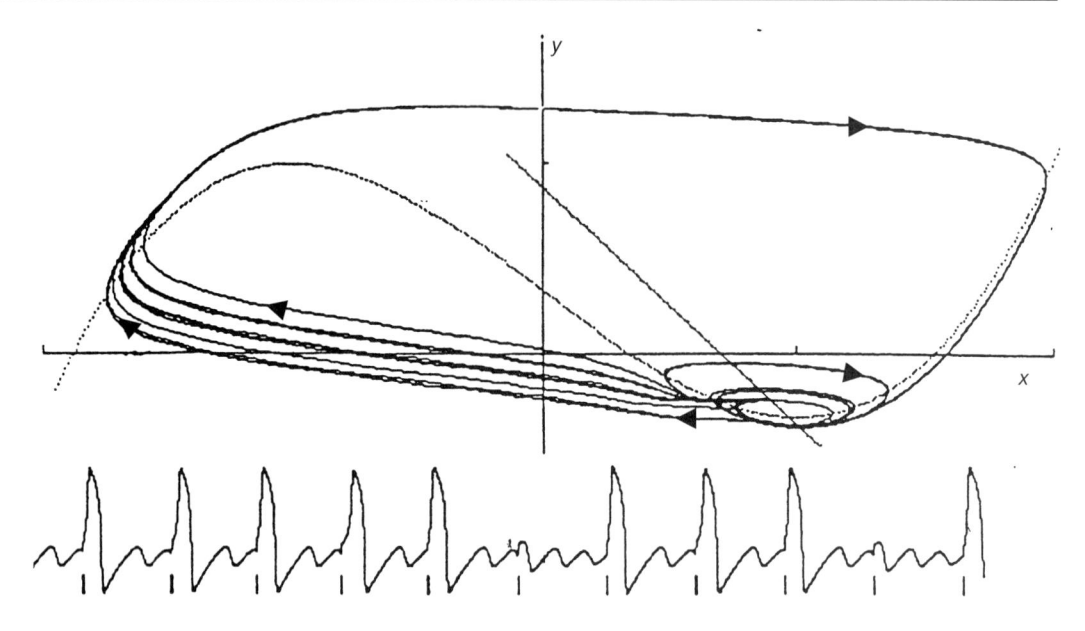

Fig. 5.16.
The response of an excitable system to the periodic excitations (at the vertical lines), $SN(t)$ in (5.5.2), which have too high a frequency, combined with too weak a strength, to maintain a regular response.

These details are presented, not because we are going to model them, but, on the contrary, because you should be aware of the great simplification that is represented by the following model. It should be abundantly clear that the heart is an extended system. The electrical activity spreads through the surface of the heart in the form of waves of considerable complexity [2]. There are, moreover, electrical connectivity problems that can occur in the ventricles, which are responsible for arrhythmias of the heart and are not addressed by the following dynamics. So what is discussed here is relatively simple.

Despite this, it has been found that some very simple models of the heartbeat can be quite instructive [3]. The present model is dynamically somewhat more complex, as it is based on Fitzhugh's dynamic model of the neuron, discussed in the last section, namely the **Bonhoeffer–van der Pol equation.** Using this model, we can explore **several aspects** of the heart's dynamics. The **first** study will concern the **reliability of a pacemaker** in the presence of environmental **noise,** and the **second** involves a study of the **periodic excitation** of the heart's excitable components. This latter study is a further application of the excitations considered in the previous section.

We consider first the dynamics of a **pacemaker** when subjected to environmental noise. The pacemaker is modeled by the Bonhoeffer–van der Pol equation, but now in its **periodic state,** and subjected to environmental **noise (NZ),** as in the case of Equation

(5.4.2). In contrast to that study, the dynamics is now **autonomously periodic,** and with **no deterministic excitations.** The condition required for **periodic dynamics** is that z be sufficiently negative, as indicated in Equation (5.5.1).

$$\frac{dx}{dt} = c\left(x - \frac{x^3}{3} + y + z + NZ\right) \quad (-0.345 \geq z)$$

$$\frac{dy}{dt} = \frac{(a - x - by)}{c} \quad (a = 0.7, b = 0.8, c = 3)$$

(5.5.1)

Explore 5.5A

The exploration here involves studying the possible influence of noise on a pacemaker, which might influence its **reliability.** Here one can study this for **different pacemakers,** as modeled by different values of **z,** and for **different** maximum magnitudes of the **noise, NZ.** This model is not intended to be based on any physical properties of the sinus node, but is used simply to show how the pacemaker dynamics need not be as simple a system as one might think. To illustrate this, **initially set z = –0.37,** so that the fixed point is **unstable** ($-0.345 \geq z$), and all solutions tend to a **stable limit cycle.** This will yield the desired **regular periodic motion** of the pacemaker dynamics. However, **the stability of this dynamics** to external noise depends on the value of z. The **smaller** it is than –0.345, the **more stable** is the periodic motion. To see what happens for a small amount of noise (maximum amplitude 0.1), **initiate the noise (<N>),** and see if the dynamics in this case is influenced **over a long time.** Repeat this for **other pacemakers,** by slowly increasing z (e.g., **z = –0.36, –0.355, –0.35,** and **–0.347**), and see the difference in the influence of the noise. Do you **understand the cause of this difference?** You can **explore further,** by changing the magnitude of the noise in the program on the line ######, and seeing how this **correlates with z.** At this stage, you might like to **automate** these **changes in z and NMAX** from the keyboard (see Program 5-4 for some help).

Now we turn from the dynamics of a pacemaker cell under the influence of noise to that of **an excitable system** under the influence of a **periodic excitation.** We use this as a simple model of the heartbeat produced by the steady sinus node periodic excitations. The equations are similar to (5.4.2) except that the excitation is now produced by the **periodic sinus node, SN(t),** which has **a variable frequency and strength.**

$$\frac{dx}{dt} = c\left(x - \frac{x^3}{3} + y + z + SN(t)\right) \quad (z = -0.33)$$

$$\frac{dy}{dt} = \frac{(a - x - by)}{c} \quad (a = 0.7, b = 0.8, c = 3)$$

(5.5.2)

Explore 5.5B

In **Program 5-5B,** the **model (5.5.2)** is used to **roughly** model the **strong pulsations** of the ventricular chambers when they are excited by the sinus node (through the av node). To do this, the value of $z = -0.33$ is used, which ensures that the **fixed point** is now a **global attractor** ($z \geq 0.333$). In this case, the periodic excitations from the sinus node **(SN)** are required to keep the "heart" beating properly (one hopes!—any irregularities in the heartbeat are called **arrhythmias**). The program starts with the **SN frequency = 0.25,** which can be **increased/decreased** using `<I>/<D>`, and its **strength = 0.35,** which can be made **stronger/weaker** using `<S>/<W>`. The program also starts with **a sound at both the SN pulse and** the **maximum excitation** of the "heart" (the minimum value of x). These sounds can be turned **on/off** using `<N>/<F>`. Begin by **decreasing the strength** slightly and seeing when the synchronized excitation is maintained by the SN. Return the strength to **0.35** and slowly **increase the frequency** to likewise see when there is no arrhythmia. Then set the frequency **near 0.4** and see what **strength is necessary** to ensure no arrhythmia. You can also discover other periodic forms of arrhythmias (excitations nonsynchronous with the SN period) and can **refine** these searches with **smaller step sizes** of the strength and frequency. The possible dynamics of the model (5.5.2) is quite rich and interesting.

An illustration of the complicated dynamics that can occur from the dynamics **(5.5.2)** in **Program 5-5B** is shown in **Fig. 5-16.** In this example, there is a combined effect from **too large a frequency** and **too weak an excitation strength** to maintain a regular response from this excitable system, illustrating an **extreme "arrhythmia."**

A few general references to the history of the dynamic studies of the heart are given in reference **[4]**, and some references related to various arrhythmias and **related diseases** can be found in reference **[5]. But also note the following Special Resource.**

Special Resource

There is now an impressive and very welcomed **free open-source software** called **PhysioNet** available on the web at **http://www.physionet.org.** This site offers **free access** to large collections of **recorded physiologic signals** and **related open-source software.** It is a public service of the **Research Resource for Complex Physiologic Signals,** funded by the National Center for Research Resources of the National Institute of Health.

References

1. R. D. Bauer, R. Busse, and E. Wetterer. Biomechanics of the Cardiovascular System. p. 619 in *Biophysics,* W. Hoppe, W. Lohmann, H. Markl, and H. Ziegler, eds., Springer-Verlag, New York, 1983.
2. A. T. Winfree. Electrical Turbulence in Three-Dimensional Heart Muscle. *Science, 266,* 11 Nov. 1994, pp. 1003–1006.

    ——— The Geometry of Excitability, in *1992 Lectures in Complex Systems* (Vol. 5 in Santa Fe Institute Studies in the Science of Complexity, L. Nadel and D.L. Stein, eds.), Addison-Wesley, Reading, MA, 1993.
3. L. Glass. Dynamics of Cardiac Arrhythmias. *Physics Today,* Aug. 1996, pp. 40–45.
4. **History:**

    V. I. Arnold. *Chaos, 1,* 20–24 (1991) (originally in Moscow University diploma dissertation, 1959); L. Glass, *Chaos 1,* 13–19, 1991 (reference to Builder and Roberts in 1939).
5. **Arrythmias and Diseases:**

    A. Garfinkel, et al. Quasiperiodicity and Chaos in Cardiac Fibrillation. *J. Clinical Investigation, 99,* 305–314 (1997).

    R. L. Verrier, B. D. Nearing, and E. G. Lovett. Complex Oscillatory Heart Rhythm: A Dance Macabre. *J. Clin. Invest., 99*(2), pp. 156–157, 1997.

    A. L. Goldberger, B. J. Bhargava, B. J. West, and A. J. Mandel. Some Observations on the question: is ventricular fibrillation "chaos"? *Physica 19D,* 282–229, 1986.

    J. M. Davidenko, A. M. Pertsov, R. Salomonsz, W. Baxter, and J. Jalife. Stationary and Drifting Spiral Waves of Excitation in Isolated Cardiac Tissue. *Nature* (Lond.) *355,* 349–351 (1991).

6
Temporal Chaos and Fractal Structures

One of the turning points in understanding dynamics was the discovery by Poincaré that **deterministic mathematical equations,** such as those of Newton's, **predicted** that, in general, **empirical observations** are **nondeterministic,** regardless of the accuracy of measurements. This is the implication of the Poincaré–Birkhoff theorem, as discussed in **Sections 2.1 and 2.2.** It is related to the property of the exponential sensitivity of dynamics to their initial conditions (now referred to as **"sensitivity to initial conditions"**). This phenomena has already been seen in the **chaotic dynamics** of the various map dynamics in **Chapter 3.** In this chapter, we will explore a method to characterize the "strength" of such chaotic dynamics, which was introduced by the Russian mathematician, A. M. Lyapunov (1857–1918) around the beginning of the twentieth century. Following this we will consider a type of dimensional characterization of structures that are not simple smooth sets of points (**"fractals"** with **"fractal dimensions"**). Although this is most readily demonstrated mathematically, we will see how **such structures can arise from dynamics,** such as the random-plus-rule dynamics in Section 2.4. We begin with the characterization of chaos.

6.1 THE COMPUTATIONAL LYAPUNOV EXPONENTS OF MAPS

In this section we will see how some **properties** of this **chaotic motion** can be **quantified.** As has been seen, chaotic motion has **two important characteristics.** The **first** is that nearby states tend, on average, to separate exponentially fast, exhibiting **"sensitivity to initial conditions."** In other words, two initial states that are initially nearly identical rapidly develop into different behaviors. The second essential feature of chaotic motions is that the **dynamics is bounded** in the phase space (it does not go to infinity). This **requires nonlinear dynamics,** as was seen in the simple banking dynamics of Sections 1.3 and 1.5. So let us first look into the issue of **how fast nearby states tend to separate on average.** To do this we begin by investigating the dynamics of a **general map**

$$x(t + 1) = F[x(t)] \quad (6.1.1)$$

and seeing how rapidly two nearby solutions diverge from each other, on average, over long periods of time. Let us denote a **second solution** by

$$y(t + 1) = F[y(t)] \quad (6.1.2)$$

which differs from (6.1.1) only in the fact that their **initial conditions** are nearly equal, $y(0) \cong x(0)$. We then set

$$y(t) = x(t) + d(t) \quad (6.1.3)$$

where $d(t)$ is the **"distance" of separation** between their solutions. In the following we **assume** that $|d(t)| \ll |x(t)|$ **for all the values of t that are used.** For any finite time, this can always be accomplished by making $d(0)$ sufficiently small. If we substitute (6.1.3) into (6.1.2), and keep the **first-order approximation** in the small quantity, $d(t)$, we obtain

$$x(t + 1) + d(t + 1) = F[x(t)] + \left(\frac{dF}{dx}\right)_{x(t)} d(t) \quad (6.1.4)$$

which, with the help of (6.1.1), reduces to

$$d(t + 1) = \left(\frac{dF}{dx}\right)_{x(t)} d(t) \quad (6.1.5)$$

Note that the derivative must be evaluated at $x(t)$, which corresponds to the local value of t. In other words, (6.1.5) simply gives us the change in the distance between the two solutions in **one time step,** starting at $x(t)$.

In the next time step, by substituting $(t + 1)$ in for t in (6.1.1) and (6.1.4), and finally using (6.1.5), we obtain

$$d(t+2) = \left(\frac{dF}{dx}\right)_{x(t+1)} d(t+1) = \left(\frac{dF}{dx}\right)_{x(t+1)} \left(\frac{dF}{dx}\right)_{x(t)} d(t)$$

This now contains **the product** of the **two derivatives** evaluated at the successive locations, $x(t)$ and $x(t + 1)$, given by the dynamics (6.1.1). If we take yet another time step, substituting $(t + 1)$ in for t in the last expression, we readily find that

$$d(t+3) = \left(\frac{dF}{dx}\right)_{x(t+2)} = \left(\frac{dF}{dx}\right)_{x(t+1)} \left(\frac{dF}{dx}\right)_{x(t)} d(t)$$

So in general, if we determine the separation distance between the two solutions after **N steps,** starting with the distance $d(0)$, we find that it is given by the product of **N derivatives**

$$d(N) = \left(\frac{dF}{dx}\right)_{x(N-1)} \left(\frac{dF}{dx}\right)_{x(N-2)} \cdots \left(\frac{dF}{dx}\right)_{x(0)} d(0) \quad (6.1.6)$$

taken at the N locations, $x(0) \ldots x(N - 1)$. Note that if $d(0)$ is sufficiently small, then $d(N)$ can still satisfy the condition $d(t)| \ll |x(t)|$ for some finite range of t.

If the distance $d(N)$ increases on **average as an exponential function of the time**

$$d(N) \approx d(0) \exp(L \cdot N) \quad (6.1.7)$$

for **large values of N,** the dynamics is called **sensitive to the initial condition $d(0)$, provided that $L > 0$.** Comparing (6.1.7) with (6.1.6), we find that this exponential factor must be related to $F(x)$ by

$$\exp(L \cdot N) = \left(\frac{dF}{dx}\right)_{x(N-1)} \left(\frac{dF}{dx}\right)_{x(N-2)} \cdots \left(\frac{dF}{dx}\right)_{x(0)}$$

If we take the logarithm of both sides, and use the fact that $\log(A \cdot B) = \log(A) + \log(B)$, we arrive at an expression for the exponent L:

$$L(N) = \left(\frac{1}{N}\right) \text{sum } (k = 0, \ldots, N-1) \left(\frac{dF}{dx}\right)_{x(k)} \quad (6.1.8)$$

"sufficiently large N, and small $d(0)$"

This exponent will be referred to as the **computational Lyapunov exponent of the dynamics (6.1.1)**. In mathematical treatments, the expression "sufficiently large N" in (6.1.8) would be replaced by the limit of **N going to infinity** [preceded by $d(0)$ tending to zero]. This would yield the **mathematical Lyapunov exponent, L**. However, in all experiments and computations, N is necessarily **finite**, as is the difference **$d(0)$**, so N is simply taken large enough for $L(N)$ to remain **"reasonably constant"** as N changes. These considerations show that the mathematical L is an abstraction of scientific observations or computations. This again relates to issues that are similar to those discussed in **Section 4.1**, and quite **dramatically** in the **host–parasite interaction models, Section 5.2.** For a more general treatment of computational Lyapunov exponents in higher dimensions, see reference [1], and references therein.

While $L(N)$ is a useful characterization of sensitive dynamics, such as chaos, it should not be forgotten that it only expresses the **average separation rate** of two nearby solutions of (6.1.1), and not their instantaneous rate of separation at any particular time. This fact is important, as we will see later.

Explore 6.1A

In this discussion, nothing has been said about what value is to be used for $x(0)$. Now use **Program 6-1A** to examine the values of the computational Lyapunov exponent for the **logistic map, with $c = 4$,** but for different values of $x(0)$. Do you find "similar values" of $L[x(0)]$ for all $x(0)$? What are the "sufficiently large" values of N? Notice that the values of L are taken to be **"quite close"** if they differ by **less than 0.0001.** You can, of course, change this condition in the program. **Go into the program** and locate the line containing `defdbl`. The variables that follow this are computed in **double precision** of accuracy, so they are computed to 16 places of accuracy. You should see what happens if you return to the usual 7 places of accuracy, by removing this command. (placing an apostrophe at the beginning of the line) There is a **big** difference!

Now you can go on to explore the values of $L(c)$ for **different values of c.** Here there are dramatic effects, related to the character of the chaos, and to whether the dynamics is chaotic.

Think 6.1

What should be a **common feature** of all values of $L(c)$ when the dynamics is **periodic?** Similarly, what should be the common feature of all values of $L(c)$ if the dynamics is chaotic? From these results, what special values of c should yield $L(c) = 0$? What chaotic case should yield the largest $L(c)$?

Explore 6.1B

To illustrate these points, use **Program 6-1B** to consider the logistic map for **various values of c.** You select at least three values of c that you expect will yield very different values of $L(c)$, and obtain the computed value of $L(c)$. In particular, select one for which $L(c)$ should be "nearly zero," within our accuracy. To assist in this **selection of c,** refer back to Figs. 3.8, 3.9, and 3.10 in **Section 3.5.** In each case, determine how $L(c, N)$ varies with N (record some representative examples), thus obtaining a specific characterization for **"sufficiently large N,"** as it depends on c. If this $L(c, N) > 0$, the **dynamics is chaotic** for that **value of c.**

The **dependence of $L(c)$** [here denoted as $\lambda(c)$] **on c** is illustrated in **Fig. 6.1.** This was obtained by Shaw [2], using $N = 100{,}000$, and values of c differing by only 0.002. The negative values of $L(c)$ correspond to attractive periodic orbits, including the windows" with periods, 3, 5, and 6, and the "microcosm" 3 × 3 window inside the period 3 window. The scale used for $\lambda(c)$ was log to the **base 2,** rather than the natural logarithm.

It should again be emphasized that what was introduced here is the **computational Lyapunov exponent,** which is obtained from a **known function, $F(x)$,** in (6.1.1). When the dynamics is

Fig. 6.1.
The dependence of the Lyapunov exponent, $L(c)$, on c, for the logistic map.

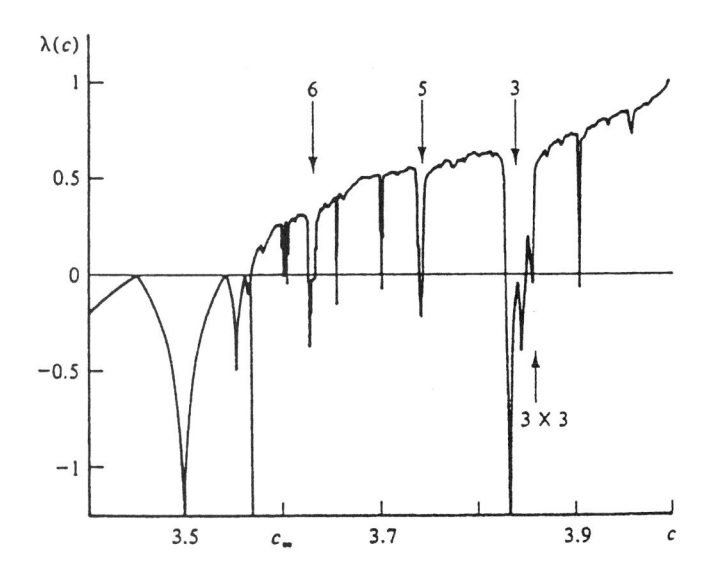

modeled by ordinary differential equations, the process involves the determination of $d(t)$ on a "continuous" range of time, which is computationally time consuming, since $d(0)$ must be very small, if t gets large. An interesting example of this is to determine **if our planetary system is chaotic (L > 0)** over **millions of years.** Examples of these heroic calculations can be found in reference [3].

The issue of determining a quantity like ***L* from empirical data** is much more difficult, and often impossible. This is because the "adjacent" data may not be obtainable, or tracked with sufficient accuracy for sufficient periods of time. Other **empirical approaches** to the concept of chaos is discussed in the fine reference **[4].**

References and Notes

1. E. A. Jackson. *Perspectives in Nonlinear Dynamics,* Vol. 2, Section 7.10. Cambridge University Press, Cambridge, 1990.

2. R. Shaw. Strange Attractors, Chaotic Behavior, and Information Flow, *Z. Naturforsch, 36a,* 80–112 (1981).

3. Studies of chaos in our **planetary system** have been made with careful numerical calculations. Here the integration traces the dynamics for 200 million years, and the Lyapunov exponent is only around 2×10^{-5}/year (that is, the so-called Lyapunov time is around 5×10^{-6} years). Some examples are:

   G. J. Sussman and J. Wisdom. Numerical Evidence that the Motion of Pluto is Chaotic, *Science, 241,* 433 (1988).

   G. J. Sussman and J. Wisdom. Chaotic Evolution of the Solar System, *Science, 257,* 56 (1992).

   J. Laskar, A Numerical Experiment on the Chaotic Behavior of the Solar System, *Nature, 338,* 237 (1989).

   J. Wisdom, M. Holman, and J. Touma. Symplectic Correctors: Integration Algorithms and Classical Mechanics, *Fields Inst. Commun., 10,* 217–244 (1996).

   N. Murray and M. Holman. The Origin of Chaos in the Outer Solar System. *Science, 283,* 1877, 1999; to obtain the Lyapunov exponent, this study reported following **two dynamic solutions** of Jupiter, Saturn, and Uranus in which **the initial conditions of Uranus differed by only 1.5 mm.**

4. For a clear introductory discussion of evaluations of the Lyapunov exponent in many situations, and other **empirical issues,** see F. C. Moon, *Chaotic and Fractal Dynamics*, Wiley, New York, 1992.

6.2 STRANGE SETS CALLED FRACTALS, WITH STRANGE DIMENSIONS

When we think of everyday structures around us, we typically think of representing them in terms of lines, planes, tubes, spheres, blocks, or what have you. Nothing particularly exotic here! All of these objects have simple dimensions associated with them. By common practice, a point is said to have dimension zero (it takes up no space!—now this is all **mathematics** that we are talking about here—we will get back to physical reality soon). Similarly, a line has dimension 1, a plane dimension 2, and **solid** tubes, spheres, blocks, etc. all have dimension 3. We live in a three-dimensional space, and we can't fit in anything that is four-dimensional in this space. So if we have a four-dimensional phase space, we can't represent it in our humble three-dimensional space.

Although this all seems pretty reasonable, the mathematical definition of these dimensions was not settled until around 1920, and they are now referred to as the **topological dimension** of a set of points, D_t. The basic idea behind the definition was due to Henri Poincaré, and if you are interested in his clever approach to the problem see reference [1]. Note that all of these dimensions have **integer values,** and the above set of points are all "smooth" and continuous collections of points.

We are going to consider a **more general definition of dimensions,** which allows us to characterize the structural features of **sets of points that do not have the smooth or continuous characteristics** of the above-mentioned sets. On the other hand, if the set of points turns out to have these properties, we want the new dimension to be the same as the topological dimension, D_t. What will be presented here is a fairly general and precise definition of these new dimensions that are associated with structures that are called **fractals.** There are other ways of introducing this concept and the related concept of chaos. The practical issues of how the concepts of chaos and fractal structures arise in **applied sciences and engineering** can be found in reference [2]. A very broad and simple introduction to the concepts of fractals and chaos, as they arise in the **life sciences,** can be found in reference [3]. A variety of references are given, which touch upon such diverse topics as science, architecture, musical inspirations, patterns from nature, and computer dynamics, etc., and a general mathematical and computational viewpoint.

So let us begin with these nice smooth and continuous sets, but define a **capacity dimension, D_c,** which will numerically equal D_t. Consider first a line of points on a plane, enclosed with little squares that have **sides of length ε.** Let

$N(\varepsilon)$ = **the minimum number of "cubes" with sides of size ε that is required to enclose all the points of a set S** (6.2.1)

Here the general term "cubes" is being used to refer to any dimensional equal-sided enclosure needed to enclose the points of the set S. Moreover we are interested in **the limit when ε is very small** (mathematics would take it going to zero, but physically we will settle for "very small," compared to the extent of the set).

Next let us determine $N(\varepsilon)$ for some **line of length L.** Clearly, it will not take more than $(L/\varepsilon) + 1$ such cubes to enclose all of the points of the line. Therefore, for $\varepsilon \ll L$

$$N(\varepsilon) \cong \left(\frac{L}{\varepsilon}\right) \quad \text{(a line)} \qquad (6.2.2)$$

Similarly, if we consider some plane set of points that lie within a rectangular area with sides L_1 and L_2, then the number of squares needed to cover this set is less than $(L_1/\varepsilon)(L_2/\varepsilon)$, say some fraction $\alpha(<1)$. Therefore in this case

$$N(\varepsilon) \cong \left(\frac{\alpha L_1 \cdot L_2}{\varepsilon^2}\right) \quad \text{(a rectangular area)} \qquad (6.2.3)$$

In these two cases the corresponding topological dimensions are

$D_t = 1$ (a line)   and $D_t = 2$ (a rectangular area)

Comparing these with the expressions (6.2.2) and (6.2.3), we see that the values of D_t equals the values of the exponents on ε. This suggests that if

$$N(\varepsilon) \cong \left(\frac{A}{\varepsilon^k}\right) \quad \text{(as ε becomes small)} \qquad (6.2.4)$$

then we can **identify the new capacity dimension as $D_c = k$.** Note that if we take the logarithm of both sides of (6.2.4), we obtain

$$\log[N(\varepsilon)] \cong \log(A) + k \log\left(\frac{1}{\varepsilon}\right)$$

so as ε **becomes very small,** the Log(A) term is negligible compared to the last terms, and we obtain by substituting $k = D_c$

$$D_c = \frac{\log [N(\varepsilon)]}{\log (1/\varepsilon)} \quad \text{("very small } \varepsilon\text{")} \quad (6.2.5)$$

This is the general expression for the **capacity dimension.** The issue of what **"very small ε"** means will now be discussed.

Mathematical Fractals

1. **There are an infinite number of points in the set S.**
2. **"Very small ε"** means that one takes the limit $\varepsilon \to 0$.

The construction of an infinite set of points, S, can be most easily achieved by mathematical **operations that are recursive** in character. Two famous examples will now be discussed.

S = The Cantor Middle-Third set

This set of points is generated by a repeated application of a rule to an existing set of points. One starts out with the set of points, **$S(0)$,** which is the line element $0 \le x \le 1$. The **repeated rule** that generates the Cantor set says that one should **remove (and discard) the middle-third set of points from any line element.** In the present set, $S(0)$, this means that we should **discard the points** $\frac{1}{3} < x < \frac{2}{3}$ from $S(0)$, leaving the set of points $S(1)$

$$0 \le x \le \tfrac{1}{3}; \quad \tfrac{2}{3} \le x \le 1 \qquad S(1)$$

One proceeds by applying the same **middle-third rule** to each of the segments of $S(1)$. This generates the set of points

$$0 \le x \le \tfrac{1}{9}; \quad \tfrac{2}{9} \le x \le \tfrac{3}{9}; \quad \tfrac{6}{9} \le x \le \tfrac{7}{9}; \quad \tfrac{8}{9} \le x \le 1 \quad S(2)$$

which now consists of **four line segments,** as compared to the two line segments of $S(1)$. This is illustrated in **Fig. 6.2.** Moreover, the first two line segments have an **identical structure to $S(1)$,** if we **scale their values of x** by multiplying x by 3. Similarly, the last two segments of $S(2)$, if shifted down by $\tfrac{6}{9}$, have this same scaling relationship to $S(1)$. When a set of points, $S(k)$, consists of subsets

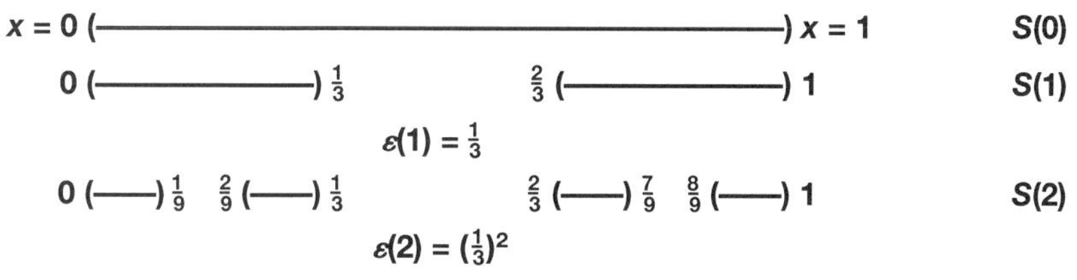

Fig. 6.2.
The first two steps in constructing the **Cantor middle-third set** of points, which consists of the **end points** of all the line segments of length $\varepsilon(k)$ in $S(k)$, as $k \to \infty$.

```
x = 0 (——————————————————) x = 1              S(0)
    0 (—————) 1/3        2/3 (—————) 1         S(1)
              ε(1) = 1/3
    0 (—) 1/9  2/9 (—) 1/3   2/3 (—) 7/9  8/9 (—) 1   S(2)
              ε(2) = (1/3)²
```

that all have the **same scaling relationship** to $S(k-1)$, we say that $S(k)$ **is self-similar to** $S(k-1)$.

▪ Think 6.2A

Construct the **set** $S(3)$, and show that it consists of **eight line segments,** and that suitable subsets of $S(3)$ are self-similar to $S(1)$ for a suitable scaling of x. Determine the subsets and the scaling factor. ▪

The **Cantor middle-third set** is the infinite set of points that remain after applying this rule an infinite number of times. It is not difficult to see, starting with **Fig. 6.1,** that the **only points** that remain after an infinite number of applications of this rule are **the end points** of the line segments that are produced in this process.

What is the capacity dimension of this set? Because of the self-similarity of the set, the answer is very easy to obtain. We see that we can cover all the points of this set by using $N[\varepsilon(k)] = 2^k$ line elements of size $\varepsilon(k) = (\frac{1}{3})^k$. If we substitute these into (6.2.5), **for any value of k,** we obtain,

$$D_c = \frac{\log(2^k)}{\log(3^k)} = \frac{k \cdot \log(2)}{k \cdot \log} = \frac{\log(2)}{\log(3)}$$

which is **independent of k because of the self-similarity of the Cantor set.** Thus

$$D_c = \frac{\log(2)}{\log(3)} \cong 0.63 \quad \text{(Cantor middle-third set)}$$

We see that the dimension of this infinite set of points is **less than the line segment** $S(0)(D_t = 1)$, **and greater than any finite set of points** $(D_t = 0)$.

Koch Triadic Curve

The second example of a fractal is the **Koch triadic curve.** In contrast to this infinite set of points, the **Koch triadic curve** provides an example of a **continuous fractal curve.** It likewise enjoys a **self-similar structure,** which makes its capacity dimension simple to evaluate. The **generator** of this set is the curve illustrated in **Fig. 6.3(A)**.

Beginning with the line segment $0 \leq x \leq 1$, one repeatedly applies the rule to change any straight line segment into this shape, by scaling the size of the generator (in both directions) so that its

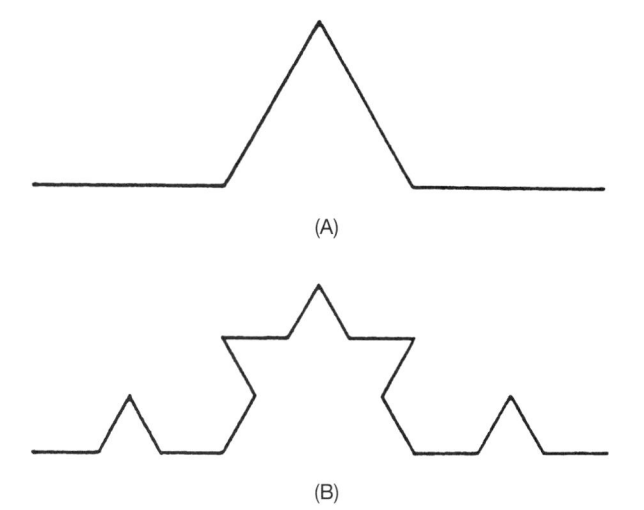

Fig. 6.3.
(A) The **generator** of the Koch triadic curve consists of this structure, which has line segments of length $\frac{1}{3}$ and a base length of unity. (B) The **first step** in constructing the Koch curve, where all line segments are now of length $(\frac{1}{3})^2$.

base equals that of one line segment. Thus, the first application of this rule simply yields **Fig. 6.3(A)**. The next application of the rule produces the line segment shown in **Fig. 6.3(B)**.

We will refer to the curves in **Fig. 6.3 (A and B) as $K(1)$ and $K(2)$** and, in general, after k applications of the rule, the set will be referred to as $K(k)$. To cover this set of points, we can use $N(k) = 4^k$ line segments, each of length $\varepsilon(k) = (\frac{1}{3})^k$. Substituting these into (6.2.5), we obtain

$$D_c = \frac{\log(4^k)}{\log(3^k)} = \frac{\log(4)}{\log(3)} \cong 1.26 \quad \text{(Koch triadic curve)}$$

The independence from the value of k is again a consequence of the **self-similarity** of this curve. This curve has a **dimension greater than any smooth curve ($D_t = 1$), but less than any planar region ($D_t = 2$).**

Explore 6.2B

Refer to **Program 6-2A,** which generates a structure that might be referred to as a **"Cantor comb," shown in Fig. 6.4.** Determine the capacity dimension of this comb. **Is it a fractal?** [Hint: assume that the length of the first level is 1, and that each vertical step is $(\frac{1}{9})$, which is about right. Now **cover this comb** with rectangles of height $(\frac{1}{9})$, and widths of length $\varepsilon(k) = (\frac{1}{3})^k$, for each k. What is $N(k)$ for the **entire comb** up to that k? Begin with $k = 0, 1, 2,$ and 3, then see if you can generalize this result. Even if

Temporal Chaos and Fractal Structures

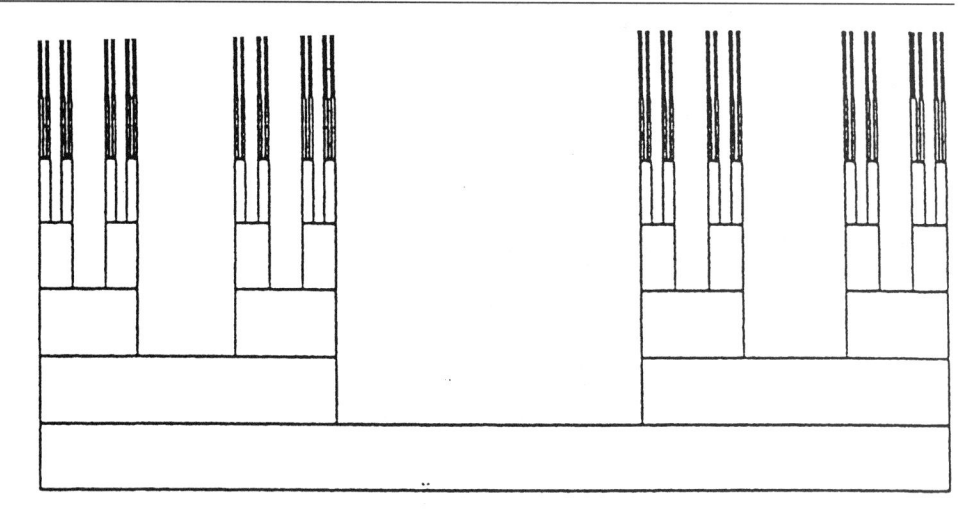

Fig. 6.4.
This "Cantor comb" is generated by **Program 6-2A.** Is it a fractal?

you can't do this, you may be able to reason whether the Cantor comb is fractal.]

A "Dynamically Generated Fractal"

Although fractals are associated with sets of points, and in this sense are geometric in character, nonetheless, such sets of points can be generated by dynamic processes. A nice example of a "fractal" that is generated by a **dynamic process** is the one considered in **Explore 2.4B** and discussed further in **Explore 2.4C.** This dynamics is repeated in **Program 6-2B,** but now with an **added color correlation effect.**

Explore 6.2B

Run **Program 6-2B,** which is a colorful version of **Program 2-4C.** Even though the number of dynamic points that can be computed is necessarily finite, a **reasonable capacity dimension** can be attributed to this set of points, because of the **self-similarity** of the set of points (within the limits of computational accuracy). This (**uncolored**) set of points is illustrated in **Fig. 6.5.**

Think 6.2B

Determine **the generator and the rule** that can be used to cover this set of points generated in Program 6-2B when applied an

6.2 Strange Sets Called Fractals, With Strange Dimensions

Fig. 6.5.
A set of points generated by a **random-plus-rule form of dynamics in Program 6-2B.** The **mathematical** form of this structure is known as **Sierpinski's sieve.** But **how can this structure come about from a dynamics with a totally random component?** See the color "correlations" in Program 6-2B.

"infinite" number of times. From this, infer the capacity dimension of this set. Although this may be clear from the figure in Program 6-2B, you may find it useful to refer back to **Explore 2.4C** and the associated **Program 2-4C**.

Think 6.2C

Explain why the result of Think 6.2B need not depend on the limitations of a computer, by referring back to the general theoretical idea raised in **Think 2.4A.** That is, establish that the mathematical (infinite precision processes) capacity dimension can be inferred from theoretical reasoning. Refer again to **Explore 2.4C.**

Think 6.2D

In contrast with the black and white Fig. 6.5, the display in Program 6-2B makes uses of **colors** to demonstrate **one type of correlation between where successive values of Z end up in the triangle**. Can you think of how to use this method to uncover a **"more interesting" type of correlation?**

Physical Fractals?

When it comes to making physical measurements, one is always limited in the possible precision of the measurement. Hence, by a **"physical fractal,"** it can only mean that the relationship (6.2.4) holds for a fixed value of k when ε varies over a limited range of small values, say $\varepsilon = a$, $a/10$, $a/100$ (which would be **two orders of magnitude**), where a is some suitable small number. It has been pointed out [4] that the claim that some physical structures are "fractal" may have little or no empirical support, because the range of ε may not even be one order of magnitude, much less two. This limitation stands in contrast to the theoretical concepts of the originator of this fractal view of nature. [5].

Random References

Formal and Applied Basics

1. E. A. Jackson. *Perspectives of Nonlinear Dynamics,* Vol. 1, Section 2.6, and Appendix B. Cambridge University Press, Cambridge, 1990.
2. F. C. Moon. *Chaos and Fractal Dynamics: An Introduction for Applied Scientists and Engineers.* Wiley, New York, 1992.

Life Sciences

3. L. S. Liebovitch. *Fractals and Chaos: Simplified for the Life Sciences.* Oxford University Press, Oxford, 1998.

Science and "Fractals"

4. D. Avhir, O. Biham, D. Lidar, and O. Malcai. Is the Geometry of Nature Fractal? *Science, 279,* 39, 1998. Made a seven year survey of all (96) experimental papers in *Phys. Rev.* A to E and *Phys. Rev. Lett.* reporting fractal analysis of data, and found the mean number of orders that were used was only 1.3.
5. H. E. Stanley and N. Ostrowsky (Eds.). *On Growth and Form: Fractal and Non-fractal Patterns in Physics.* Martinus Nijhoff, Boston, 1986.
6. B. B. Mandelbrot. *The Fractal Geometry of Nature.* Freeman, New York, 1983.
7. B. B. Mandelbrot. Fractal Geometry: What Is It and What Does It Do? In *Fractals in the Natural Sciences,* M. Fleischmann, D. J. Tildersly, and R. C. Ball (Eds.), Princeton University Press, Princeton, 1990, pp. 3–16.

Architecture

8. R. Eglash. Fractals in African Settlement Architecture. *Complexity,* 4(#2), 21–29, 1998. Remarkable similarities between some African settlement formations and known fractal formations.

Musical Inspirations

The concept of chaos and fractals have inspired a number of musical explorations. A few are:

9. Z. Davids. Heartsongs: Musical Mappings of the Heartbeat. Ivory Moon Recordings, 22 Rutgers Road, Wellesley, MA 02181. Compositions

whose melodic line is obtained directly from the time intervals between heartbeats (averaging out fluctuations due to movement or breathing), which are mapped onto integers ranging from 1 to 18, corresponding to the diatonic scale, to which the composer is free to choose the rhythm and harmonic accompaniment for each melody.

10. M. Gardner. *Fractal Music, Hypercards and More.* Freeman, New York, 1992.
11. A. T. Scarpelli. 1/f Random Tones: Making Music with Fractals. *Personal Computing, 3,* 17–27 (1979).

Patterns from Nature and Computer Dynamics

12. E. Porter and J. Gleick. *Nature's Chaos.* Viking Penguin, New York, 1990.
13. J. Briggs. *Fractals: The Patterns of Chaos.* Simon & Schuster, New York, 1992.
14. M. Fleischmann, D. J. Tildersley and R. C. Ball (Eds.). *Fractals in the Natural Sciences.* Princeton University Press, Princeton, 1990.
15. H. Lauwerier. *Fractals: Endlessly Repeated Geometrical Figures.* Princeton University Press, Princeton, 1991. A fine book, with numerous simple mathematical and computational examples, and **a collection of basic programs.**
16. M. Barnsley. *Fractals Everywhere.* Academic Press, Boston, 1988. A more advanced book that focuses on how fractal geometry can be used to simulate some structures in the physical world. The results are often impressive, but **shed no light on the physical dynamics** responsible for these structures.

Imagination Galore!

17. J. Briggs and F. D. Peat. Turbulent Mirror: *An Illustrated Guide to Chaos Theory and* **the Science of Wholeness.** Harper & Row, New York, 1989. A mind-stretching encounter, with wonderful illustrations by Cindy Travernise.

6.3 "SYSTEMS" AND "ENVIRONMENTS": ROLES OF "CHAOS," FRACTALS, AND FUNCTIONAL AND ADAPTIVE INTERACTIONS

To discuss the topics listed above, it is important to introduce a new level of description of what is involved in the dynamics of nature. Up to now, and quite commonly in discussions of dynamics, one uses the term **"system"** in what seems be a rather obvious sense, with "their" dynamics perhaps represented by some recursive equations. We have also described in a similar manner the interaction between different systems, or species. In order to describe the concept of **functional** and **adaptive** forms of **dynamics** or physical **structures,** we need to refine our ideas about what we mean by a **"system"** and its **"environment."** So let us look into these ideas briefly, before focusing on examples of today's functional or adaptive forms of dynamics.

All physical systems arise out of some environmental context, and then interact with their future environment in various ways. A **"system"** is a collection of matter that has some type of "regular" dynamics in a **domain, D.** Whatever is **outside D** is typically referred to as the system's **"environment."** But a system is always influenced by its environment, to one degree or another, and the type of influence can vary greatly, depending on the time scales that are considered. From an extended point of view, over sufficiently long time scales, **all physical systems** are **ephemeral structures.** Even mountains only persist for a finite amount of time, as will the Earth. The reason for this is that all systems are strongly influenced by their environment over **"sufficiently long"** periods of time. So **"systems"** and their **"environment"** are **time-scale dependent concepts.** Appreciating this coupling is essential for the future of the world, as was briefly noted in issues related to the **human population of the world** and finite natural resources **(Section 1.5).**

What is most enlightening in **appreciating the world in which we exist** is to reflect upon the remarkable and **very special** sequence of dynamic processes, extending over **billions of years,** that produced the **nature** that we now experience. Many of the presently known details involved in these remarkable evolutionary events should surely give anyone an appreciation of our very special planet, with its higher forms of life.

As a **thumbnail sketch, consider the following [1].** In the early stages of the universe this involved such evolutionary sequences as the formation of "stars" from gaseous nebula (their "environment"); the formation of the "planets" and "asteroids" from the "stellar nebula" (their "environment"); the initial formation of the "early Earth," benefiting from **chaotically induced collisions** with specially "mineral-enriched asteroids," all associated with the stellar nebula of this "particular star" (the early "environment," see References and Notes). From this followed the slow cooling of the Earth, maintained in part by meteoroid impacts, one near-impact being thought to have ripped out the "moon system." This was followed by a sequence of dynamic processes that produced the formation of the "ocean and our atmosphere," all contingent on their "earthly mineral environment." Following this were the tectonic plate motions, generating divergent continents on which different forms of life could originate, possibly first near hot thermal vents in the oceans, all being coupled to the "moon environment," generating massive "ocean tides," which in turn are part of the "moon's environment," causing it to recede from the earth (presently at around 4 cm per century). After this followed the generation of "simple forms of life (bacteria)," and the **"different forms of living systems,"** all dependent on their **"local environments,"** which, of course, had to be provided by the overall **"earth environment."** For many more details that will make you appreciate the title of the book, see reference [1].

Following the initial stage of inanimate evolution, a profoundly different form of dynamics arose, giving rise of **living systems** (around 3.8 billion years ago). While there is no general consensus as to how this came about, a few theorists credit the great variability of the above **"local environments"** (e.g., see extremophiles [1]). The dynamics of living systems is fundamentally different from **inanimate systems** because all **living systems** are **far from in equilibrium** with their "environment." This means that **living systems** must **exchange matter and energy** with their **environment** in order to sustain their structure and dynamics. **Inanimate examples** of this type of energy and/or matter exchange with the environment is the basis of the **autocatalytic oscillators** discussed in **Section 5.3.** In **living systems** this exchange sustains **metabolic processes** involving **chemical reactions** of widely different ranges of complexity [2, 3]. The details of this need not concern us here, but the basic requirement of energy/matter **exchange with the environment, and internal metabolic processes,** are the first **essential features of all living systems** [4].

As evolution progressed, and **cognitive capabilities** of increasing complexity arose in some living systems, their interactions with their "environment" (now including other cognitive systems, of course) required the development of entirely **new forms of interactions** and related dynamics. Roughly, the interactions initially involved acquiring **information,** then **internally processing this information** (consciously or other wise), to **respond** in some manner to what was happening in the environment. What the nature of the information acquired, the processing performed, and manner of response actually involves, varies immensely over the realm of cognitive systems (e.g., think of the differences in the environmental accomplishments of ants, honeybees, birds, tigers, whales, chimpanzees, and humans). For a wonderfully broad introduction to the vast diversity of life and its accommodations to its environments, see references [5, 6]. For a spectrum of ideas concerning the cognitive activities of humans, see the sampling in reference [7].

To describe some ideas about the **connection of living systems with their environment,** two useful concepts are:

> **Functional Dynamics and Structures:** there are two forms of these:
>
>> **Internal:** internal features that aid in sustaining a living system in a state far from equilibrium with its environment. There is no more dense and diverse collection of functional dynamics than that **within a human cell [8, 9].**
>>
>> **External:** features external to the system that aid in keeping the system alive in its environment.
>
> **Adaptive Dynamics:**
>
>> **Internal:** a form of internal dynamics that adjusts a functional dynamics so that it remains functional in a randomly changing environment, within specific limits, of course.
>>
>> **External:** a change in dynamics **relative to the environment** that accommodates to new life-threatening environmental changes.

In this section are a few explicit examples that certain multicellular living systems appear to employ either **fractal structures** for **metabolic** functional dynamics, and, possibly, **chaotic dynamics** for adaptive purposes in **informational interactions.** More generally, the importance of fractal structures in the living world has now been documented over scales ranging from **bacteria to whales,** which is **over 21 order of magnitude. There is nothing like this in the inanimate world.**

The first example is the "fractal" structure of the **human lung.** This can reasonably be referred to as a fractal, since it contains **over twenty "generations" of branching.** This is illustrated in **Fig. 6.6 (a).** Figure 6.6(b) is a computer generated fractal model of the blood circulation of vertebrates.

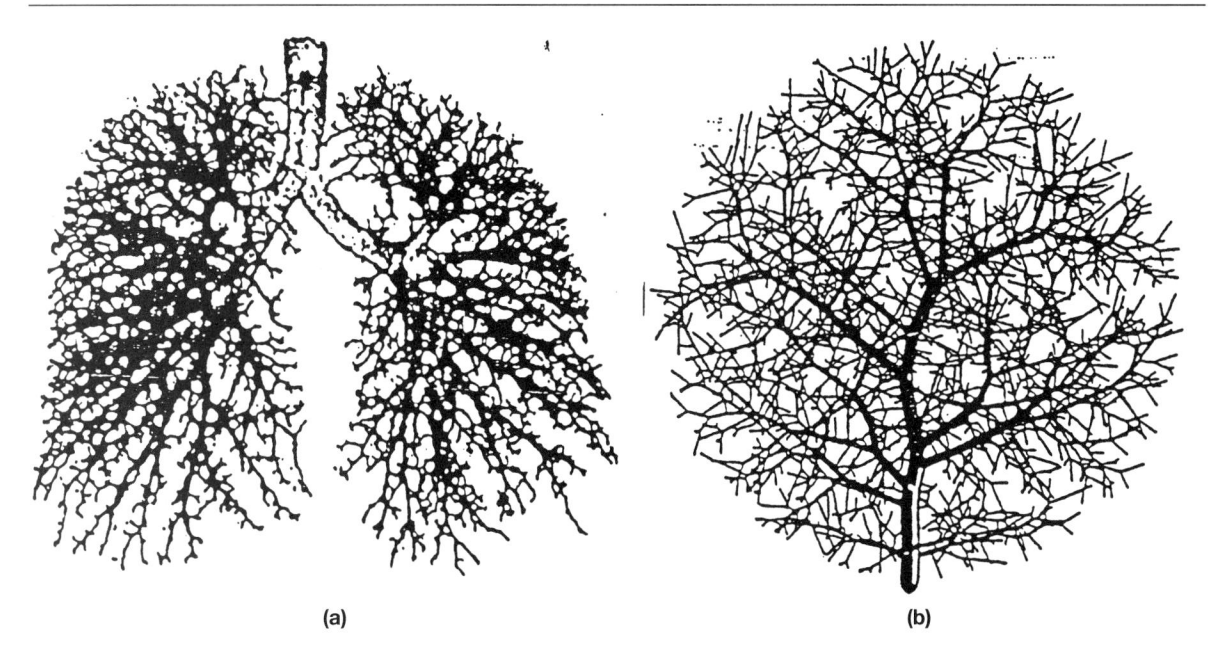

Fig. 6.6.
(a) The structure of a human lung, with over twenty generations of branching. (b) A computer generated model of a vertebrate vascular system, using fractal geometry.

Fractal Structures of Metabolic Interactions with Environment

There is no reason in the world that one would look into the idea that fractals have anything to do with metabolism, unless there were some initially perplexing observations that led to this connection. These observations come from **allometric scaling features** that have long been observed in biology.

A nice example of functional structures that are related to fractals are the circulatory systems of plants and animals. The functional character of the dynamics within some circulatory systems has been related to energy transport through the fractal pathways of transportation [10]. Three requirements that were considered were:

1. The network must reach all parts of the three-dimensional body.

2. The amount of energy that is expended in the transportation of resources should be minimized.

3. The ends of the network, such as the capillaries in a circulatory system, must all be the same size for all species, because their cells are roughly similar in size.

"The dependence of a biological variable Y on body mass M is typically characterized by an allometric scaling law of the form $Y = Y_0 \cdot M^b$, where b is the scaling exponent and Y_0 is a constant that is characteristic of the kind organism." **Examples** of this are: for the metabolic rates of entire organisms, $b = \frac{3}{4}$; for rates of cellular metabolism, heartbeat, and maximal population growth scale as $b = -\frac{1}{4}$; times of blood circulation, embryonic growth and development, and life-span scale as $b = \frac{1}{4}$; the cross-sectional areas of mammalian aortas and of tree trunks, scale as $b = \frac{3}{4}$ **[10]**. An example of the distribution network of a plant is illustrated in Fig. 6.7.

After considering these few specific examples, you are invited to explore the following "smorgasbord" of readings.

References and Notes

1. For a detailed study in support of the thesis that higher forms of life are very unlikely in the universe see P. D. Ward and D. Brownlee, *Rare Earth: Why Complex Life is Uncommon in the Universe* (Copernicus; Springer-Verlag, New York, 2000). For another general review of the sequential evolution of many physical structures in nature (not addressing issues concerning the dynamics of "systems" and their "environments") see C. Emiliani, *Planet Earth: Cosmology, Geology, and Evolution of Life and Environment,* Cambridge University Press, Cambridge, 1992.

Chaotic Planetary Dynamics

1. On meteorites, and the nonchaotic "Moon–Jupiter environment" see I. Peterson, *Newton's Clock: Chaos in the Solar System,* Freeman, New York, 1993, an interesting presentation of an account of Poincaré's error in his prize-winning article in *Acta Mathematica* and the subsequent "rectifications."

Fig. 6.7.
(a) The topology of the branching of the vascular system in plants. (b) The symbolic three-dimensional character of the conducting tubes around the nonconducting core tissue **[10]**.

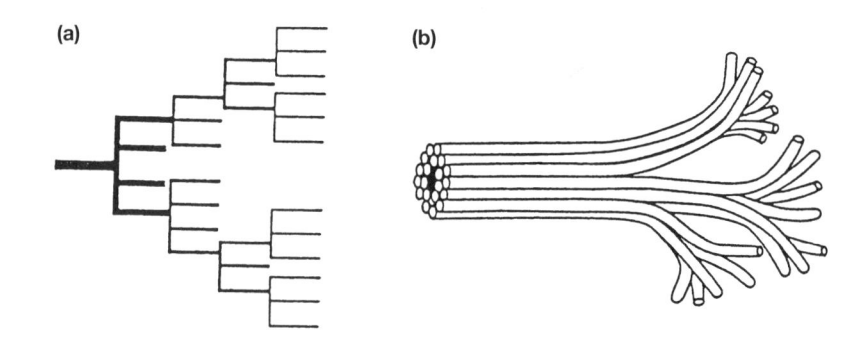

B. Parker. *Chaos in the Cosmos: The Stunning Complexity of the Universe.* Plenum, New York, 1996.

N. Murray and M. Holman. The Origin of Chaos in the Outer Solar System. *Science, 283,* 1877, 19 Mar. 1999.

G. J. Sussman and J. Wisdom. Numerical Evidence that the Motion of Pluto is Chaotic, *Science, 241,* 433 (1988).

G. J. Sussman and J. Wisdom. Chaotic Evolution of the Solar System, *Science, 257,* 56 (1992).

J. Laskar, A Numerical Experiment on the Chaotic Behavior of the Solar System, *Nature, 338,* 237 (1989).

R. Malhotra. Migrating Planets. *Sci. Amer.,* 56–63, Sept. 1999.

Murray and M. Holman. The Origin of Chaos in the Outer Solar System. *Science, 283,* 1877–1881, 19 Mar. 1999.

E. W. Thommes, M. J. Duncan, and H. F. Levison. The Formation of Uranus and Neptune in the Jupiter-Saturn Region of the Solar System. *Nature, 402,* 635–638, 9 Dec., 1999.

R. Malhotra. *Chaotic Planet Formation. Nature, 402,* 599, 9 Dec., 1999.

Metabolism and the Evolution of Life

2. H. J. Morowitz. *Beginnings of Cellular Life: Metabolism Recapitulates Biogenesis.* Yale University Press, New Haven, 1992.

3. H. J. Morowitz. A Theory of Biochemical Organization, Metabolic Pathways, and Evolution. *Complexity, 4*(4), 39–53, July/Aug., 1999.

4. For some of the amazing details of the energy-harvesting methods that have evolved in living systems (such as those involving mitochondria, eucaryotic cells, and chloroplasts in plant cells), see Chapter 14 in *Molecular Biology of The Cell,* (3rd ed., B. Alberts, D. Bray, J. Lewis, M. Raff, K. Roberts, and J. D. Watson (Eds.), Garland, New York, 1994.

5. To acquire some appreciation of the **diversity of life** that has evolved on earth see E. O. Wilson, *The Diversity of Life*, W. W. Norton, New York, 1992.

6. You should not miss this classic, B. Hölldobler and E. O. Wilson, *Journey to the Ants: A Story of Scientific Exploration*, Belknap Press of Harvard University Press, Cambridge, 1994.

7. A sample of some recent studies of the **development of cognitive dynamics in the human brain:**

   T. W. Deacon. *The Symbolic Species: The Co-evolution of Language and the Brain.* W. W. Norton, New York, 1997.

   G. M. Edelman. *Bright Air, Brilliant Fire: On the Matter of the Mind.* Basic Books, New York, 1992.

   P. Kruse and M. Stadler (Eds.). *Ambiguity in the Mind and Nature: Multistable Cognitive Phenomena* Springer-Verlag, Berlin, 1995.

   J. R. Harris. *The Nurture Assumption: Why Children Turn Out the Way They Do.* The Free Press, New York, 1998.

   S. Pinker. *How the Mind Works.* W.W. Norton, New York, 1997.

J. H. Barkow, L. Cosmides, and J. Tooby (Eds.). *The Adapted Mind: Evolutionary Psychology and the Generation of Culture.* Oxford University Press, New York, 1992.

T. R. Blakeslee. *Beyond the Conscious Mind: Unlocking the Secrets of the Self,* Plenum, New York, 1996.

V. S. Ramachandran and S. Blakeslee. *Phantoms in the Brain: Probing the Mysteries of the Human Mind.* William Morrow, New York, 1998.

F. Crick. *The Astonishing Hypothesis: The Scientific Search for the Soul.* Charles Scribner's Sons, New York, 1994.

8. G. Brown. *The Energy of Life: The Science of What Makes Our Minds and Bodies Work.* The Free Press, New York, 2000.

9. B. Rensberger. *Life Itself: Exploring the Realm of the Living Cell.* Oxford University Press, New York, 1996.

Scaling in Biology

G. B. West and J. H. Brown (Eds.). *Scaling in Biology.* Oxford University Press, 1999.

Allometric Scaling and Fractals (Allometric: the relative change of a part of an organism in comparison with the whole)

10. G. B. West, J. H. Brown, and B. J. Enquist. A General Model for the Origin of Allometric Scaling Laws in Biology. *Science, 276,* 122–126, April 4, 1997.

   G. B. West, J. H. Brown, and B. J. Enquist. A General Model for the Structure and Allometry of Plant Vascular Systems. *Nature, 400,* 664, 12 Aug. 1999.

   G. B. West, J. H. Brown, and B. J. Enquist. The Fourth Dimension of Life: Fractal Geometry and Allometric Scaling of Organisms. *Science,* 1677, June 4, 1999.

   N. Williams. Fractal Geometry Gets the Measure of Life's Scales. *Science, 276,* 34, April 4, 1997.

   The fractal character of some ecosystems, produced by species of different sizes, is illustrated in **[5], ff. 207.**

Spatial Scaling and Biodiversity

B. T. Mime. Applications of fractal geometry in wildlife biology, pp. 32–69 in *Wildlife and Landscape Ecology: Effects of Pattern and Scale.* J. A. Bissonette (Ed.). Springer-Verlag, New York, 1997.

M. E. Ritchie and H. Olff. Spatial Scaling Laws Yield a **Synthetic Theory** of Biodiversity. *Nature,* 400, 557–560, Aug. 10, 1999. Another mode of "understanding"; **Synthetic** in logic: not true by the meaning of component terms alone but by virtue of observation and not resulting in self-contradiction.

Mixture of Functional Dynamics

A. L. Goldberger, D. R. Rigney, and B. J. West. Chaos and Fractals in Human Physiology. *Sci. Amer., 262,* 43–49. 1990.

Books on Fractal Structures in Nature

J. Bassingthwaighte, L. Liebovitch, and B. West. *Fractal Physiology.* Oxford University Press, Oxford, 1994.

P. M. Lannaccone and M. Khokha. *Fractal Geometry in Biological Systems.* CRC Press, Boca Raton, 1996.

H. Takayasu. *Fractals in the Physical Sciences.* Wiley, Chichester, U.K., 1992.

P. Ball. *The Self-Made Tapestry: Pattern Formation in Nature.* Oxford University Press, Oxford, 1999.

Chaotic Aspects of Heart Dynamics. In the heart, "chaotic" dynamics may be either life-threatening or an indication of health. The distinction is based partially on the time scales and whether the chaos is the dominant mode of the dynamics or a small component added to the primary functional dynamics. Much has yet to be learned about the details of these clinical observations.

Life-threatening

A. Garfinkel, P.-S. Chen, D. O. Walter, H. S. Karagueuzian, B. Kogan, S. J. Evans, M. Karpoukhin, C. Hwang, T. Uchida, M. Gotoh, O. Nwasokwa, P. Sager, and J. N. Weiss. Quasiperiodicity and Chaos in Cardiac Fibrillation. *J. Chin. Invest., 99,* 305–314, 1997.

L. Glass. Dynamics of Cardiac Arrhythmias. *Physics Today,* 40–45, Aug. 1996.

Adaptive and Precursor Dynamics

A. L. Goldberger. Non-linear Dynamics for Clinicians: Chaos Theory, Fractals, and Complexity at the Bedside. *Lancet, 347:* 1312–1314, 1996.

A. L. Goldberger. Fractal Variability Versus Pathologic Periodicity: Complexity Loss and Stereotypy in Disease. *Perspectives in Biology and Medicine, 40*(4), 543–561, Summer, 1997.

W. J. Freeman. Simulation of Chaotic EEG Patterns with a Dynamic Model of the Olfactory System. *Biol. Cybern., 56,* 139–150, 1987.

W. J. Freeman. *Societies of Brains: A Study in the Neuroscience of Love and Hate.* Erlbaum, Hillsdale, NJ, 1995.

W. J. Freeman. *A Pathway to Brain Dynamics.* Division of Neurobiology, LSA 129, University of California at Berkeley, Berkeley CA 94720, 1991.

Y. Yao and W. J. Freeman. Models of Biological Pattern Recognition with Spatially Chaotic Dynamics. *Neural Networks, 3,* 153–170, 1990.

7
Periodic Systems Coupled to Periodic Environments

7.1 THE CONCEPT OF SYSTEMS RESONATING WITH THEIR ENVIRONMENT

Periodic motion (or near-periodic [1]) is one of the most important forms of dynamics in nature. We use it in our clocks to keep track of time, and thereby motion, whether it be with the old pendulum clock, or one that relies on atomic vibrations. The music we hear consists of linear superposition of harmonic vibrations in the air. The colors we see, the radio we hear, and the television we watch are due to periodic vibrations of electromagnetic waves. Most importantly, there is the periodic motion of our planet around the Sun, and the rotation around its axis, which influence the lives of all living plants and creatures in their seasonal and daily activities. Besides many obvious influences, we are generally not aware that a large number of hormonal activities in living systems (us!) are influenced by these periodic actions. Those activities that are influenced by the daily changes in light from day to night are referred to as **circadian rhythms** and are discussed in **Section 7.4.** Longer aperiodic effects, such as those involving the Earth's motion about the sun, may have been responsible for dramatic climatic effects, such as the ice ages [2].

Periodic Systems Coupled to Periodic Environments

In the next few sections, we will consider how the periodic motion of a **system** can be influenced in different ways by a periodic activity within its **environment,** depending on how strongly the environment is coupled to the system **[3]**. To do this, we take for the **system** a **nonlinear oscillator,** which is acted on by a **periodic force** from its **environment.** So our **periodically forced nonlinear oscillator model** will be taken to be

$$\frac{dx^2}{dt^2} + \mu\left(\frac{dx}{dt}\right) + \omega^2 x + \varepsilon x^3 = \alpha \cos(\Omega t) \qquad (7.1.1)$$

If $\alpha = 0$, so that there is no forcing action, the remaining equation is the **Duffing oscillator,** discussed in Section 4.3. The term proportional to μ represents the **frictional damping** of this oscillator, and the term proportional to ε is the **nonlinear restoring force** in this oscillator. The angular frequency ω is the frequency of the harmonic motion (when μ, ε, and α are all zero), and Ω is the angular frequency of the applied force.

The simplest dynamics occurs when there is no nonlinear term, $\varepsilon = 0$. In this section, we will consider this case of a **periodically forced, weakly damped harmonic oscillator,**

$$\frac{dx^2}{dt^2} + \mu\left(\frac{dx}{dt}\right) + \omega^2 x = \alpha \cos(\Omega t) \qquad (\omega \gg \mu) \qquad (7.1.2)$$

The idea of being a weakly damped oscillator is expressed by the inequality in (7.1.2), which means that it carries out many oscillations as the magnitude of its oscillations slowly decreases.

The general solution of (7.1.2) is obtained in courses in mechanics, and is rather messy. However, because of the damping factor, the solution for long times (the **"asymptotic solution,"** when $\mu \gg 1$) is relatively simple, and it is the dynamics of greatest general interest. This "asymptotic solution" is

$$x(t) \cong \frac{\alpha \cos(\Omega t + \delta)}{[(\omega^2 - \Omega^2)^2 + \mu^2 \Omega^2]^{1/2}} \qquad (7.1.3)$$

where δ is a constant phase factor, of no importance in the following effects. Note that in this asymptotic solution there is no longer any reference to the initial conditions, $[x(0), v(0)]$. **All of the dynamics tends to the motion (7.1.3). It is the "attractor" dynamics.**

▪ Think 7.1

If you want a challenge, show that (7.1.3) is an exact solution of

(7.1.2), by direct substitution, provided that $\sin(\delta)/\cos(\delta) \equiv \tan(\delta) = -\mu\omega/(\omega^2 - \Omega^2)$. You will need to review some trigonometry! Notice the sign of the last term, and determine whether the force reaches its maximum value **before or after** $x(t)$ reaches its maximum value. This is their **"phase relationship."**

The **maximum value of $x(t)$** is called the **amplitude** of the oscillation. This maximum value of $x(t)$ occurs when $\cos(\Omega t + \delta) = 1$, so we have

$$\text{amplitude of oscillation} \equiv A(\Omega) \cong \frac{\alpha}{[(\omega^2 - \Omega^2)^2 + \mu^2\Omega^2]^{1/2}} \quad (7.1.4)$$

$A(\Omega)$ depends on the value of the applied frequency, Ω, and the strength of the force α. In this section, we will consider **α to be a fixed value**, but **take Ω to be variable**. If Ω is varied around the value of the **frequency** ω, then **$A(\Omega)$ becomes largest** when $\Omega \cong \omega$ ($\omega \gg \mu$), and this maximum value is approximately

$$\text{maximum } A(\Omega) \cong \frac{\alpha}{\mu\omega} \quad (\text{when } \Omega \cong \omega) \quad (7.1.5)$$

The dependence of $A(\Omega)$ on Ω is shown in **Fig. 7.1**. This **maximum response** of the oscillator occurs when the environmental frequency is near the natural frequency of the oscillator, that is, **when $\Omega \cong \omega$**. This maximum response of an oscillator to an applied periodic force is known as a **resonance effect.**

This concept of a "resonant response" appears to extend well

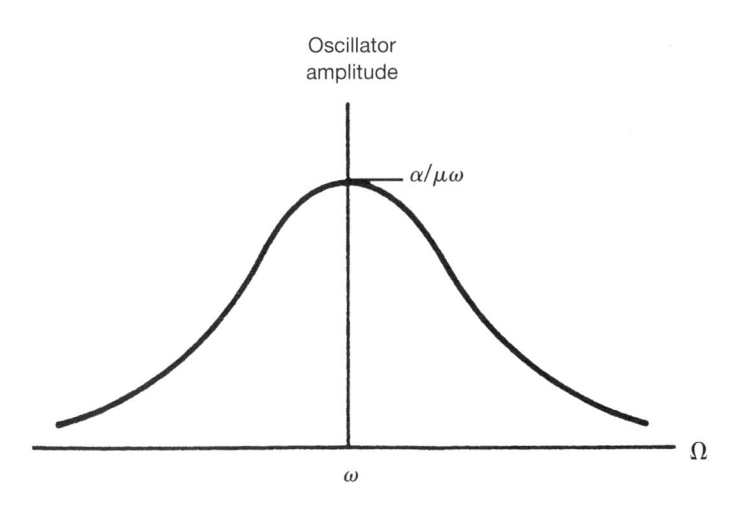

Fig. 7.1.
The dependence of the amplitude, $A(\Omega)$, on Ω yields this famous bell-shaped curve. This illustrates **the "resonance effect,"** namely that a **system responds the most** when the environmental periodicity is near its natural frequency, $\Omega \cong \omega$.

beyond this oscillator example to phenomena that **produce "resonant responses"** when the environmental actions are "natural" to a **person's cognitive dynamics.** This shows that when a new dynamic concept is discovered in one context, it should **stimulate your imagination** to look for the possible extension to nonoscillatory forms of **"resonant effects"** in very different contexts. One of these is in the **field of education,** where the responsiveness of students can be enhanced it they are presented material in terms of concepts that they are already familiar with (**"resonating"** with their **"natural insights"**) [5]. Hopefully, you "resonated" to the example of the **banking dynamics** at the beginning of this book, and therefore found it **easier to understand** such concepts as recursive relationships, exponential growths, the reality of limited resources, the resulting nonlinear effects, and implicit and uncertain assumptions involved in all models.

Explore 7.1

Use **Program 7-1** to see how the **theoretical amplitude,** $A(\Omega)$ in **(7.1.4),** compares with the **actual amplitude** of the asymptotic dynamics, given by the equations of motion, **(7.1.2).** This **amplitude** is given by the maximum value of $x(t)$, which can be seen in the phase space and compared with the theoretical value, marked by **blue circles.** Also shown is the amplitude versus Ω, as in **Fig. 7.1.** You can supply the coordinates and scales (copying from **Program C3**).

Think 7.2

There are several features about **Program 7-1** that you should think about. First of all the **maximum value** of $x(t)$ has to be "captured" during the dynamics. See if you can improve on the method used in the program. Second, you may note that as the amplitude decreases, the orbits are not identical with those on the low-frequency side. Is this an error, or some real physical effect?

The results of this section apply to the case when the nonlinear effects can be ignored [since we have set $\varepsilon = 0$ in (1.1.7)], or, in other words, the case of weak periodic forcing. **Next,** we will consider what happens when there is a **"moderate"** force (**values of** α). In this case, $A(\Omega)$ **becomes larger,** (7.1.4), so that the first **nonlinear effects in (1.1.1) influence the dynamics.** In particular, the nonlinear effect will **dramatically modify** the resonance effect, illustrated in **Fig. 7.1,** introducing what is called a **hysteresis phe-**

nomenon. So let us study this rather amazing change in the response of the oscillator to a moderate force.

References and Notes

1. As was discussed in **Section 6.3,** when it comes to functional and adaptive processes in natural systems, as contrasted with engineered systems (e.g., clocks, radios, television, etc.), perfect periodic motion is not desirable. It appears that some degree of "chaos," dependent on the **time scales,** must often be present in these dynamics to accomplish these purposes. If one extends the concept of "functional" to include psychological forms of enjoyment, then the vibrato is an essential component of musical tones, as are overtones, etc., none of which are based on simple periodic motion.

2. An example of this is the theory, due to Milankovitch (1924), that the **ice ages** may be due to certain long-period aspects of the Earth's dynamics about the Sun (e.g., 41,000 years and 22,000 years). The fascinating history of this scientific debate can be found in J. Imbrie and K. P. Imbrie, *Ice Ages: Solving the Mystery*, Enslow, Hillside, NJ, 1979. These studies continue to be an active area of research, e.g., R. A. Kerr, Why the Ice Ages Don't Keep Time, *Science,* 23 July 1999, p. 503; J. A. Rial, Pacemaking the Ice Ages by Frequency Modulation of Earth's Orbital Eccentricity, *Science,* 23 July 1999, pp. 564–568; D. P. Schrag, Of Ice and Elephants, *Nature,* 2 March, 2000, pp. 23–24; G. M. Henderson and N. C. Slowey, Evidence from U-Th dating against Northern Hemisphere Forcing of the Penultimate Deglaciation, *Nature,* 2 March, 2000, pp. 61–66.

3. The concepts of a **"system"** and its **"environment"** are discussed in some detail in **Section 6.3.** As the references in **[2]** indicate, the variable effects due to the "Earth system" coupling to the "sun environment" can have can have profound impacts on life on Earth over **those time scales.** This coupling is also attributed to the **snowball Earth theory** of the ice ages during the past **200 million years,** and its **possible** connection with the famous **Cambrian explosion** in the evolutionary development of life on Earth **[4].**

4. P. D. Ward and D. Brownlee. Rare Earth: *Why Complex Life is Uncommon in the Universe.* Copernicus: Springer-Verlag, New York, 2000; also see **Section 6.3.**

5. This is a **pedagogical perspective** that has been emphasized by Alfred Hübler of the University of Illinois, Urbana–Champaign.

7.2 MODERATE PERIODIC FORCES CAN PRODUCE BISTABLE OSCILLATORS

In the last section, we derived the classic study of the response of a harmonic oscillator system to a weak periodic force, and the resultant resonant response when the driving frequency, Ω, is near the natural oscillator frequency, ω. In this section, we will see the first **dramatic difference** that comes about from **nonlinear effects.** To do this, we will consider the case of a "moderate" strength force, α, in the Equation (7.1.1), namely

$$\frac{dx^2}{dt^2} + \mu\left(\frac{dx}{dt}\right) + \omega^2 x + \varepsilon x^3 = \alpha \cos(\Omega t) \qquad (7.2.1)$$

If $\alpha = 0$ and $\mu = 0$, then (7.2.1) is the classic **Duffing oscillator**, discussed in **Section 4.3** (with suitable changes in the coefficient notation).

In that section, it was shown that the nonlinear Duffing oscillator has a natural frequency, $\omega_n(A)$, which **depends on the amplitude of the oscillation, A,** as estimated in **Appendix M5**. This **nonlinear frequency** is given by

$$\omega_n(A) \cong \omega + \frac{3\varepsilon A^2}{8\omega} \qquad (7.2.2)$$

The present idea of a **"moderate" force, α,** in (7.2.1) is that we can try to **approximate** the behavior of the nonlinear **Duffing oscillator** by use of a **"harmonic oscillator,"** but **with the nonlinear frequency, (7.2.2), which depends on the amplitude** of the oscillation. It is not at all clear under what conditions this is a reasonable approximation, except if α is "sufficiently small," so that A is likewise "small." Nonetheless, we proceed by using this idea, and see later how well it represents the dynamics of (7.2.1). We are thereby led to the **"linear" approximation of (7.2.1)**, with an **unknown $\omega_n(A)$**

$$\frac{dx^2}{dt^2} + \mu\left(\frac{dx}{dt}\right) + \omega_n(A)^2 \cdot x = \alpha \cos(\Omega t) \qquad (7.2.3)$$

This looks like the forced harmonic oscillator example in the last section, (7.1.2), except for the **all-important feature** that $\omega_n(A)$ appears in (7.2.3), rather than the linear frequency, ω, and the **amplitude A is only determined after (7.2.3) is solved.** In other words, one has to establish a **self-consistent** relationship between A in $\omega_n(A)$ and the amplitude generated by the dynamics of (7.2.3)

To do this, we reapply the resonant behavior found in the harmonic case (7.1.2), except now we will use the model (7.2.3) with the frequency (7.2.2). The new asymptotic solution of (7.2.3) is now, using **(7.1.3),**

7.2 Moderate Periodic Forces Can Produce Bistable Oscillators

$$x(t) = \frac{\alpha \cos(\Omega t + \delta)}{[(\omega_n^2 - \Omega^2)^2 + \mu^2\Omega^2]^{1/2}} \quad (7.2.4)$$

and, using (7.1.4), the new amplitude is

$$A = \frac{\alpha}{[(\omega_n(A)^2 - \Omega^2)^2 + \mu^2\Omega^2]^{1/2}} \quad (7.2.5)$$

In order **to determine $A\Omega$),** we see that (7.2.5) leaves us with a new type of problem to solve, because the amplitude is expressed as a function of itself through the frequency (7.2.2). The new problem is algebraic in character.

First of all, in order to remove the square root factor in (7.2.5), square both sides to obtain

$$A^2 = \frac{\alpha^2}{[\omega_n(A)^2 - \Omega^2]^2 + \mu^2\Omega^2}$$

To simplify this further, multiply by the denominator on the right side, to yield

$$A^2[(\omega_n(A)^2 - \Omega^2)^2 + \mu^2\Omega^2] = \alpha^2 \quad (7.2.6)$$

It can be seen from (7.2.2) that $\omega_n(A)$ only **depends on A^2,** so that (7.2.6) also only **depends on A^2.** In other words, (7.2.6) is an **algebraic equation** that **determines $A(\Omega)$.** However, this equation only involves A^2, so one is left with the problem of **solving a cubic equation in A^2.** Of course, nobody remembers the solution of a cubic equation, but it can be found in many sources, so one simply has to look it up.

▪ Think 7.2A

Before we proceed with some simple algebra, it is important to think about **what such an equation means.** One needs only to recall some basic facts about **cubic equations.** Such equations can have either **three real solutions** (i.e., ordinary numbers), **or one real solution** and a pair of "complex conjugate" numbers. In the present case, A must be a **real number,** so these are the only roots that interest us. But, in contrast to the single solution of a linear equation, the **nonlinear equation** can now have **three different amplitudes for the same values of Ω and α.** That is a remarkable new possibility. But all solutions $[x(t), v(t)]$ are uniquely determined by their initial conditions $[x(0), v(0)]$, where $v = dx/dt$. So what *might* this say about the initial conditions in different regions of phase space? Read further on for more insight into this matter. ▪

Now let us return to (7.2.6) and simplify matters by using the fact that the **most important frequencies Ω are near** the natural frequency ω. We therefore use this idea to set

$$\Omega = \omega + \varepsilon\Delta \qquad (7.2.7)$$

so **$\varepsilon\Delta$ is the new applied frequency variable** and ε is the **small** parameter. Using this in one of the factors of (7.2.6), yields

$$(\omega_n^2 - \Omega^2)^2 \cong (\omega_n^2 - \omega^2 - 2\omega\varepsilon\Delta)^2$$

where we always retain only terms up to the **first order in ε**. Next we substitute $\omega_n(A) \cong \omega + 3\varepsilon A^2/8\omega$, which eliminates the factor ω^2 and also allows us to factor out the common term $(2\omega\varepsilon)$, so the above expression becomes (keeping only the lowest term in ε)

$$(\omega_n^2 - \Omega^2)^2 \cong \left(\frac{3A^2}{8\omega} - \Delta\right)^2 (2\omega\varepsilon)^2 \qquad (7.2.8)$$

Referring back to (7.2.6) we see that we must add to this the term $\mu^2\Omega^2$. Since **(7.2.8) is proportional to ε^2**, we must **restrict** μ^2 to be of this same order of smallness. So we now simplify the factor $\mu^2\Omega^2$ in (7.2.6) by requiring that the damping factor **μ to be of order ε**. If we keep only the largest factor that involves ε, then $\mu^2\Omega^2 \cong \mu^2\omega^2$, and substituting this and (7.2.8) back into (7.2.6) yields the desired **cubic equation for A^2**

$$A^2\left[\left(\frac{3A^2}{8\omega} - \Delta\right)^2 (2\omega\varepsilon)^2 + \mu^2\omega^2\right] = \alpha^2 \qquad (7.2.9)$$

That takes care of all the algebra, but it is "neater" if one introduces some notation to clean up all the factors. This is left for you do work out the details.

Think 7.2B

Show that if $\beta = 3A^2/(8\omega)$, and $F = 3\alpha^2/(32\varepsilon^2\omega^3)$, then (7.2.9) becomes

$$\beta\left[(\beta - \Delta)^2 + \left(\frac{\mu}{2\varepsilon}\right)^2\right] - F = 0 \qquad (7.2.10)$$

Recall that μ is of the order of ε, so (μ/ε) is of order 1, as are the other factors, provided that **α is also of order ε**. This requirement then gives the formal definition of a **moderate periodic force**. It is really more of a mathematical ordering device than any physical statement about the magnitude of the force. But ultimately, of course, this whole approximation scheme will need to be judged by

7.2 Moderate Periodic Forces Can Produce Bistable Oscillators

how closely it resembles the actual dynamics of (7.2.1). Its main advantage (or raison d'être, if you like) is that it exposes the fact that there can be three real solutions to the (7.2.1).

Before proceeding further with the analysis of (7.2.10), it is useful to **graph the general behavior of β versus Δ for different strengths of F**. These results are illustrated in **Fig. 7.2**.

Recall that the **physical meaning** of these terms are ($\leftrightarrow$ means "related to," not equal to):

$\beta \leftrightarrow$ **amplitude of oscillation**

$\Delta \leftrightarrow$ **the amount the applied frequency is above the natural frequency, $(\Omega - \omega)$**

$F \leftrightarrow$ **the strength of the applied force**

It can be seen that for a **small applied force**, there is no nonlinear effect, and the amplitude response is the simple **resonance response** shown in **Fig. 7.1**. As F is increased, the **first nonlinear effect** is to "tip" this amplitude curve so that the **maximum amplitude** now occurs at **larger values of $(\Omega - \omega)$**, in other words, when the **applied frequency is larger than the natural frequency**. If the magnitude of **F is increased further**, there comes a **particular value** where this "tipping" structure becomes like a breaker of a water wave, and there are **three values of β**; in other words, there are **three possible amplitudes** of the oscillator. Of these three, **only the** solutions with the **largest and smallest amplitudes** are stable solutions, and represent **dynamic attractors**. The **middle-amplitude** solution represents an unstable **(repellor)** solution. This is a **new aspect** of nonlinear dynamics: **An oscillator can have two stable periodic states, and is referred to as a bistable oscillator.**

Bistable oscillators can also occur in autonomous systems (not acted on by periodic forces), as will be seen in **Section 8.2**.

Fig. 7.2.
A schematic illustration of the solutions $\beta(\Delta)$ of (7.2.10) for increasing values of F, showing the possible generation of **more than one amplitude for a given Δ**. This is shown for a "hard" nonlinear oscillator, $\varepsilon > 0$. If $\varepsilon < 0$, replace Δ by $-\Delta$, so the "tipping" is toward the left.

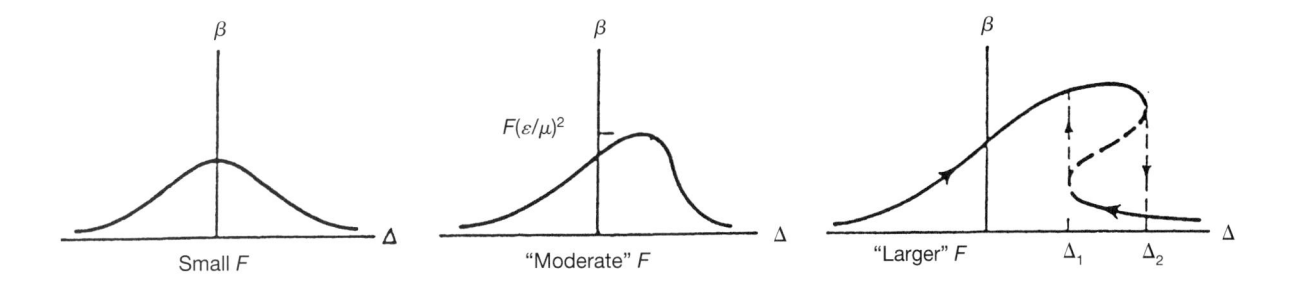

Small F

$F(\varepsilon/\mu)^2$

"Moderate" F

"Larger" F Δ_1 Δ_2

Many other forms of bistability occur in **biological systems,** as illustrated in note **[1]**.

Now it is time to begin to check out this **approximate theory** and to determine some **quantitative results,** using the dynamics **(7.2.1).**

Explore 7.2

Program 7-2 will be of help in determining whether all the theoretical predictions of this **approximation** are in fact true, and if so, for **what numerical values** of the frequency Ω and strength of the force α in **(7.2.1).** The theoretical results are at best only a guide to such numbers, assuming that the theory is correct! There are two useful stages to explore:

1. See if you can recover the **middle figure in Fig. 7.2.** The program has been set up to **automatically increase omega** (O in the program) in steps of **STP = 0.02.** These steps only occur after the **asymptotic state** of the dynamics is reached, because it is this amplitude that is plotted in Fig. 7.2. You should study the program to see how this is done. Begin this part with the suggestions in the program. The time has been normalized so that the **natural frequency** of the harmonic oscillator **is 1,** and **mu = 0.15, epsilon = 0.1.** As you will see below, these quantities can have **no influence on the qualitative behavior** of the solutions. You will see that the nonlinearity not only produces the form in the central figure of Fig. 7.2, but **also a more dramatic behavior.** Note down for what values of α and Ω (A and O in the program) this behavior occurs.

2. Next you will see if there is indeed a **hysteresis effect.** In part (1) you should have noticed some sharp changes in the asymptotic amplitude when α became larger than some value, say α^* (or A and A* in the computer program). Moreover, this occurs for some specific value of omega (O in the program). Call this value O*. Now select a value of **A = A* + 0.2,** and **turn off the automatic omega-increase line in the program,** because you will need to **both increase and decrease O, carefully.** Select **O = O* − 0.2,** and set the STP (omega step) at **0.01** (or less). **Slowly increase O, waiting or the asymptotic state (when MAXX stops changing) before changing O again.** In this process you are going from left to right in the third figure of Fig. 7.2, and

7.2 Moderate Periodic Forces Can Produce Bistable Oscillators

the question is, **can you find values of o, like the Δ_1 and Δ_2 in that figure?** Notice that you can now **both increase and decrease o.** Also note that you may need to adjust the graphics magnification, and screen location (e.g., **at z:**)

What follows is a further analysis, indicating how one can determine what this theory says about the **important question: Under what conditions does (7.2.10) have three real solutions?**

Well, first of all, one can go to some reference book and find out that an equation of the form

$$\beta^3 + a\beta^2 + b\beta + c = 0 \qquad (7.2.11)$$

has three real roots whenever

$$q^3 + r^2 < 0 \qquad (7.2.12)$$

where q and r are the unlikely looking combinations

$$q = \frac{b}{3} - \left(\frac{a}{3}\right)^2 \quad \text{and} \quad r = \frac{ab - 3c}{6} - \left(\frac{a}{3}\right)^3 \qquad (7.2.13)$$

Moreover, **at least two** of the three real roots **are equal if**

$$q^3 + r^2 = 0 \qquad (7.2.14)$$

Finally, there is only **one real root** if

$$q^3 + r^2 > 0$$

To relate all this to the model results (7.2.10), a quick comparison between this and (7.2.11) yields

$$a = -2\Delta, \quad b = \Delta^2 + \left(\frac{\mu}{2\varepsilon}\right)^2, \quad c = -F \qquad (7.2.15)$$

Then there remains the messy substitution into (7.2.13) before the results on the number of solutions can be identified with the original model.

Think 7.2C

However, all is not lost! Note that r is the only parameter containing $c = -F = -3\alpha^2/(32\varepsilon^2\omega^3)$. In other words, **r is the only parameter dependent on α, which is the strength of the applied force.** Notice also that the equality **(7.2.14)** represents the

217

boundary between the **regions of one solution and three solutions.** This makes (7.2.14) a particularly interesting feature of this dynamics. Since this **equality** only involves r^2, which is **quadratic in** c and hence F in **(7.2.14) is a quadratic equation for F.** We all know (or can find out) how to **solve** this quadratic equation **for F** (with the requirement that $F > 0$; **see Think 7.2B**). For those with the fortitude to check out the accuracy of the following, it is claimed that the answer to all this is

$$F = \left(\frac{2}{27}\right)\Delta^3 + \left(\frac{2\Delta}{3}\right)\left(\frac{\mu}{2\varepsilon}\right)^2 \pm \left(\frac{2}{27}\right)\left[\Delta^2 - 3\left(\frac{\mu}{2\varepsilon}\right)^2\right]$$

and if one plots out the two curves given by this for which $F > 0$, **one obtains Fig. 7.3.**

In part (2) of **Explore 7.2,** the changes in O involved **moving up and down in this figure.** When one includes the information of the asymptotic amplitude β of (7.2.10), this yields a three-dimensional representation of all possibilities, illustrated schematically in **Fig. 7.4**

In the next section we will see new dynamic features that arise when the force is even stronger.

Fig. 7.3.
The two roots of (7.2.14) as a function of Δ and F (7.2.16), scaled for any value of $|2\varepsilon/\mu|$. This set of points is known as a **cusp catastrophe set, K.**

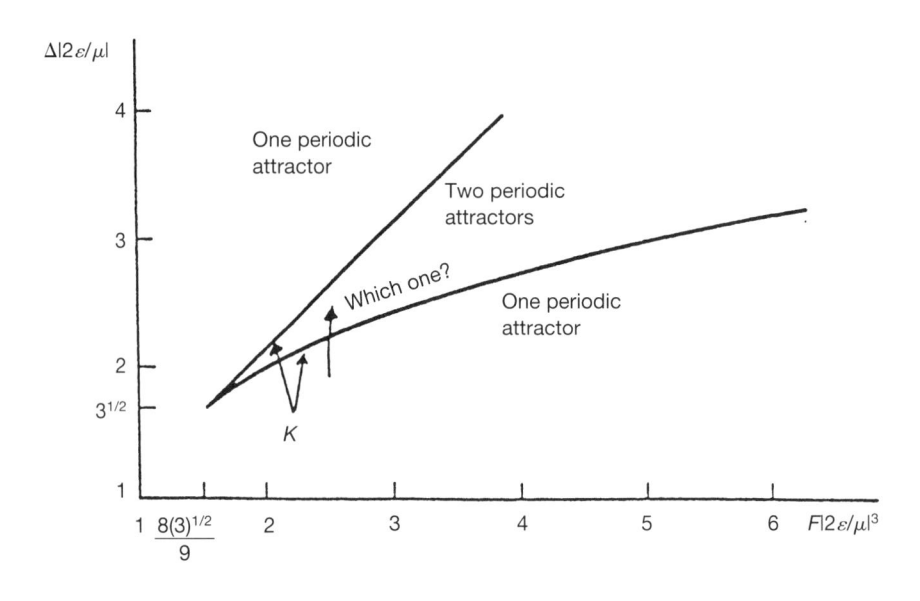

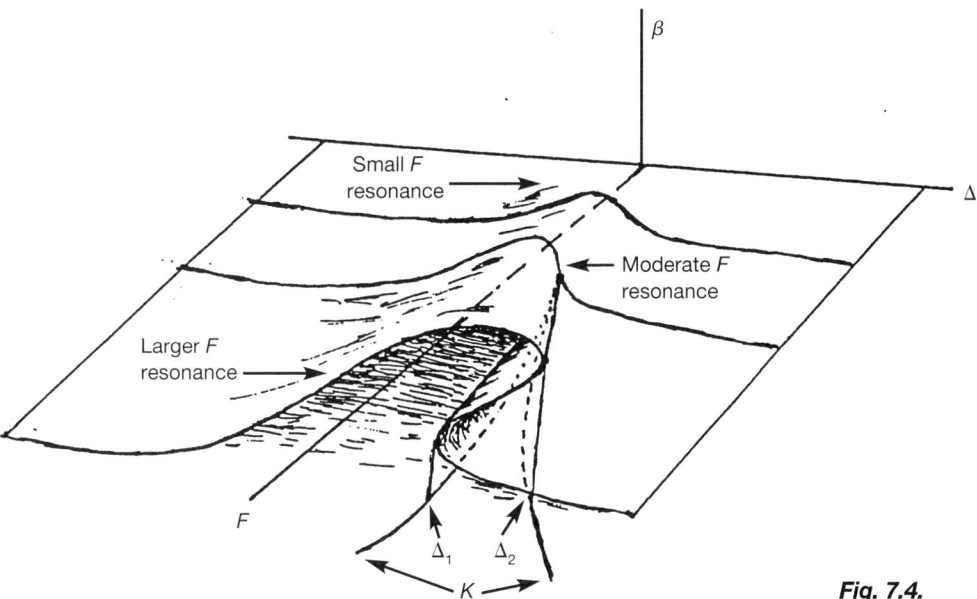

7.3 Symmetry-Breaking and Chaos From Strong Forces

Fig. 7.4.
The general structure of the asymptotic amplitudes, β, of (7.2.10), as a function of Δ and F. **Relate this to Figs. 7.2 and 7.3.**

References and Notes

1. The importance of **bistability** (and multiple stability) extends into many aspects of **living systems,** ranging from genetic to cognitive dynamics. An example is a synthetic, **bistable gene-regulatory network,** explored in (with additional references): T. S. Gardner, C. R. Cantor, and J. J. Collins, Construction of a Genetic Toggle Switch in *Escherichia coli, Nature, 403,* Jan. 20, 2000, pp. 339–342. Their (autonomous) model is

$$\frac{dU}{dt} = \frac{\alpha_1}{1 + V^\beta} - U; \quad \frac{dV}{dt} = \frac{\alpha_2}{1 + U^\gamma} - V \quad (\beta, \gamma > 1)$$

   which has a very simple V-shaped catastrophe set when plotted in the $\log(\alpha_1)$ versus $\log(\alpha_2)$ plane (with slopes β and $1/\gamma$, respectively, analogous to Fig. 7.3). Explore this model, and see if you agree with their results.

   Multistability occurs in a variety of **cognitive phenomena,** as illustrated in P. Kruse and M. Staler (Eds.), *Ambiguity in Mind and Nature: Multistable Cognitive Phenomena,* Springer-Verlag, New York, 1995.

2. **"Catastrophes"** are large sudden changes in a system's behavior, representing a discontinuous response of a system to a smooth change in some environmental conditions (here, α and Ω in (7.2.1)). The **catastrophe set, K,** is where this discontinuity takes place in the environmental parameter space (e.g., Δ, F in Fig. 7.3). René Thom introduced the termi-

nology "catastrophe" in 1972, and it soon became famous in the public press, and infamous among many for its lack of details, much less rigor. For Thom's wide-ranging visions, see: R. Thom, *Structural Stability and Morphogenesis* (Benjamin/Cummings, Reading, MA, 1975). For many early examples of its envisioned applications, see: E. C. Zeeman, *Catastrophe Theory : Selected Papers 1972–1979*, Addison-Wesley, Reading, MA, 1977.

For an enjoyable, insightful, and authoritative account of these matters, see V. I. Arnold, *Catastrophe Theory*, Springer-Verlag, Berlin, 1984.

7.3 SYMMETRY-BREAKING AND CHAOS FROM STRONG FORCES

In the last section, we saw that even a "moderate" periodic force can produce some dramatic behavior in a nonlinear oscillator. In Sections 4.4 and 4.5, it was found that autonomous nonlinear systems can have more than one stationary (fixed point) attractor. In the last section, with the addition of a periodic force it was seen that there could be **two** periodic attractors **(bistability)**. These attractors are **both symmetric about the origin** in the phase space, but with different amplitudes. By this is meant that **if (x, v) is a point on the orbit** in the phase space, $(-x, -v)$ **is also a point on the same orbit.** Which of these two oscillations actually occurs depends on either the initial conditions or on the way that the driving frequency was changed in the past, giving rise to **the hysteresis effect.**

In this section, we will see that when the **periodic force becomes stronger** a number of new dynamic phenomena can occur, depending on the strength, C, of the periodic force in (7.3.1)

$$\frac{dx^2}{dt^2} + \mu\left(\frac{dx}{dt}\right) + \omega^2 x + \varepsilon x^3 = C \cos(\Omega t) \qquad (7.3.1)$$

or, in terms of the original two variables of the phase space,

$$\frac{dx}{dt} = v$$

$$\frac{dv}{dt} + \mu v + \omega^2 x + \varepsilon x^3 = C \cos(\Omega t) \qquad (7.3.2)$$

As C is increased in magnitude, the **first effect** that is found is that the periodic orbits in phase space no longer are symmetric

7.3 Symmetry-Breaking and Chaos From Strong Forces

about the origin. This change is illustrated in **Fig. 7.5.** In Fig. 7.5(a) the orbit satisfies the symmetry condition [(x', v') is on the same orbit as $(-x', -v')$]. The change from symmetric orbits to asymmetric orbits is called a **symmetry-breaking bifurcation,** and although this dynamics is still periodic, it is more complicated and interesting in several respects.

In particular, if there is **one asymmetric periodic orbit** on which there is a point (x, v), there must be a **second asymmetric periodic orbit** on which there is the point **$(-x, -v)$.**

▪ Think 7.3A

See if you can give a proof of this statement. It will be explained soon, but first give it a try. [Hint: consider a solution at the time $t + p$, where $p = \pi/\Omega$. How does it relate to $[x(t), v(t)]$? ▪

Now, because the first orbit is not symmetric, neither will be the second orbit, and hence they are **distinct orbits** (not just one orbit at two different times).

▪ Think 7.3B

Use this fact to sketch the second asymmetric periodic orbit corresponding to the one in **Fig. 7.5(b).** Do these two orbits intersect in this phase space representation? But **all solutions of (7.3.2) are unique,** so how can two orbits occur at the same point in the phase space, or self-intersect, as in Fig. 7.6? ▪

Fig. 7.5.
Symmetric (a) and asymmetric (b) periodic orbits in the phase space, (x, v).

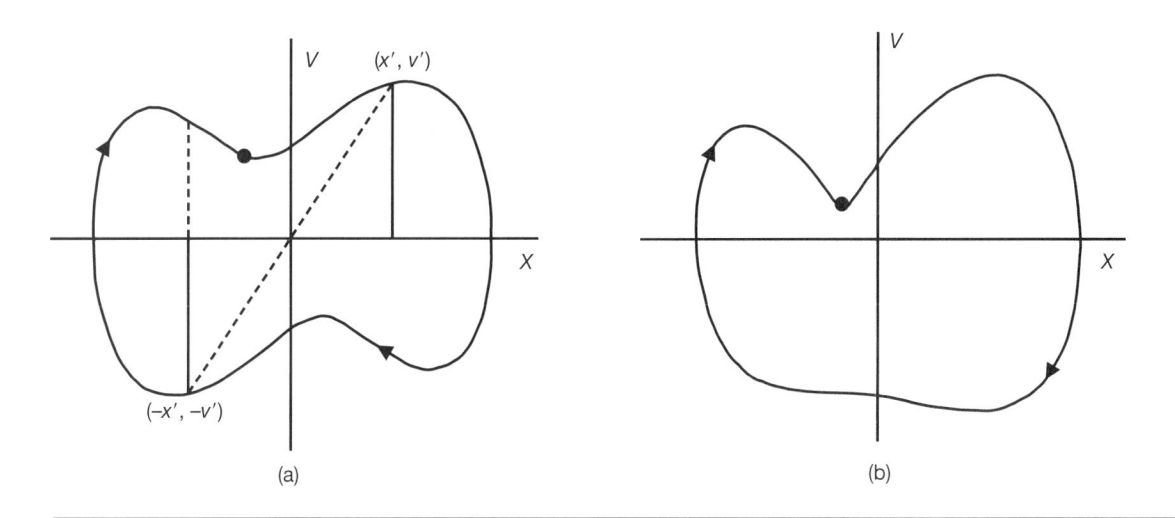

The above fact is due to the following feature of the equations (7.3.2). Note that **if [$x(t)$, $v(t)$] is a solution** of these equations, then so is **[$-x(t + p)$, $-v(t + p)$]** a solution, provided that **$p = \pi/\Omega$** [so that cos $(\Omega(t + p)) = -\cos(\Omega t)$). This conclusion follows because if one substitute the variables [$-x(t + p)$, $-v(t + p)$] in place of [$x(t)$, $v(t)$] into (7.3.2), and change t to $t + p$ in cos (Ωt) (so that the time agrees with that of the new variables on the left), one finds that the minus signs cancel out everywhere in (7.3.2). This proves that [$-x(t + p)$, $-v(t + p)$] satisfy the same equations as do [$x(t)$, $v(t)$]. If [$x(t)$, $v(t)$] is a symmetric solution, this result is trivial, since both solutions are on the same orbit, just at a half-period difference in time. But if [$x(t)$, $v(t)$] is an asymmetric orbit, then the two solutions must represent distinct solutions.

This last statement is important, because it makes no reference to whether the asymmetric orbit is periodic or not. That is irrelevant to the general conclusion. So we have the general fact that:

In any symmetry-breaking bifurcation of (7.3.2) two distinct orbits are generated, either both periodic or both aperiodic.

Explore 7.3A

Before proceeding to the dynamics related to "very large" values of C, use **Program 7-3A** to study the **symmetry-breaking bifurcation** of this dynamics that occurs first as C is increased.

The parameter values in (7.3.2) have been set at **$\varepsilon = 1.0$, $\omega = 1$**,

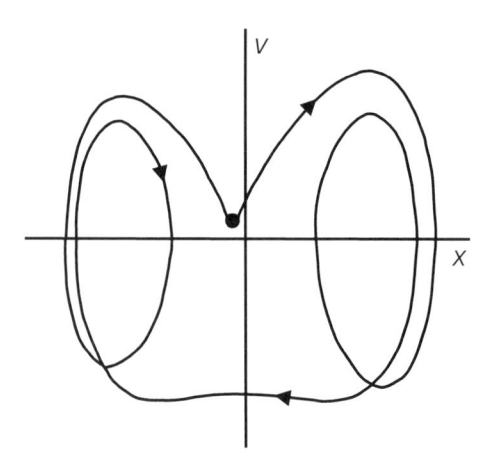

Fig. 7.6.
A nonautonomous orbit can **intersect itself** in the phase space (x, v), where **time is not accounted for.** This is a period-one orbit, **marked** by a circle at the times $t_n = n2\pi/\Omega$.

7.3 Symmetry-Breaking and Chaos From Strong Forces

and $\mu = 0.15$ in Program 7-3A, and $\mu = 0.10$ in Program 7-3B, so this is a weakly damped, very nonlinear system. The details of the dynamics are strongly influenced by the values of these parameters.

The graphics contain **blue circles** attached to the orbits, which designate the periodic times $t_n = n2\pi/\Omega$ of the forcing oscillation. In periodic cases, you can then determine if the orbit is period-one, or period-two, etc. with respect to the forcing oscillation. These times are also basic to a **Poincaré map** to be introduced later.

There are two factors that can be changed from the **keyboard** in this program. You can **increase/decrease the value of C,** using `<I>/<D>`, and you can **raise/lower the value initial condition,** $x(t = 0)$, using `<R>/<L>`. For all selections, the other initial condition is taken to be $v(t = 0) = 0$. The purpose of this second option will become clear shortly.

(A) Begin with a low value of $C = 1$, and increase it slowly (giving time for the dynamics to approach "asymptotics"); **use the `<C>` to clear the screen** of old data. Determine the **lowest value of C** for which you see symmetry breaking occur. You will see **one orbit appear,** similar to **Fig. 7.5(b).**

(B) But there must be another asymmetric orbit. Leaving the **value of C fixed,** now **increase** the location of the initial condition, $x(0)$. It will be in the **basin of attraction** of the first orbit, and then enter the basin of attraction of the other asymmetric orbit. Determine the location of this **boundary between the two basins.** Having established the existence of the two attractors, what dynamic phenomena exits? This process can be extended to larger values of $x(0)$ to find points in the original basin of attraction, and therefore another boundary point between these two

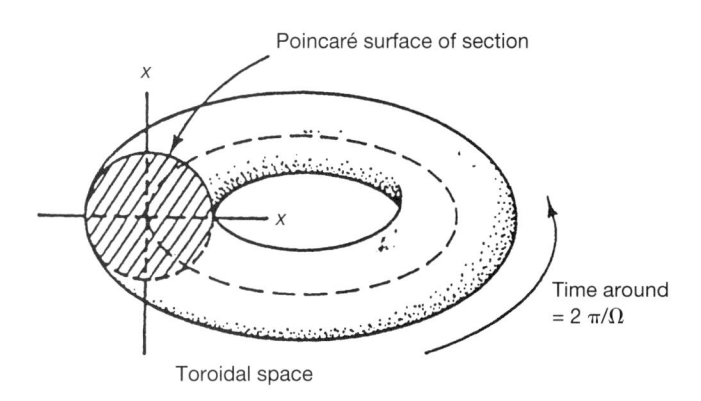

Fig. 7.7.
The dynamics in the **extended phase space (EPS),** $(x, v, t),$ in which all **orbits are unique.** These orbits intersection a plane $(x, v),$ known as a **Poincaré surface of section,** at the times $t_n = n2\pi/\Omega,$ generating a **Poincaré map of any orbit.**

basins. It can be seen that the dynamics is getting pretty complicated, even for these modest values of C.

In order to recapture a unique representation of the dynamics it is necessary to account for the time by introducing the **time as a third coordinate,** in addition to (x, v). This makes a new type of phase space, which is referred to as an **extended phase space (EPS).** In this EPS, the orbits are unique, and do not intersect each other. Unfortunately, this is difficult to show in a graphical representation. However, because the **force is periodic,** one can represent the dynamics uniquely in an **EPS** that is constructed like "doughnut," or more formally, a **solid torus,** as illustrated in **Fig. 7.7.** The interior of the solid torus is simply a **disk, D,** with its origin swept around on a perpendicular "time axis" circle in three dimensions, returning to the original origin. Several possible **periodic orbits** in this space are shown in **Fig. 7.8.**

The "time axis" in this solid torus has a length of the period of the applied force, $P = 2\pi/\Omega$. Up to that time, the representations of all solutions in the torus are unique, and hence they are also unique at that time. But since the equations have entirely returned to the same form (because the periodic force also returns every $2\pi/\Omega$ **seconds**), the same reasoning applies to any number of circulations around this torus, and hence the **orbits are uniquely represented in the EPS space for all time.** It is important to realize that **this holds for aperiodic (chaotic) dynamics, as well as periodic orbits.** Hence, this "doughnut" EPS gives a space of unique orbits (and the time axis does not run off to infinity!).

The **Poincaré map** now consists of **plotting** these particular values of $[x(t_n), v(t_n)]$ ($t_n = nP$, $n = 0, 1, 2, \ldots$) **in the phase space.** Thus, while an orbit in the **phase space** may intersect itself or other orbits, **this Poincaré map generated by any orbit in the EPS will be unique.**

An important point here, and in **Section 7.4,** is that in using these programs to Explore, you will **need to show a lot of pa-**

Fig. 7.8.
This illustrates some very simple orbits in the EPS. The Poincaré map of one of these has **period one,** and two have **period two. Visualize any surface of a section to determine which is which.**

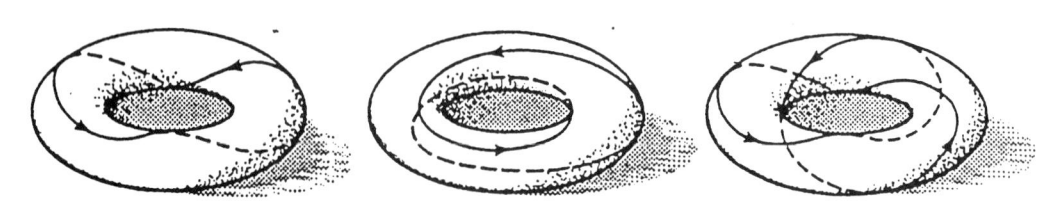

7.3 Symmetry-Breaking and Chaos From Strong Forces

tience, when you study **"asymptotic dynamics."** Computers are fast, but only so fast, and **as the applied force is increased,** the dynamic system will take **increasingly longer times** to reach its "asymptotic" state, whether it is periodic or chaotic. A **valuable tool** in this case is to use the **"clear screen" key, <C>,** which does not change the dynamics but only **erases the earlier "nonasymptotic" dynamics** (or whatever you may not be sure about).

With this rather lengthy, but necessary introduction, you are now in the position to try out new forms of exploration.

Explore 7.3B

This makes use of **Program 7-3B,** which is a modification of 7-3A. Notably, now $\mu = 0.1$, to make things interesting in this larger range of C (you might set $\mu = 0.15$, as in Program 7-3A, and see the difference). There are many things to explore, but you can begin with:

(A) select C = 20, X0 = 0, and see what you get. Is this orbit **symmetric or asymmetric** (recall what this means from **Fig. 7.5**)? What is the period of this dynamics relative to the applied period (note the circles on the orbit and **understand their meaning**). Now decrease C slightly until you see a change in the periodicity, then increase C back to 20. What do you find, and what does it mean? Why is the period not the same as you first obtained? You can then go on to explore the **basins of attraction** by changing the initial X0.

(B) Explore around the values **24 < C < 33,** for a beginning. Find the **"self-intersecting"** (?) orbit in Fig. 7.6. Here you **need patience;** wait for **asymptotic results** and use <C> to record the asymptotic properties. Can you find **three attractors** at the low end? Are they **period-2 attractors**? What is the **lowest value of C that yields chaos**? Use the **Poincaré map** to study the structure of the chaotic attractors. Is the **chaotic attractor symmetric?** Does the attractor change with different initial values of X0? What happens at the higher end of this range of C values? An example of the Poincaré map is shown in Fig. 7.9.

Think 7.3C

If two attractors are not symmetric, what is the ("holistic" or global) relationship between their structures in the EPS? If they

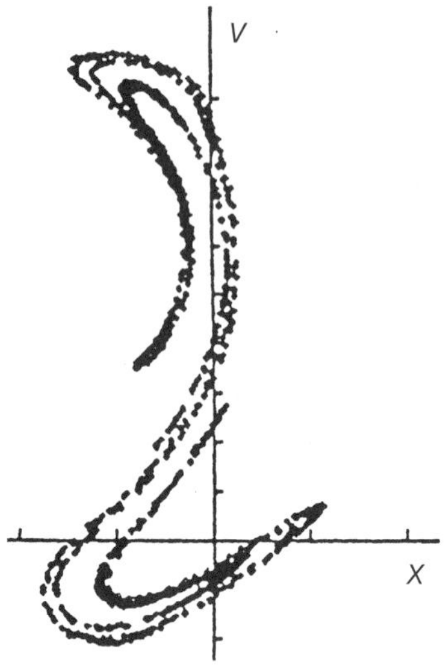

Periodic Systems Coupled to Periodic Environments

Fig. 7.9.
A Poincaré map in the EPS consisting of 2000 "points" (circles of radius 0.6, hence a crude indication of the structure). The scale is twice as large along the x axis as along the v axis.

are aperiodic, do you think they have a **fractal structure?** Why? Issues similar to this will be discussed in Section 8.2.

Think 7.3D

As a final check of your understanding of solutions that have asymmetric orbits, consider the three orbits in **Fig. 7.10.** Two of these orbits are solutions of (7.3.2), for the same parameter values but different initial conditions, and one is not a solution. **Add the necessary orientation arrows,** and decide **which orbit is not a solution (identifying all sections).** If you are not sure of your answer, the means for identifying these orbits can be found in Program 7-3B.

7.4 PACEMAKERS, ZEITGEBERS, CIRCADIAN RHYTHMS, AND ENTRAINMENT

In the course of evolution, nature has developed a variety of methods by which various physical processes within living systems are adjusted to meet changing environmental needs. Sunflowers rotate faithfully to face the sun; our heart, respiratory rates, and eye-blinking rates vary, depending on physical or emotional activities.

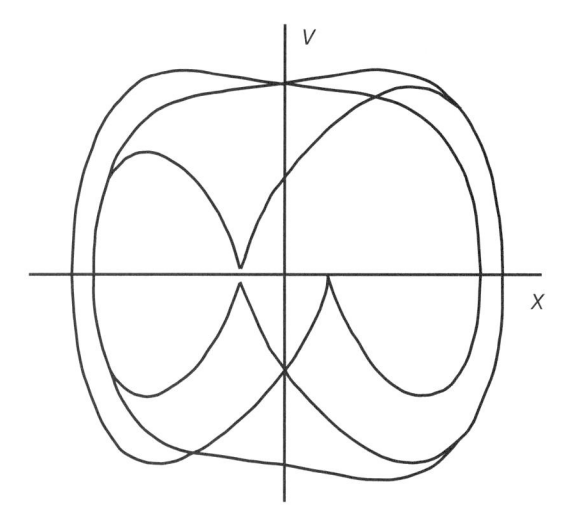

Fig. 7.10.
This shows three "orbits" (requiring orientations), **only two** of which **are solutions of (7.3.2),** associated with different initial conditions. Which is not a solution for these parameter values?

The **base levels** of other processes such as growth hormones, insulin and hemoglobin in the blood, blood pressure, heart rate, body temperature, urinary flow rate, activity of EEG waves, etc. are also **modulated on a daily basis [1, 2].**

In contrast to the **modulation** of these base levels, some of which relate to the **periodic rates** of processes that are **relatively fast** compared with daily periodic changes, there are processes that are periodic on environmental time scales that are **relatively slow** compared to 24 hours. Many of these are associated with **reproductive processes,** such as the menstrual cycles of women **[1, p. 135],** the many examples of breeding in marine animals, the periodic return of salmon to their particular breeding rivers or sea turtles to their specific beaches **[3, 4],** and the periods of reproduction of different species of locusts, to mention only a few. Some of these processes are believed to be associated with the phases of the moon and stages of the tides, whereas others are obvious only over longer time scales, and are more dynamically complex in their physiological origin.

There are a number of internal **"biological clocks"** in living systems which measure out time intervals over some fairly **regular period in the absence of environmental influences. Any** such biological clock is referred to as a **pacemaker.** In the **early 1960s,** Jürgen Aschoff and colleagues showed that volunteers who lived in isolation bunkers, with no natural light or other clues of time, maintained a normal sleep–wake pattern of roughly 24 hours (more precisely, 24.18 hours). Any pacemaker that has a **period near to 24 hours** is now referred to as having a **circadian rhythm** (*circa diem* means "about a day"). Since Aschoff's study, there has been a great deal of interest in the **coupling of circadian rhythms to daylight periodicity,** resulting in a number of books related to this subject **[6]**. In this section, we will focus on how these circadian rhythms can be influenced by an environmental periodicity, frequently related to daylight.

In recent times, there has been an explosion in the identification of such pacemakers in a wide variety of living systems **[2, 5]**. Some of these pacemakers are directly influenced by the periodicity of the daylight, whereas others are coupled to this information by other light-influenced pacemakers. The dynamic process by which a pacemaker is directly influenced to become **synchronized ("locked-in') with some environmental periodicity** is referred to as the **entrainment of that pacemaker.** The environmental periodicity that produces this entrainment is often referred to as a **zeitgeber ("time giver").** We will explore a simple model of this dynamic entrainment process, in order to understand some of the factors that are required to produce such a synchronization.

Of great interest has been the discovery of the group of nerve cells in the region of the our, as well as other mammal's, brains called the suprachiasmatic nucleus (SCN) **[2, 5]** that controls the production cycles of a multitude of biological processes. It has been found that a special set of "clock" genes within the SNC are turned on and off by the proteins that they encode, which act back in a feedback loop, producing the above circadian rhythm **[2, 4]**. It appears that these clock genes are regulated by light-sensing cells in the eyes, which in turn influence other biological clocks in various organs (such as the liver), but produce entrainment over longer time scales, leading to longer jet-lag reactions. Thus, there are multiple forms of entrainment in nature, and the present study involves only the simplest example of this dynamic process.

The **periodic environmental influences** are referred to as **zeitgebers** (time-givers), and those **pacemakers** that have periods **near 24 hour** are referred to as **circadian rhythms.** Some examples of this are the insulin content of the blood, the level of hemoglobin in the blood, heart rate, body temperature, and respiratory rate, each of which tends to peak at different times of the day **[1]**. The details of the biological processes that give rise to these **circadian clocks** are under active investigation, and in recent times many cases been identified with individual autocatalytic cells, or groups of cells, within living systems. Many of these ap-

7.4 Pacemakers, Zeitgebers, Circadian Rhythms, and Entrainment

pear to be under the control of a "Clock" gene [2], while others are directly influenced by light [5].

To get some idea of what is required for an environmental periodicity to be able to entrain a biological clock of a different periodicity, we will consider a very simple model. In this model, the **pacemaker** will be represented by an **autocatalytic oscillator,** specifically the van der Pol oscillator. The influence of an **environment, which has a different period (the zeitgeber),** will be represented by a periodic force acting on the van der Pol oscillator. Thus the present model is

$$\frac{dx}{dt} = y \qquad (7.4.1a)$$

$$\frac{dy}{dt} = -x + K(1-x^2)y + A\cos(\Omega t) \qquad (7.4.1b)$$

where A is the **strength** of the **zeitgeber action** and Ω is its **angular frequency.** The harmonic portion of the autocatalytic oscillator has been scaled to unity. **The period** of the autonomous ($A = 0$) **pacemaker, $P(K, A = 0)$,** depends on the magnitude of **K.** For example (as can be found in Explore 7.4A), $P(1, 0) \cong 6.67$, $P(2, 0) \cong 7.62$, etc. The larger the value of K, the longer the period, and **all exceed** the period of the zeitgeber, $P(\Omega) \cong 2\pi/\Omega$, **if $\Omega > 1$.** At the same time, the larger the value of K, the stronger is the nonlinearity, and the more resistant is this pacemaker to any outside influence. So larger values of K **increase the difficulty** of achieving **entrainment,** which is **defined by** the condition that $P(\Omega) = P(K, A)$ **(zeitgeber period "equals" the pacemaker period).**

In general, the system (7.4.1), with three independent parameters (K, A, Ω), is a very complicated dynamic system [7]. However, the present interest concerns the requirements needed to **entrain** the pacemaker to the zeitgeber, whose period, $P(\Omega) = 2\pi/\Omega$, **is not greatly different from that of the pacemaker.** This simplifies the range of dynamics considerably. Note also that all variables in (7.4.1) are **dimensionless.**

Explore 7.4A

Program 7-4 displays the dynamics of Equation (7.4.1) [the blue orbit in the **(x, y) phase space**], as well as the dynamics of the **$A = 1$ zeitgeber (yellow** orbit in the **same space,** where now "x" = cos (Ωt), "y" = $-\sin(\Omega t)$; (the minus sign makes d"x"$/dt$ = "y"). Also packed into this display are the periods of the zeitgeber, $P(\Omega)$

229

$= 2\pi/\Omega$, and the period of the pacemaker, $P(K, A)$, determined from the dynamics (you should **see how this is done** within the program). On the blue orbit (the pacemaker influenced by the zeitgeber), you will find **small yellow circles,** which are placed there every time the zeitgeber dynamics **crosses the $x > 0$ axis.** If these **yellow circles move along the blue orbit,** that indicates that the two dynamics do not have the same period (the pacemaker is **not entrained**). You can also see this information quantitatively. **Finally,** at the bottom of the screen you will find a display of $x(t)$ and "x" = $\cos(\Omega t)$ versus t, color coded as above. This will show that even when there is entrainment, there will be a **phase difference** between these two functions. The following figures were obtained from this program, with suitable modifications.

First take **$K = 1$, $\Omega = 1$, $A = 0$** (no interaction with the zeitgeber). You will find that the yellow circles move along the blue orbit **(why in that direction?),** and that the two curves at the bottom shift with respect to each other, similar to **Fig. 7.11.**

Now increase the coupling strength, A, until you find that **entrainment occurs,** at some value $A^*(\Omega = 1, K = 1)$. You can either **see or hear the entrainment** phenomena, by noting the stationary location of the yellow circle on the pacemaker orbit, or by **turning on the sound, <N>. Sounds are made** when the pacemaker or zeitgeber cross the $x > 0$ axis (**high and low tones,** respectively). Use <F> to turn **off the sound.** This entrainment is seen by the fixation of the yellow circle on the blue orbits, as in **Fig. 7.12.**

Explore 7.4B

Now determine how this entrainment is influenced by the strength of the pacemaker's nonlinearity, by increasing the value of K (recall from above how this also influences the periodicity of the pacemaker). Specifically, determine at least the required coupling strengths for entrainment, $A^*(\Omega = 1, K = 2)$, and $A^*(\Omega = 1, K = 3)$. Put these ideas together to see what you might conclude about how nature arranged biology to give the results found by Jürgen Aschoff (above).

Explore 7.4C

Another issue is how the entrainment is modified if the zeitgeber frequency differs more from that of the autonomous $K = 1$ pace-

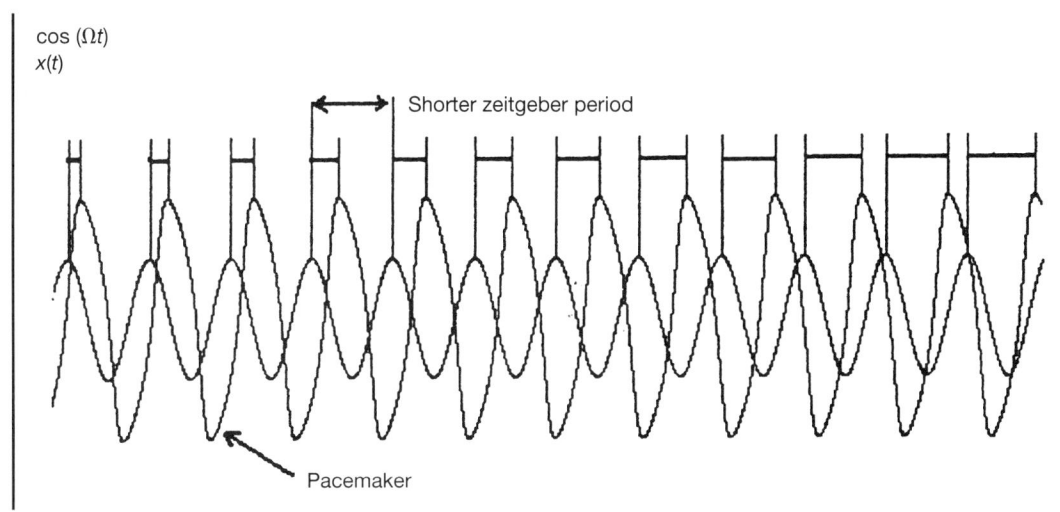

7.4 Pacemakers, Zeitgebers, Circadian Rhythms, and Entrainment

Fig. 7.11.
This shows the pacemaker, $x(t)$, and the zeitgeber function, $\cos(\Omega t)$, in (7.4.1) versus t, when there is no coupling, $A = 0$. The zeitgeber has a shorter period and "drifts" behind the pacemaker's curve.

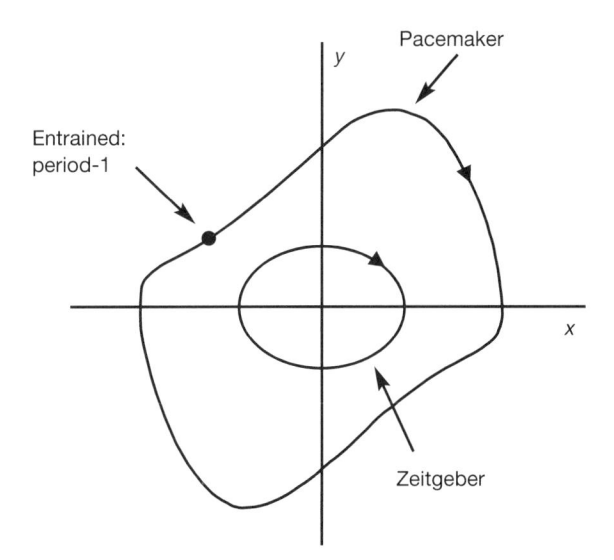

Fig. 7.12.
The joint "phase space" representation of the zeitgeber and the pacemaker, and the fixed reference circle with entrainment.

maker. So determine how $A^*(\Omega, K = 1)$ changes as a function of the difference $D = [\Omega - P(K = 1, A = 0)]$ when **Ω is increased, and when Ω is decreased. Is there a difference?** Can you give an explanation? Once again, one may consider what Jürgen Aschoff found to be the case.

References and Notes

1. S. Binkley. *The Clockwork Sparrow: Time, Clocks, and Calendars in Biological Organisms*. Prentice-Hall, Englewood Cliffs, NJ, 1990.
2. M. W. Young. The Tick-Tock of the Biological Clock, *Scientific American,* 64–71, March, 2000.

Adaptations to phases of the moon and tidal stages:

3. Rachel Carson. *The Sea Around Us*. Oxford University Press, Oxford, 1961, 1991. Also see the beautiful extract, *Tides,* in *Eyewitness to Science: Scientists and Writers Illuminate Natural Phenomena from Fossils to Fractals*, J. Carey, Ed., Harvard University Press, Cambridge, MA, 1995. In particular, see p. 348: "The most curious and incredibly delicate adaptations, however, are the ones by which the breeding rhythm of certain marine animals is timed to coincide with the phases of the moon and stages of the tide. . . ." It is also fascinating to realize how this has changed over billions of years, as the moon receded from the earth, and the horrendous "tides" have dwindled to their present state.
4. D. Palmer. Time, Tide and the Living Clocks of Marine Organism. *American Scientist, 84,* pp. 570–578 (1996).
5. **Circadian Clock Everwhere!**

    D. Whitmore, N. S. Foulkes, and P. Sassone-Corsi. Light Acts Directly on Organs and Cells in Culture to Set the Vertebrate Circadian Clock. *Nature, 404,* 87–91, 2 Mar., 2000.

    U. Schibler. Heartfelt Enlightenment. *Nature, 404,* 25–28, 2 Mar., 2000.

    S. M. Reppert. A Clockwork Explosion! *Neuron, 21,* 1–4 (1998).

    J. C. Dunlap. Molecular Bases for Circadian Clocks. *Cell, 96*(2), 271–290, Jan. 22, 1999.

    N. Barkai and S. Leibler. Circadian Clocks Limited by Noise. *Nature, 403,* 267, Jan. 20, 2000.

    M. W. Young. The Molecular Control of Circadian Behavioral Rhythms Their Entrainment in *Drosophila. Ann. Rev. Biochemistry, 67,* 135–152 (1998).

    A. Goldbeter. *Biochemical Oscillations and Cellular Rhythms*. Cambridge University Press, 1996.

    W. J. Schwartz. Internal Timekeeping. *Science and Medicine,* 44–53, May/June, 1996.

6. **Basic Books**

   A. T. Winfree. *The Timing of Biological Clocks.* Scientific American Library, New York, 1986.

   A. T. Winfree. *The Geometry of Biological Time.* Springer-Verlag, New York, 1990.

   L. Glass and M. C. Mackey. *From Clocks to Chaos: The Rhythms of Life.* Princeton University Press, Princeton, 1988.

   L. N. Edmund. *Cellular and Molecular Basis of Biological Clocks.* Springer-Verlag, New York, 1988.

   M. Young. *The Metronomic Society: Natural Rhythms and Human Timetables.* Harvard University Press, Cambridge, 1988.

   S. H. Strogatz. *The Mathematical Structure of the Human Sleep-Wake Cycle.* Lecture Notes in Biomathematics, Vol. 69, Springer-Verlag, Berlin, 1986.

   M. C. Moore-Ede and C. A. Czeisler (Eds.). *Mathematical Models of the Circadian Sleep-Wake Cycle.* Raven Press, New York, 1984.

   M. C. Moore-Ede, F. M. Sulzman, and C. A. Fuller. *The Clocks That Time Us: Physiology of the Circadian Timing System.* Harvard University Press, Cambridge, 1982.

8

A Variety of Dynamics in Space and/or Time

This chapter contains a very diverse collection of examples of dynamics. They are rather abbreviated presentations of these subject matters, in large part because some of the topics are not well developed at present (and perhaps will never be!). However, a number of references are provided for further discussions on these topics. Some of what follows should therefore be recognized as rather rudimentary models that would need considerable development in order to make them of serious empirical use. Other models correspond closely with observed phenomena. The dynamic models will be in two and three dimensions, involving deterministic and random elements, and using discrete and continuous variables, so there is quite a spread of ideas. We begin with an important class of spatiotemporal dynamics.

8.1 SPATIAL PATTERN DYNAMICS OF EXCITATION–DIFFUSION SYSTEMS

There are many physical systems in nature whose dynamics are based on the fact that they are composed of **excitable elements** that interact locally with each other (**nearest-neighbor interactions**). This type of dynamics is based on the fact that these elements go from an **"excited" state** to a **"refractory" state,** in which they do not interact with their neighbors. Following this, the elements go to a **"quiescent" state,** where they remain until **"excited" neighboring elements** may cause them to once again go from the quiescent state to an excited state. It is this cycling between these three states that can generate the spatial patterns to be discussed in this section. The details of some of the various dynamic rules that involve these three states will be illustrated shortly.

The other feature of this dynamics is that the **nearest-neighbor interactions** mean that any excited element can only pass on this excitation to an adjoining element. This is characteristic of **diffusive dynamics.** [On the other hand, **neurons,** which likewise **have excited, quiescent, and refractory stages** (Section 5.4), **can** interact with other neurons over large distances, in which case the dynamics is not diffusive in character. However, if a set of neurons interact only with their neighbors, then this set will have a dynamics similar to the one discussed in this section.]

There are interesting and important examples of this type of dynamics, which occurs in such diverse systems as **chemical oscillations** (Belousov– Zhabotinksy reactions [7]), as part of the remarkable **dynamics of slime molds [8],** in spatial **ecological interactions [6],** and in critical aspects of **cardiac arrhythmias [2, 3].** Some generalizations of the present dynamics may even offer insights into how epidemics spread over regions of space. The references at the end will give a few ideas about where you can begin to learn about these diverse topics.

Here, however, our objective will simply be to study the dynamics of a simple model of a **reaction–diffusion system.** This dynamics will be demonstrated on both a square and hexagonal spatial lattice of adjacent cells. Groups of cells in these lattices are illustrated in **Fig. 8.1,** with the numbers indicating the state of each cell. The square dynamics is easier to program, whereas the

Fig. 8.1.
A group of cells in a square (a) and hexagonal (b) lattice, interacting with **four and six adjacent neighbors, respectively.** Each cell is in one of three states: **excited (2), quiescent (1), or refractory (0).**

(a)

(b)

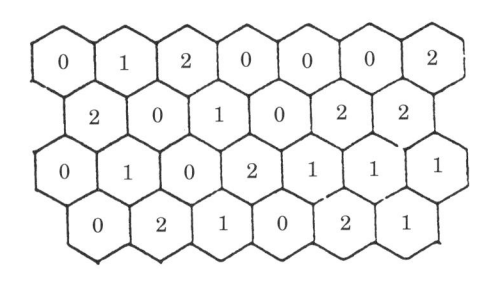

hexagonal spatial patterns are more realistic, and the dynamic options are also greater.

The **simplest dynamics** is based on cells that can have only three states, as indicated in **Fig. 8.1**. A cell in a **quiescent state (1)** remains in that state unless it is excited by **excited adjacent** cells, that is, cells in **excited states (2)**. After a cell is excited, it next passes into a **refractory state (0),** in which it is not responsive to its neighbors. From that state it next returns after some period of time to the quiescent state, in which it will become responsive to the excited state of its neighbors (it is excitable). Thus **excitement** is only passed along to an adjoining quiescent cell. The **diffusion feature** of this dynamics is due to the fact that the **interactions** are only between **adjacent neighbors [four and six in (a) and (b), Fig. 8.1]**. In the simplest case, it will be assumed that a quiescent cell will be excited if any of its neighboring cells are excited. **Generalizations** of these rules are given in **Explores 8.1B and 8.1C.**

Thus, the **simplest dynamics** is defined by the **next-step rules:**

$$0 \to 1, \quad 2 \to 0$$

$1 \to 1$ if there is **no excited** adjacent neighbor (8.1.1)

$1 \to 2$ if there is **any excited** adjacent neighbor

Think 8.1A

Referring to **Fig. 8.1,** use the above rules to determine the state of all of the cells in the next time step. The answer does not depend on the state of any of the adjacent cells not shown in the figure. ∎

Think 8.1B

Fig. 8.2(a) shows a sequence of patterns obtained from **Program 8-1A.** Because only shades of gray can be used in these figures, the **black** cells represent the excited state **(2),** the gray cells represent the refractory state **(0),** and the white cells represent the quiescent states. Given the sequence of patterns, and the rules (8.1.1), determine which are the excited and refractory cells. The answer will be found in the program, but **figure it out first.** ∎

Explore 8.1A

The dynamics in **Program 8-1A** follow the rules (8.1.1) in the square lattice, **Fig. 8.1 (a),** with **four neighbors** for each cell, except the **boundary cells,** which have three neighbors. The state in

A Variety of Dynamics in Space and/or Time

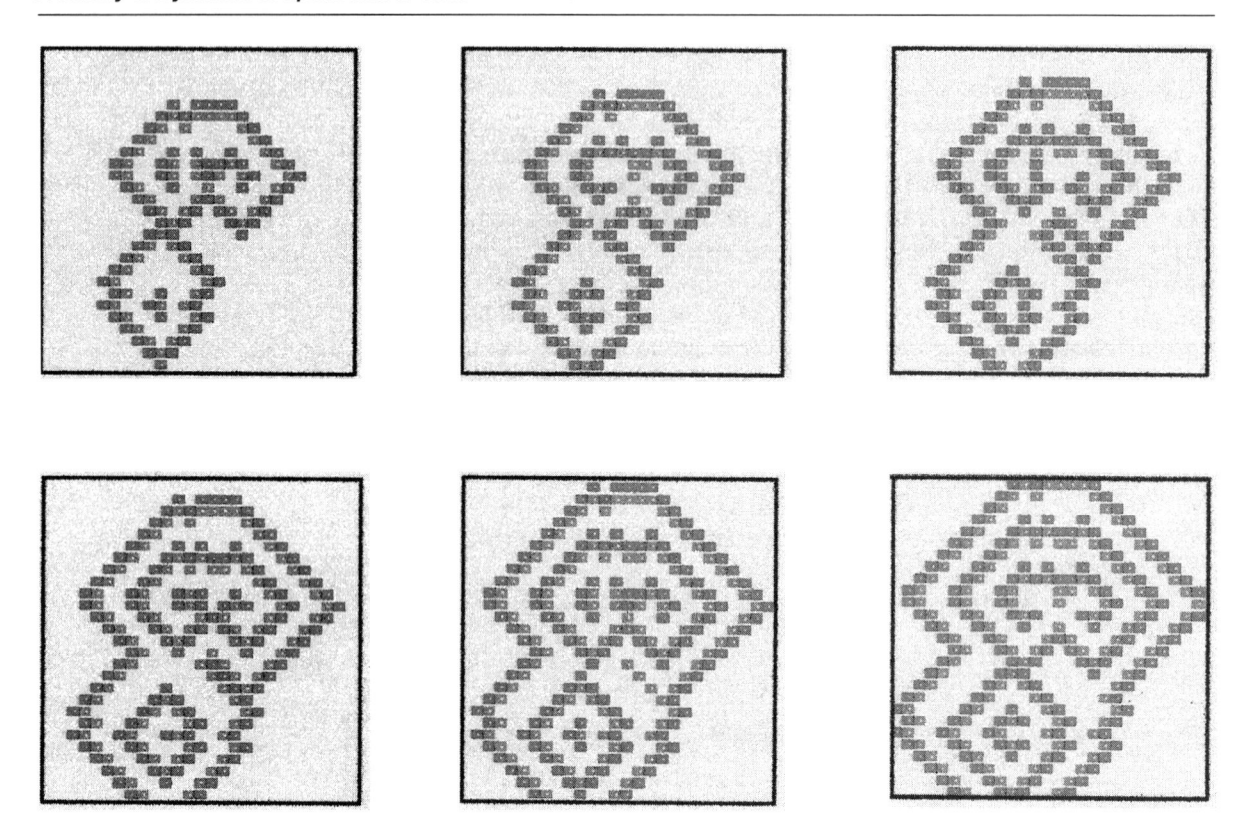

Fig. 8.2.
(a) An early sequence (left to right on two rows) of patterns from **Program 8-1A,** using $RT = 1$. Notice the periodic character near the center. What is the period? **Note:** Since we cannot show colors in this book, the colors on the computer screen (red = excited, light blue = quiescent, and dark blue = refractory) are represented here as dark cells, lighter cells, and white cells.

each cell is **color-coded** (0 = dark blue, 1 = light blue, 2 = red), which is illustrated in the program, in contrast to the shades of gray used in **Fig. 8.2.** The **boundary** of this region is **inactive** (it has a permanently refractory environment). The explorations involves **changing the initial conditions** in the program, which contain a number of **possibilities** for you **to activate, deactivate** (with apostrophes at the beginning of lines), or better yet, **ignore and entirely modify.** The program has a place for you to do this easily. The possibilities include a small percentage of **random initial states** (use <N> to survey these effects). You can **modify all of this,** and do much better!

Think 8.1C

Beginning with all cells in a quiescent state, how can you modi-

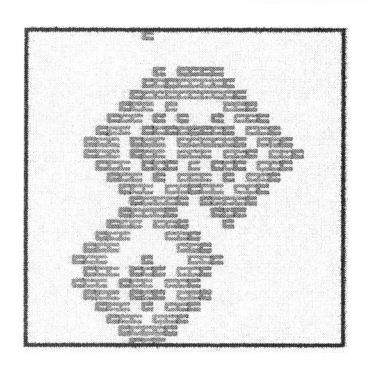

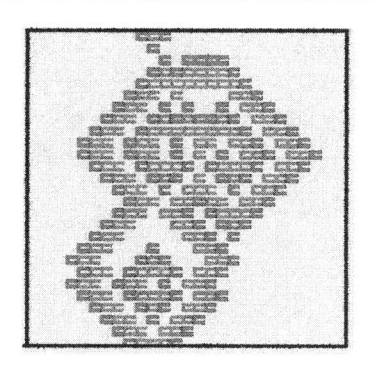

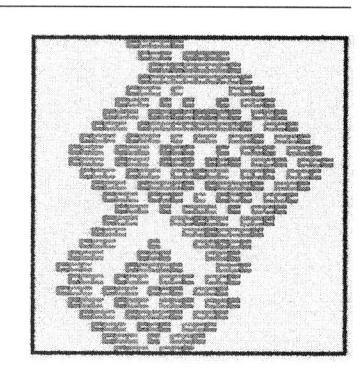

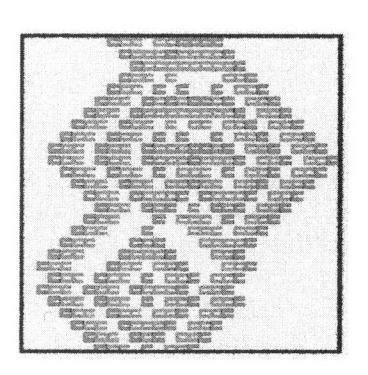

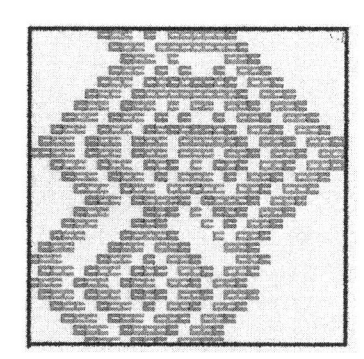

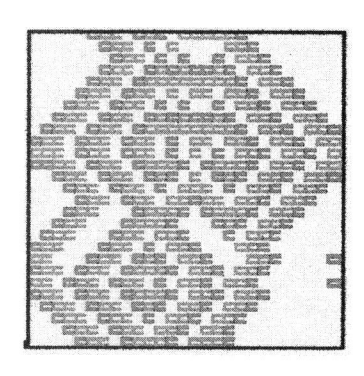

8.1 Spatial Pattern Dynamics of Excitation–Diffusion Systems

Fig. 8.2. *(continued)*
(b) An early sequence (left to right on two rows) for **Program 8-1B**, using $RT = 2$. This also differs from Fig. 8-2(a) in that the **boundary conditions are now periodic.** Note the "invasion" from the bottom to the top, and left to right edges.

fy the states of **only two cells** so as to generate a **continual production** of **only** concentric "rings" [the diamond patterns in Fig. 8.2(a)]. These are the famous **"target patterns"** observed in many excitable systems **[1, 3–8]**.

Explore 8.1B

Program 8-1B contains a number of generalizations of Program 8-1A. You can **explore the following:**

(a) The rules (8.1.1) have been generalized by replacing the **rule 2 → 0** by the **rule 2 → 1 − *RT*,** where ***RT*** is a **"refractory time,"** and **also appending** a new rule for a **cell in state *S***

  if $S < 0$ then → $S + 1$ (in the next step)

239

This means that **RT is the number of steps** it takes for the cell **to return to the quiescent state, 1.**

(b) Change the number of interacting "neighbors" to **eight** (including "corner" cells). This can yield new, "unnatural," patterns.

(c) Introduce a **variable threshold number,** *TH,* for the number of adjacent excited cells that is required to excite a quiescent cell. This threshold is *TH* = 1 in Program 8-1A.

(d) Introduce **"periodic" boundary conditions,** the results of which are illustrated in Fig. 8.2(b). Periodic boundary conditions means that the cells along the **top edge** are considered to be **adjacent to** those along the **bottom edge** (which would form a tube in physical space), and the cells along the **left edge** are **adjacent to** those along the **right edge.** Putting these together means that the dynamics occurs in a rather exotic **physical region** that is **the surface of a torus.** (A toroidal phase space was introduced in **Section 7.3.** Whereas that is a volumetric region, the present region is only the outside surface of such a torus). As can be seen from **Fig. 8.2(b),** this type of excitation dynamics is much **more complex** than what occurs in **Fig. 8.2(a).**

These modifications, and other suggestions, are easily accomplished by **activating appropriate lines** that already exist in the program by **removing apostrophes** or **modifying at locations** indicated by ####.

When the excitation–diffusion dynamics takes place in **three-dimensional regions,** the forms of dynamics can be dramatically different, as was discovered by Winfree **[2]**. In particular, he found **linked structures of excitations.** These are topologically the same as the periodic attractors that occur in the **phase space** of the Lorenz system (next section), as will be discussed in **Section 8.3.** This, however, is only a mathematical similarity, in two very different spaces with no physical connection.

Think 8.1D

If the refractory time is increased [modification **(a)** above], how would this change the **periodicity** seen in **Fig. 8.2(a)?**

Think 8.1E

Is it possible to change the boundary conditions on this square

8.1 Spatial Pattern Dynamics of Excitation–Diffusion Systems

region so as to represent the dynamics on the **surface of a sphere,** which might be more like a region of the heart? (Think about geographicsl maps of the Earth's surface.)

For a more "continuous" representation of excitable systems, one can use systems composed of hexagonal cells, each of which have six neighbors **[Fig. 8.1(b)]**. To appreciate some of the differences in this representation, see the following Explore.

Explore 8.1C

Program 8-1C illustrates the dynamics in a space of **hexagonal cells,** each of which has **six nearest neighbors.** The graphics is "nicer" but more complicated to program. (It is **not** done very efficiently in this program! Improve upon it.) The patterns in **Fig. 8.3** now more clearly illustrates a spiral pattern that is widely observed in nature **[1, 3–8],** interacting with a self-sustaining target pattern, seen in the lower parts of Fig. 8.2. The **"turbulent" initial conditions** give a more general sense of the dynamics. See if you can find any target patterns in this case. You can likewise make extensive explorations by using the **alternative dynamics (a), (c),** and **(d),** listed in **Explore 8.1B.**

Fig. 8.3.
(a) An example of a spiral pattern interacting with a target pattern, found in Fig. 8.2; (b) turbulent dynamics generated by random initial conditions. $RT = 2$ in both cases.

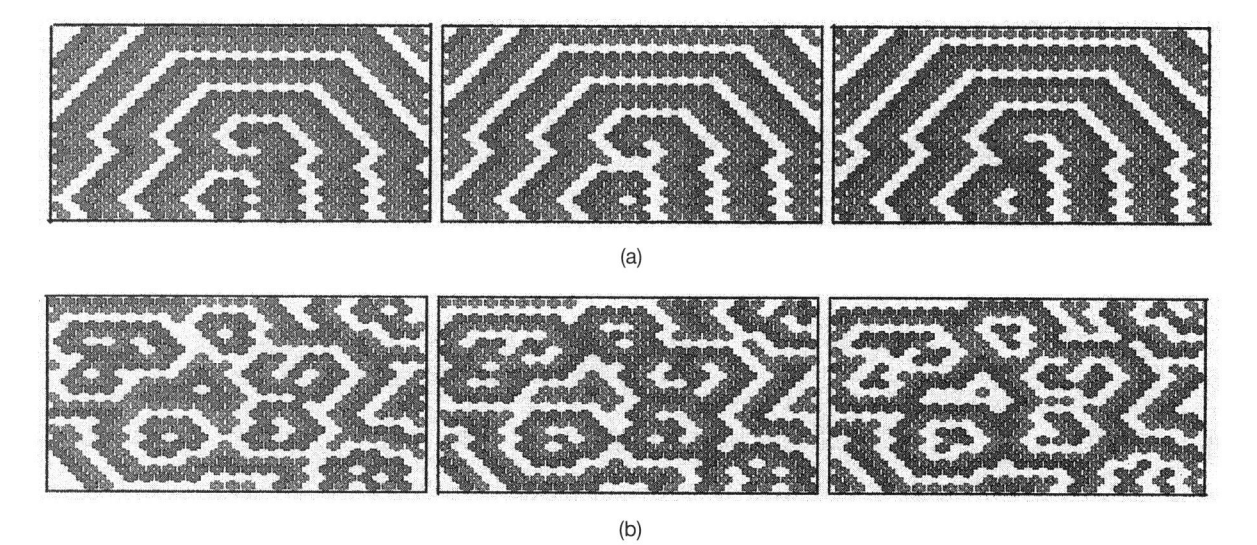

(a)

(b)

References and Notes

The spatial aspects of cardiac dynamics are discussed in:

1. L. Glass, Dynamics of Cardiac Arrhythmias. *Physics Today,* 41–45, Aug. 1996. Contains many references.

2. A. T. Winfree, *When Time Breaks Down: The Three-Dimensional Dynamics of Electrochemical Waves and Cardiac Arrhythmias.* Princeton University Press, Princeton, NJ, 1987.

    ———, Persistent Tangled Vortex Rings in Generic Excitable Media. *Nature, 317,* 233–236, Sept. 1994.

    ———, Electrical Turbulence in Three-dimensional Heart Muscle. *Science, 266,* 1003–1006, Nov. 1994.

For spatial dynamics in more general contexts:

3. L. Glass and M. C. Mackey, *From Clocks to Chaos: The Rhythms of Life.* Princeton University Press, Princeton, NJ, 1988. In particular, read the discussion in **Section 8.3:** Waves and Spirals in Two Dimensions, **and following sections,** with references to the famous Belousov–Zhabotinsky chemical spiral oscillations, and a number of living systems, including slime molds, which aggregate toward a chemically emitting pacemaker cell, and the relationship to complex cardiac arrhythmias. **Also read** the very interesting discussion of **dynamic diseases** in **Chapter 9** of this book. This concerns the breakdown of any normal temporally organized unit in the human body, being replaced by some abnormal dynamic behavior.

A few references for biological models of spatial pattern dynamics:

4. E. Mosekilde and O. G. Mouritsen (Eds.), *Modelling the Dynamics of Biological Systems: Nonlinear Phenomena and Pattern Formation,* Springer-Verlag, Berlin, 1995.

A number of useful articles can be found in:

5. S. A. Levin (Ed.). *Frontiers in Mathematical Biology.* Lecture Notes in Biomathematics 100. Springer-Verlag, Berlin, 1994.

6. R. M. May. Spatial Chaos and Its Role in Ecology and Evolution. In S. A. Levin (Ed.), *Frontiers in Mathematical Biology,* pp. 326–344. Lecture Notes in Biomathematics 100. Springer-Verlag, Berlin, 1994. May discusses a cellular model, as in this section, discussed in **section 5.2.** That dynamics is thereby generalized to include **spatial propagation** and **chaotic possibilities.**

7. J. J. Tyson. What Everyone Should Know About the Belousov–Zhabotinsky Reaction. In S. A. Levin (Ed.), *Frontiers in Mathematical Biology,* pp. 569–587. Lecture Notes in Biomathematics 100. Springer-Verlag, Berlin, 1994. A further discussion of scroll and spiral wave dynamics.

8. A number of examples can be found Chapter 3 "Life the Excitable Medium," in B. Goodwin, *How the Leopard Changed Its Spots: The Evolution of Complexity* (Charles Scribner, New York, 1994).

8.2 THE DISCOVERY OF CHAOTIC ATTRACTORS IN METEOROLOGY

One of the many important discoveries that has been made, thanks to the digital computer, is the **chaotic attractor** (or "strange attractor"). This discovery was made by Edward Lorenz, a professor of meteorology at MIT, who had long been struggling with the vagaries of weather dynamics (see reference **[1], Chapter 3,** "Our Chaotic Weather"). He had worked with models involving twelve or more variables, searching for parameters that would produce the nonperiodic dynamics found in real weather patterns. Having found such equations, he then studied their dynamics over longer time periods. To quote the circumstances of his important discovery of the **dynamic sensitivity to initial conditions** and chaos (related to the strange attractor, to be discussed):

> At one point I decided to repeat some of the computations in order to examine what was happening in greater detail. I stopped the computer, typed in a line of numbers that it had printed out a while earlier, and set it running again. I went down the hall for a cup of coffee and returned after about an hour, during which time the computer had simulated about two months of weather. The numbers being printed were nothing like the old ones. I immediately suspected a weak vacuum tube or some other computer trouble ... [however] instead of a sudden break, I found that the new values at first repeated the old ones, but soon afterward differed by one then several units in the last decimal place.... The differences more or less steadily doubled in size every four days or so.... This was enough to tell me what had happened: the numbers that I had typed in were not the exact original numbers, but were the rounded-off values that had appeared in the original printout. The initial round-off errors were the culprits; they were steadily amplifying until they dominated the solution. In today's terminology, **there was chaos.**

This discovery by **Lorenz,** through the use the **computer,** brought to the attention of the scientific community a basic dynamic fact that had been discovered **mathematically** at **the beginning of the 20th century** by **Poincaré,** as recounted in **Section 2.1.** This fact is that **all nonlinear dynamics are sensitive to both the initial conditions and environmental perturbations** (see the reference to **Duhem in Section 2.1**). Other earlier mathematical **[2a, 2b],** as well as a concurrent computational studies **[2c],** also contributed to an appreciation of the richness of nonlinear dynamics, reflected in this feature of dynamic sensitivity. However, Lorenz's computational study was able to uncover the entirely new fact that **sensitive dynamics can also occur in dynamic attractors.** Prior to this, all known dynamic attractors were either fixed points, limit cycles, or a chaotic nonautonomous system **[2a].**

Lorenz's original publications are found in references **[3, 4].** To appreciate more fully the issues involved in this research, and the history of his personal discovery, read Lorenz's own account (see, **p. 130 ff, in [1]**). This includes an account of how, in 1961, he came to extract a simplified model of three variables from a seven-variable model of B. Saltzman, which had shown the desired aperiodic dynamics. It is this three-variable model that has become known as the **Lorenz equations.** Much later, it was shown how these equations could be approximated from hydrodynamic equations that are associated with only vertical air circulation.

The simple idea is to consider a fluid ("air") between a horizontal hot plate ("Earth") and a second parallel cold plate ("stratosphere"), as illustrated in **Fig. 8.4.** If the temperature difference, ΔT, is not too large, the heat is simply conducted upward to the top plate without causing any motion of the fluid, because it is held in place by the downward gravitational force. However if ΔT is increased, there comes a point when the hot fluid expands enough to become sufficiently buoyant and it rises, displacing the cooler denser air downward (like what happens with a hot air balloon). This produces a constant vertical uplifting of hot air and a countercirculatory downward flow of cool denser fluid, as illustrated in **Fig. 8.4.** This transition from the static heat conduction to the stationary circulatory fluid flow was understood in the 19th century. What could not be determined mathematically was what happens if ΔT is further increased. This leads to **time dependent motion** of this circular air flow, which can only be dynamically understood with the help of a computer.

In 1972, Lorenz gave a lecture entitled "Predictability: Does the Flap of a Butterfly's Wings in Brazil Set Off a Tornado in Texas?" Since then, the only idea that has often been extracted from this lecture is the characterization of dynamic sensitivity as "the butterfly effect." His insight went much further than this.

Think 8.2A

To appreciate Lorenz's deeper insights into this issue, it is **very beneficial to read the first publication of this lecture** in Appendix I **(pp. 181ff) in reference [1].** Here you will find the

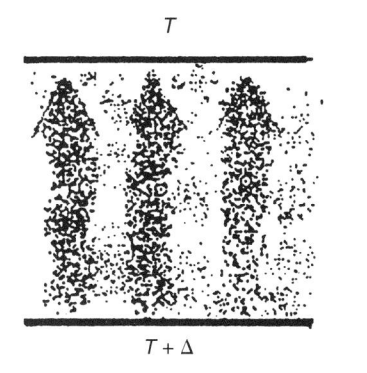

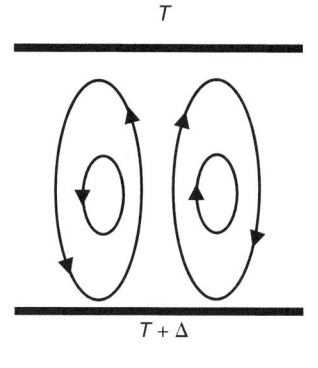

Fig. 8.4.
The heat transfer in a fluid between a lower hot surface ("earth"), at a temperature $T + \Delta T$ and an upper cooler surface ("stratosphere"), with temperature T. The fluid only moves in a steady circulatory fashion if ΔT exceeds a certain "moderate" value. For "larger" values of ΔT, aperiodic circulatory motion may occur.

true significance of Lorenz's insights revealed, in contrast to the inaccurate popularization of the "butterfly" characterization. In particular, reflect upon his statement: **"If the flap of a butterfly's wings can be instrumental in generating a tornado, it can equally well be instrumental in preventing a tornado."** What about many "butterflies," at all times?

Lorenz made a drastic simplification of a set of hydrodynamic equations, based on his appreciation of the most important asymptotic features of this dynamics. The resulting **Lorenz model** is given by the system of three differential equations [5]

$$\frac{dx}{dt} = \sigma(y - x)$$
$$\frac{dy}{dt} = rx - y - xz \qquad (8.2.1)$$
$$\frac{dz}{dt} = -bz + xy$$

The values $\sigma = 10$ and $b = 8/3$ are commonly assigned to these constants. This leaves the **one variable parameter** to be considered,

$$r \sim \Delta T g H^3 \left(\frac{\text{thermal expansion}}{\text{heat conductivity and viscosity}} \right) \qquad (8.2.2)$$

which is proportional to ΔT, the gravitational constant, g, the distance between the plates, H, as well as the other indicated influences [5, 6]. The physical parameter (environmental influence) that is of **greatest interest** is the change in the temperature difference ΔT, and because of the **relationship (8.2.2)**, the **behavior of the solutions** of equations (5.2.1) are studied as a **function of r**.

The variables $x(t)$, $y(t)$, **and** $z(t)$ **in (8.2.1)** are dynamic variables, **not coordinates in space.** They are the coefficients of certain spatial functions that describe how the fluid's circular flow behaves in time, and how the temperature changes in the vertical direction [5, 6]. Thus, the time behavior of these variables is related to the time behavior of the circular flow of the fluid and its vertical temperature variation. If $[x(t), y(t), z(t)] = (0, 0, 0)$ then there is **no fluid flow** (no convection), but only heat conduction, as illustrated on the **left in Fig. 8.4.** If $[x(t), y(t), z(t)] = (A, B, C)$, which are **three nonzero constants**, then the fluid flow is time independent, and the fluid has a **stationary convection,** as on the **right in Fig. 8.4.**

However, when ΔT **is sufficiently large,** there are **nonstationary asymptotic** ($t \to \infty$) **solutions of (8.2.1).** The most dramatic effect of this is that the **rotational sense of the convection** now becomes a function of time. This sense of rotation is mathematically connected with **sign of** $x(t)$, which is illustrated in **Fig. 8.5** [note that in this model **all** the adjacent rotations have a similar contrary sense, which change accordingly when $x(t)$ changes sign].

Think 8.2B

Determine the **three possible fixed points** of the equations **(8.2.1)** (i.e., the solutions for which there is no time dependence). For **what values of r** are three fixed points possible? Identify which of these correspond to **heat conduction** and which to **heat convection.** Using the information in **Fig. 8.5,** explain **physical-**

Fig. 8.5.
This illustrates how the **sign of the variable** $x(t)$ is related to the circulatory flow of adjacent convective circulation in space. If two **adjacent rotations have the senses shown when** $x(t) > 0$, then when $x(t) < 0$ their flow patterns both change their sense of rotation, as do all the others.

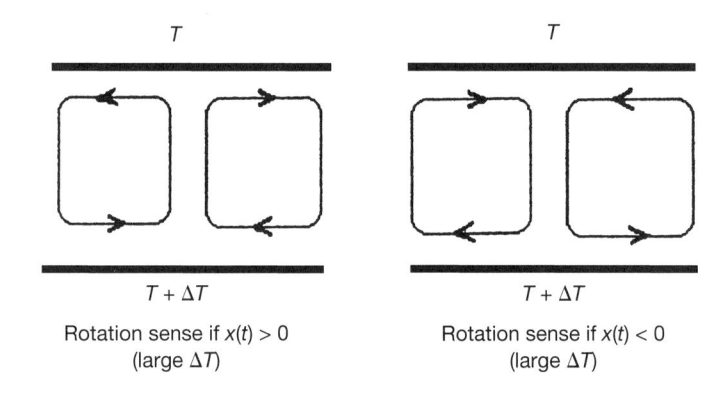

Rotation sense if $x(t) > 0$
(large ΔT)

Rotation sense if $x(t) < 0$
(large ΔT)

ly why there are **three,** rather than just two **fixed points.** Knowing these facts will make the **dynamic results in Explore 8.2A** much more meaningful to you.

If $x(t)$ **erratically flip-flops** back and forth **between positive and negative values,** then the corresponding **flow flip-flops** between the two **patterns in Fig. 8.5,** producing a highly **chaotic character of fluid motion.** This **chaotic dynamics,** when represented in the **three-dimensional** phase space, (x, y, z), is attracted to a **fractal structure.** A particularly useful characterization of this attractor is shown in **Fig. 8.6,** which was obtained by Lanford [7] (see Explore 8.2A). This type of structure was at first called a **"strange attractor,"** as it is indeed strange, but now is sometimes given the more precise name **"chaotic attractor."**

In order to **explore the various types of dynamics** that occur in (8.2.1) as the temperature difference, $r \sim \Delta T$, is increased, it is necessary to use computer programs; so now begin your explorations with the following Explore.

Fig. 8.6.
The Lorenz **"strange attractor"** dynamics of (8.2.1), in the **three-dimensional** (x, y, z) **phase space.** This chaotic dynamics only occurs if r **(hence ΔT) is sufficiently large,** as illustrated in **Explore 8.2A.**

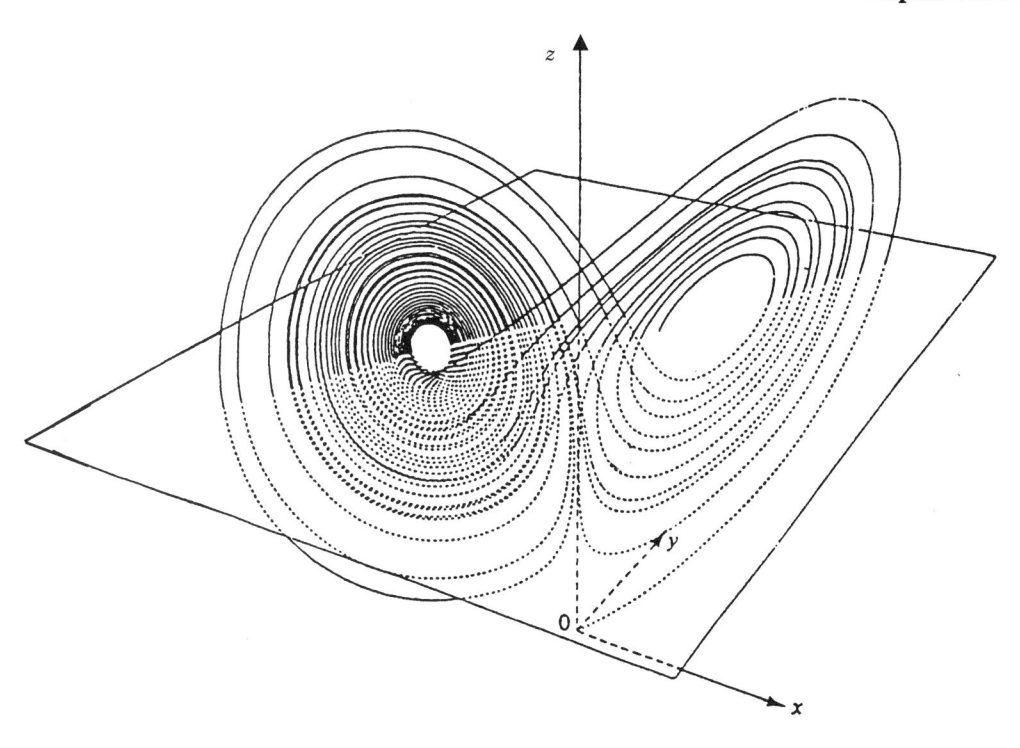

A Variety of Dynamics in Space and/or Time

Explore 8.2A

In **Program 8-2A,** you will be able to discover the various types of dynamics that are generated by (8.2.1), **depending on the value of r.** It makes use of a Euler approximation, which makes the computations faster, at the expense of some unnecessary accuracy (the qualitative features are what are important here). This dynamics is displayed in the **three-dimensional phase space, (x, y, z),** the axes of which can be rotated in any direction. To appreciate how the dynamics, illustrated in **Figs. 8.5 and 8.6,** is represented in this phase space, study the following stages of dynamics as r (=R in the program) is increased.

Note first that r has been **normalized** such that **convective flows** (which are the only dynamics of interest here) only occur if $r > 1$; so do not take smaller values of r or you will get an error statement. As r is increased, the dynamics goes through several transitions (bifurcations). One of these transitions is **very surprising,** and rarely discussed in the literature (ever? see Figs. 7.49 and 7.51 in [0]), so this will be your first discovery (with some clues!). Begin with **$r = 15$** and study the dynamics, using **a number of initial states.** What do these **two asymptotic states** represent physically? Increase the value to $r = 16$ and again try a number of initial conditions, using **<N>**. Repeat this process of increasing r, **looking for a change in the dynamics,** as indicated in the program (also noting the **change in** behavior of the **nonasymptotic portion** of the dynamics). In each case, determine what these asymptotic dynamics represent physically. Does something strike you as quite remarkable? If not, you missed something **quite unusual,** so try again! You can also **rotate the coordinates about the z and x axes** (changing "theta" and "phi"), to see the **structure** of *any* **strange attractor in three dimensions.**

Figure 8.6 shows the strange attractor passing through a plane, **below which the orbits** are represented by **dotted curves.** This is a very special plane, which is **parallel to the (x, y) plane** at $z = z_0 \equiv (r - 1)$. It is in this plane that **two fixed points** are located, at $x = y = \pm \sqrt{b(r-1)}$ (as you should have discovered in **Think 2.8B**), which each correspond to one of the stationary convective flows in **Fig. 8.5**. When r becomes sufficiently large, these **fixed points become unstable,** as is the one at the origin ($r > 1$). However, **all solutions** of (8.2.1) are always **attracted into a bounded region about the origin** (see the proof on p. 145 of **PND**, Vol. 2, [0]). Thus, they must either approach a **stable fixed point,** a periodic motion **(stable limit cycle),** or else be aperiodic (a chaotic attractor). So **when all of the fixed points are unstable,** the dynamics must become **asymptotically chaotic,** for all initial conditions (reconsider the surprise in Explore 8.2A).

Now the **solutions of (8.2.1) are uniquely defined** by their initial conditions. Therefore, as they **pass through the plane at** $z = z_0 \equiv (r-1)$ (say in a downward direction), they have **unique intersection points. If two points are the same,** it means that the **solution is periodic,** otherwise all intersection points will be distinct. In any case, **the map of successive intersections**

$$[x(t), y(t), z(t) = z_0] \to [(x(t'), y(t'), z(t') = z_0] \quad (\text{and } dz/dt < 0) \tag{8.2.3}$$

is an example of a **Poincaré map.** It has the great advantage of reducing the **complicated dynamics in three dimensions** into a **unique two-dimensional representation.** With the help of this, one can see that the **strange attractor** does not fill out a region in three dimensions, nor does it lie on a two-dimensional surface, but **is a fractal** with a capacity dimension greater than 2.

Explore 8.2B

Use **Program 8-2B** to see the **Poincaré map** of the Lorenz dynamics for **any value of** $r > 1$. This program uses the Runge–Kutta iteration method and hence gives the **accurate bifurcation values of** r for the transition between different types of flows. The **blue circles** are the **two fixed points** on the plane $z = z_0 \equiv (r-1)$. They may be either stable or unstable, depending on the value of r. Begin with $r = 20$, and **using many initial conditions,** see how the dynamics changes as r is increased, using `<I>/<D>`. A number of suggestions and questions are given along the way. **When $r > 27$,** the graphics becomes **more refined** (using `PSET` instead of `CIRCLE`, you can modify this as you like), to make it clearer that **each point is unique** (to within the graphics resolution). If you want a count of the **number of mapped points,** activate the line before `I`.

Think 8.2C

From Fig. 8.6, why do you think that it is important to use the particular plane $z = z_0 \equiv (r-1)$ to construct this map? What if you use another plane, say $z = z_1 > (r-1)$? **Try it out, if you have doubts.**

A Variety of Dynamics in Space and/or Time

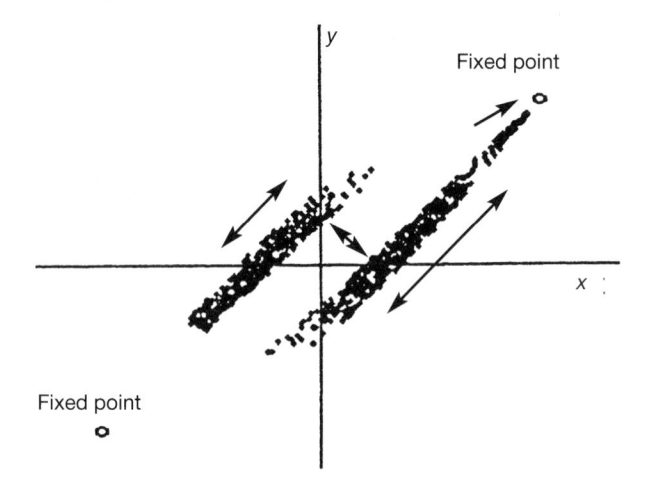

Fig. 8.7.
The Poincaré map of a **"not quite chaotic"** dynamics.

Fig. 8.8.
(a) The Poincaré map of the Lorenz chaotic attractor ($r = 30$), and (b) that of a **nonstandard chaotic attractor of (8.2.1)** ($r < 25$).

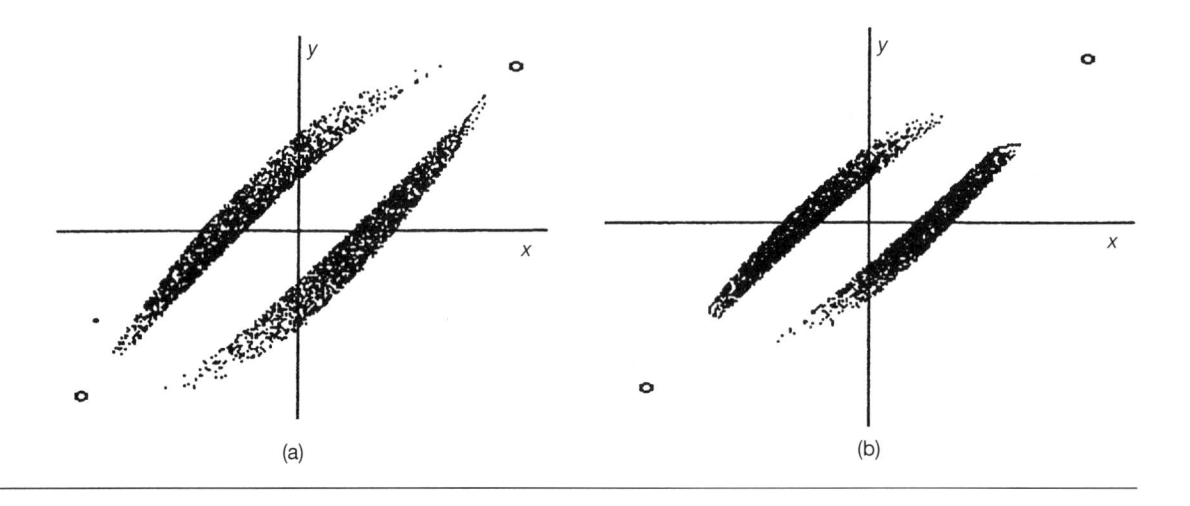

Among the types of dynamics you should discover is the **"not quite chaotic"** dynamics, illustrated by the Poincaré map in **Fig. 8.7,** in which the dynamics behaves chaotically for a limited period of time, before tending to a stationary convective state.

A more subtle form of dynamics is illustrated in **Fig. 8.8.** In this case, the Poincaré maps of two types of **chaotic dynamics are illustrated. The one on the left is** the map of **Lorenz's chaotic attractor (see Fig. 6),** whereas the one on the right is not the standard Lorenz attractor (what else exists in this case? If you do not know, then go back to Explore 8.A.). These figures both contain 5000 points, but note the distinct difference in their structure.

References and Notes

0. The references to E. A. Jackson, *Perspectives of Nonlinear Dynamics*, Cambridge University Press, Cambridge, 1991, are denoted by **PND**.

1. E. N. Lorenz. *The Essence of Chaos*. University of Washington Press, Seattle, 1993.

2. An introductory discussion of the following references can be found in in the indicated sections of **PND**.

2a. N. Levinson. A Second Order Differential Equation with Singular Solutions. *Ann. Math., 50,* 127–153, 1949. He established that chaos could be characterized (understood) by using **symbolic Bernoulli sequences,** when applied to a strongly forced van der Pol oscillator (**see Section 5.3** this volume; in **PND, Vol. 1,** see index, **and particularly p. 337**).

2b. For the Kolmogorov–Arnold–Moser (**KAM**) theorem see **PND, Vol. 2, index.** This sophisticated theorem deals with the perturbation of the dynamics of Hamiltonian systems, which are "conservative!" isolated systems. An early authoritative reference is: V. I. Arnold, Small Denominators and Problems of Stability of Motion in Classical and Celestial Mechanics, *Russian Math. Surveys, 18,* 6, 85–123, 1963. A much more accessible discussion of the ideas involved can be found in **Appendix 8** of V. I. Arnold, *Mathematical Methods of Celestial Mechanics*, Springer-Verlag, New York, 1978.

2c. M. Hénon and C. Heiles. The Applicability of the Third Integral of Motion: Some Numerical Experiments. *Astrophys. J., 69,* 73–9, 1964. As in the case of the KAM theorem, this deals with an **abstract** model of Hamiltonian dynamics (which have no attractors), computationally illustrating dynamic features related to those discussed in the KAM theorem. See **PND, Vol. 2, Section 611** and index.

3. E. N. Lorenz. Deterministic Nonperiodic Flows. *J. Atmos. Sci., 20,* 130–142, 1963.

4. E. N. Lorenz. The Problem of Deducing the Climate from the Governing Equations. *Tellus, 16,* 1–11, 1964.

5. For an elementary discussion of the basis of these equations, see **PND,** Vol. 2, Section 7.3.

6. For useful graphics and general explanatory text see L. S. Liebovitch, *Fractals and Chaos: Simplified for the Life Sciences,* Oxford University Press, New York, 1998; also see Lorenz in the index.

7. O. E. Landford. Turbulence Seminar. P. Bernard and T. Rativ (Eds.), *Lecture Notes in Mathematics,* Vol. 615. Springer-Verlag, Berlin, 1977.

8.3 A SYMMETRY-BREAKING BIFURCATION: LINKED LIMIT CYCLES

In this section, it will be shown that the Lorenz equations can go through some **interesting bifurcations** of flow as the parameter is decreased through values near $r = 160$ and 320. At these large values of r, there is **no reason to believe that this model any longer bears any physical relationship to the convective** flows discussed in Section 8.2. One might, in fact consider this dynamic phenomenon simply an abstract mathematical wonderment, except for the fact that similar topological structures have been found in the **dynamics of real space** for excitable media [1], considered in **Section 8.1**. Remember, the **present dynamics is in phase space,** so there is no physical connection with the dynamics in [1]. However, what follows illustrates how such dynamic bifurcations can be computationally discovered in a much simpler dynamic model, and their topological features investigated. The ultimate physical applications of such results remains to be discovered (by you?).

The study begins in regions of the **large values $r = 160$,** for which the **Lorenz equations** of the last section,

$$\frac{dx}{dt} = \sigma(y - x)$$

$$\frac{dy}{dt} = rx - y - xz \qquad (8.2.1)$$

$$\frac{dz}{dt} = -bz + xy \qquad (\sigma = 10, b = 8/3)$$

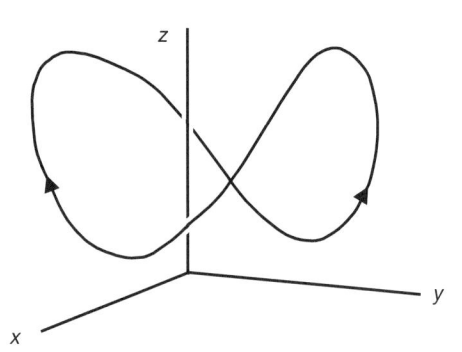

Fig. 8.9.
A symmetric stable limit cycle, satisfying (8.3.1).

8.3 A Symmetry-Breaking Bifurcation: Linked Limit Cycles

have a simple **globally stable limit cycle.** It typically looks something like the curve shown in **Fig. 8.9.** It has a "symmetric structure," in the sense that if you **change from a point on this orbit,** (x, y, z), according to the rule

$$(x, y, z) \to (x^*, y^*, z^*) \equiv (-x, -y, z) \qquad (8.3.1)$$

then (x^*, y^*, z^*) **will be another point on the same orbit.** This type of symmetry is also true for the equations (8.2.1), as one can establish by making the substitution (8.3.1) in (8.2.1) and canceling the minus signs. This is called the **symmetry property of the equations (8.2.1).** From this point of view, it seems hardly surprising that the limit cycle in Fig. 8.9 also satisfies this symmetry property. Now, whereas one might expect that **all solutions** of the Lorenz equations, (8.2.1), must **satisfy** the same symmetry property **(8.3.1),** this is **not necessarily so!** All that the symmetry property of (8.2.1) tells us is that:

(A) if $[x(t), y(t), z(t)]$ is **a solution** of (8.2.1),

(B) so is $[x^*(t), y^*(t), z^*(t)] = [-x(t), -y(t), z(t)]$ **a solution.**

Note that, **at the same time,** the solution $[x^*(t), y^*(t), z^*(t)]$ is in the "symmetric region" (a different region) of the phase space than is $[x(t), y(t), z(t)]$. This result does **not tell us** that $(x(t), y(t), z(t))$ ever goes to this location of $[x^*(t), y^*(t), z^*(t)]$ at **some different time.** However,

(S) If $[x^*(t), y^*(t), z^*(t)] = [x(t^*), y(t^*), z(t^*)]$ for some $t^* \neq t$, **there is only one solution,** because of the uniqueness of all solutions. In this case, **the solution is symmetric** in the phase space, as in **Fig. 8.6.**

Think 8.3A

If (S) is satisfied, select a location for $[x^*(t), y^*(t) z^*(t)]$ and for $[x(t^*), y(t^*), z(t^*)]$ in Fig. 8.9, and **determine the period** of the orbit **in terms of t^* and t.**

We are now going to study **symmetry-breaking bifurcations,** as was done in Section 7.3. However, in that case the dynamic equations were nonautonomous (they involved a forced oscillator), and were two-dimensional. Now we are dealing with the **three-dimensional phase space** of the **autonomous** Lorenz equations, (8.2.1). Both of **these distinctions are important.**

Now, in contrast to (S), one can have **the second possibility:**

(AS) $[x^*(t), y^*(t), z^*(t)] \neq [x(t^*), y(t^*), z(t^*)]$, **for any** (t, t^*). In this case there are two distinct solutions, and since they do not individually satisfy the symmetry property (8.3.1), they are a pair of **asymmetric solutions [2]**.

If in the course of changing the value of r one discovers that the solutions of (8.2.1) go **from a symmetric solution to a pair of asymmetric solutions,** this is referred to as a **symmetry-breaking bifurcation.**

Single Symmetric Solution ⇒ Pair of Asymmetric Solutions

This type of bifurcation apparently is quite rare, but it happens in several cases (8.2.1). To see what happens (topologically) when a symmetry-breaking bifurcation occurs, consider **Figs. 8.10(a) and (b). Figure 8.10(a)** illustrates a stable periodic orbit (the "tube" orbit), with a **particularly important surface** that contains orbits approaching this stable orbit. It has a strange structure, involving a **"full twist" around the orbit** in the course of one period of the motion (notice that this surface is not the famous Möbius strip, which involves only a "half-twist"). Moreover, this surface is such that when this periodic orbit becomes unstable it produces **two nearby new stable orbits** on this particular surface (this surface is inferred from the computations to follow). These two stable limit cycles are the **two "tube" orbits in Fig. 8.10(b),** where the dashed orbit is the now-unstable original orbit. What needs to be noticed is that, because of the "full twist" character of this surface on which the bifurcation takes place, **these two new stable limit cycles are linked** in the phase space, like two links of a chain. Topologically speaking, they are the same as shown in **Fig. 8.11(a),** whereas what is **observed in the phase space** is like **Fig. 8.11(b),** as can be found in **Program 8-3.** To "unravel" the over–under representation in Fig. 8.11(b) requires some thought,

Fig. 8.10.
A topological representation of the symmetry-breaking bifurcation of (8.2.1), from one stable symmetric limit cycle (a) to a stable pair of **linked** asymmetric periodic limit cycles (b) that lie on this "full-twist" surface.

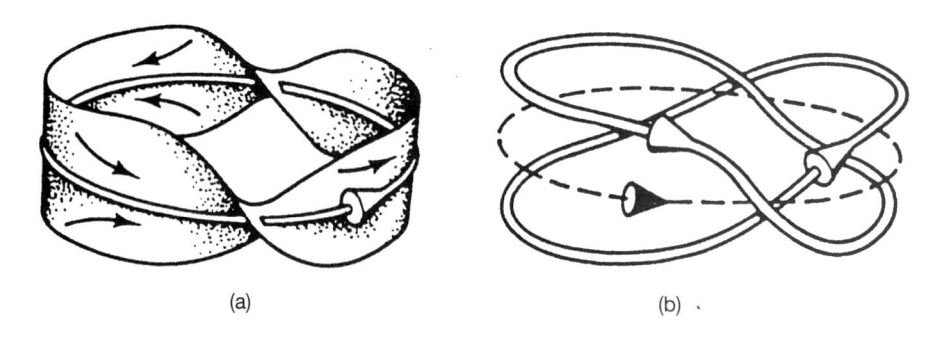

(a)   (b)

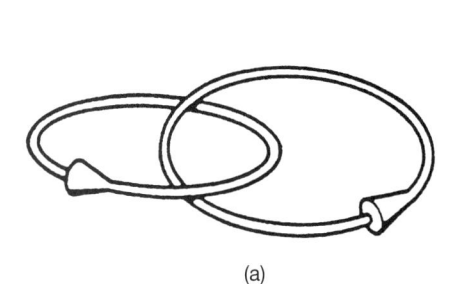

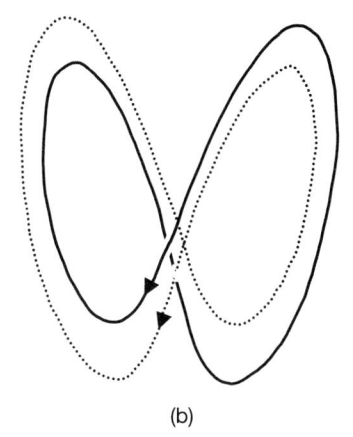

8.3 A Symmetry-Breaking Bifurcation: Linked Limit Cycles

(a)   (b)

Fig. 8.11.
A topological (a) and phase-space (b) representation of the **linked limit cycles. (a) Corresponds to Fig. 8.11(a); (b) corresponds to two limit cycles of (8.2.1), both for $r >$ 300.**

perhaps with the help of **rotations about the z axis in Program 8-3.**

Think 8.3B

Determine if the two linked orbits in (a) and (b) of Fig. 8.11 are **topologically equivalent.** That is, can you **stretch** these figures (one or both) in such a way that you can make the resulting **two orbits** from one figure **overlap** the resulting **two orbits** from the other figure? What **single change** can be made in Fig. 8.10(a) that would yield the **opposite answer?**

Now how can one **easily detect a symmetry-breaking bifurcation,** using graphical methods? Let us explore one method and see how it can reveal the above phenomena of **linked stable limit cycles.**

Explore 8.3A

Program 8-3 graphs both $[x(t), y(t), z(t)]$ **(as a red orbit)** and $[-x(t), -y(t), z(t)]$ **(as a blue orbit)** in the phase space. The perspective used is looking along the (x, y) **plane** (so the **horizontal line is this plane,** and the vertical line is the z axis). If there is a **symmetric limit cycle,** the blue and red orbits will **overlap.** If there are **two asymmetric limit cycles,** these will show up as **intersecting, due to the two-dimensional projection,** as in Fig. 8.11. Select $r = 320$ to see a **symmetric stable limit cycle,** and its symmetric image. You will need to **wait** for it to approach this asymptotic limit, **using <C> liberally** see this limiting result. Now

255

lower the value of r, **using <L>,** and determine roughly the bifurcation value, $r = r^*$, where **two** stable limit cycles occur. At this point the blue and red orbits will be slightly different, and **not overlap.** If you continue to decrease r these orbits will become more distinctly separated. If r is decreased to **around $r^* - 10$,** and moreover you change the perspective by **decreasing theta,** the **left–right projected symmetry will be lost,** and you will be able to observe the orbits looking like **Fig. 8.11(b). To see how they are entangled** requires rotating the phase space about the z axis, by changing TH (= theta). See if you can detect the linked topology illustrated in the above figures.

Explore 8.3B
Another example can be found, starting at $r = 160$, but the limit cycle is more elaborate than in the case $r = 320$. You should repeat the study in part (A) and again **decide whether these two limit cycles are linked or not.**

Explore 8.3C
In this case, you can begin with a **pair of asymmetric stable limit cycles** by taking $r = 100$. The exploration now involves discovering **what type of bifurcations** produced these limit cycles. You can study the bifurcations by both **increasing r** and decreasing r slightly **(how slightly?)** You will find quite different results in this case. The dynamics for arbitrary values of r is very complicated [3].

Think 8.3C
Run **Program 8-3** at $r = 30$ so that it "generates the Lorenz attractor." What does this mean when considered (1) **mathematically** as compared with (2) **computationally** or (3) **empirically?** Are there two **mathematically distinct** solutions [2]? Are they **computationally distinct?** Would there be any **empirical** method to distinguish them? [4]

Think 8.3C
Determine if the two linked orbits in Fig. 8.11(a) and (b) are **topologically equivalent.** That is, can you **stretch** these figures (one or both) in such a way that you can make the **two orbits** in one figure **overlap** the **two orbits** in the other figure? What **single change** can be made in Fig. 8.10(a) that would yield the **opposite answer?**

Reference and Notes

1. A. T. Winfree. Persistent Tangled Vortex Rings in Generic Excitable Media. *Nature, 317,* 233–236, 15 Sept.1994.
2. To make this point quite clear, if there exists an aperiodic solution, it cannot satisfy (8.3.1), for if it did then it would satisfy **(S),** and hence be periodic. Thus, in the case of the Lorenz system, aperiodic solutions must come in pairs. For other equations, the number of aperiodic solutions depends on their particular symmetry (invariance) properties.
3. An indication of this variety of solutions can be found in my *Perspectives of Nonlinear Dynamics,* Vol. 2, Fig. 7.61, p. 187, where the above examples can be related to more general solutions.
4. Mathematicians have **defined their "chaotic attractor"** not in terms of individual solutions but rather in terms of "dense asymptotic sets of points", which has the consequence that **all of these three concepts** of the "chaotic attractor" are **compatible concepts,** that is, noncontradictive concepts. They are not identical concepts because "dense sets" do not exist in either computations or empirical observations. This issue refers back to **basic discussions in Chapter 2.**

8.4 SELF-ORGANIZING ANT COLONY DYNAMICS: SOCIAL INSECTS

The concept of a **dynamically self-organizing system** has to do with the feature that a system, which only has **limited interactions** between its **many component parts,** may (under suitable circumstances) develop a **globally coherent form of dynamics**—it **"self-organizes"** in the sense that its globally coherent dynamics is **not dictated by long-range interactions or some external action on the system.** A living system may also self-organize its internal processes in response to **environmental changes,** often in order to sustain its viability, in which case this capability is referred to as **adaptive dynamics.** There are also some wonderfully sophisticated forms of self-organization of **groups of social insects.** Famous among these are honeybees and ants. **Honeybees** are a fascinating example of organizational behavior that is accomplished by **complex dynamic interactions [1–4].** Also, **various species of ants** are famous for their organizational properties, particularly in connection with food collection **[5–7].**

In this section we will explore a relatively simple but rather surprising example of the **dynamic organization** in some **ant colonies [8, 9].** When the density of the ants in these colonies is small, the individual **ants move** about in a **random fashion,** bumping into each other only rarely. In these circumstances, it is observed that the ants only move around for a **limited time, rest** for an **indefinite period** of time, and then at **random times** spontaneously begin to **move around again.** However, if an ant bumps into one that is at rest, it stimulates that ant to begin to move again. That is, the ant is an **excitable creature,** being stimulated by this bodily contact **[8, 9].**

Think 8.4A

What are the similarities and differences between this system and the **excitable system** considered in **Section 8.1?** Keep this in mind for some explorations below.

At **low densities** the result of this type of random individual dynamics, resting, spontaneous excitations, and "infrequent" stimulating bodily interactions is to produce a generally **random motion** within the colony of ants. This may seem to be quite reasonable, given the random character of the individual dynamics, resting periods, and occurrences of physical interactions. However, it has been observed that as the **density of ants increases** the behavior of the colony as a whole can change dramatically. It is found that the **activity** of the colony, $ACT(t)$, as measured by the number of active ants, **becomes a periodic function of time** (for details see [8]). That is, there is a collective or **self-organizing dynamic** feature established within the colony as a whole, despite only **local random behavior.**

To model this system **[9],** we take "ants" moving on a spatial lattice, in a random fashion, never occupying the same lattice point, as in **Fig. 8.12.** They have an **internal state of "excitation,"** $1 \geq E(t) \geq 0,$ which decreases with time, unless there are

Fig. 8.12.
"Ants" (circles) move randomly on a lattice (dots), while excited above a threshold. Their "excitation" decreases with time, and they may come to rest. They can become spontaneously excited and move, or may also become excited by ants entering their eight adjacent lattice points.

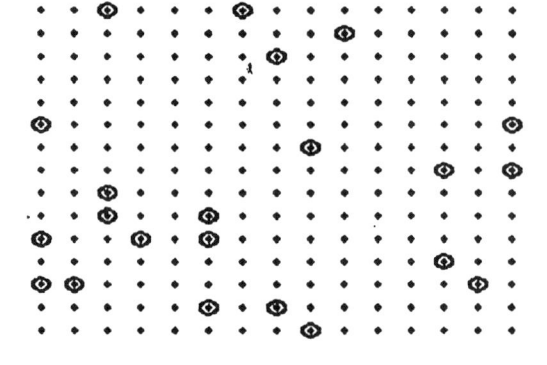

8.4 Self-Organizing Ant Colony Dynamics: Social Insects

other ants in their **eight-cell neighborhood.** These ants add to the excitation of the central ant by an amount that depends on their state of excitation.

Notice that what is going on here is the **introduction of a memory** to these dynamic components, **namely $E(t)$.** This is an **essential ingredient** in the **organizational and adaptive dynamics of living systems.**

Explicitly, closely following references [8] and [9], we will take the **time change** of the excitation $E(I, t)$ (for the ant labeled I) to be given by the hyperbolic tangent function

$$E(I, t + 1) = \tanh[G \cdot NES(I, t)] \qquad (G < 1) \qquad (8.4.1)$$

where $\tanh(z) = [\exp(z) - \exp(-z)]/[\exp(z) + \exp(-z)]$. (8.4.1) is demystified in **Fig. 8.13.** In (8.4.1), $NES(I, t)$ represents the **"neighborhood excitation sum"** for ant I, which is the **sum of the excitations,** $E(J, t)$, of **all** ants in the **nine-cell region,**

$$1 \geq |X(I, t) - X(J, t)| \quad \text{and} \quad 1 \geq |Y(I, t) - Y(J, t)| \qquad \text{(all } J\text{)}, \qquad (8.4.2)$$

which includes the central ant ($J = I$). In other words, the excitation contributions from other ants are being added to whatever the central ant possesses at that time. The map (8.4.1) is illustrated in **Fig. 8.13** for several values of G.

Fig. 8.13 illustrates the fact that, although the total neighborhood excitation can range from $9 \geq NES(t) \geq 0$ (since the number of lattice points is nine), this is always mapped into the range of **1**

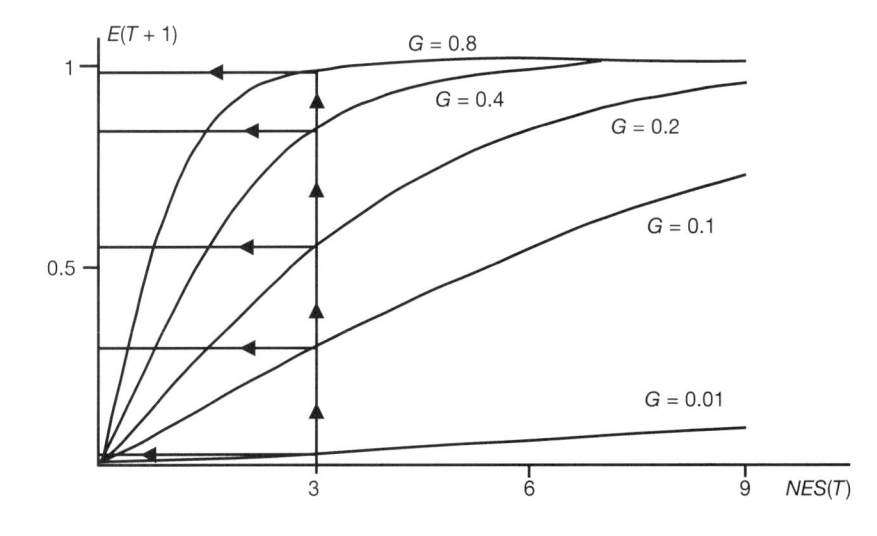

Fig. 8.13.
The maps of **NES = 3** to **$E(t + 1)$**, (8.4.1), for the indicated values of G.

$\geq E(t + 1) \geq 0$. In fact, if $G = 0.01$, then $E(t + 1)$ is squeezed into the range $0.09 \geq E(t + 1) \geq 0$.

If there are **no ants adjacent** to ant I, then (8.4.1) simply becomes

$$E(I, t + 1) = \tanh[G \cdot E(I, t)] \quad \text{("lonesome ant")} \quad (8.4.3)$$

so that $E(I)$ decays faster for smaller values of G (as should be clear from **Fig. 8.13**). The condition $G < 1$ in (8.4.1) ensures that $1 \geq E(t + 1) \geq 0$ remains satisfied. The iterative map **(8.4.3)** is illustrated in **Fig. 8.14**.

Figure 8.14 indicates the rapid decrease of $E(t)$ for a single ant, even when G is large. If $0.01 \geq G$, then $G \cdot E(t) \geq E(t + 1)$, so other actions are required to keep an ant continually excited. The ant remains active only if its excitation exceeds some **activation threshold, $E(I) > ATH$**. If $E(t) < ATH$, it comes to rest $[E(t) = 0]$, and remain at rest unless either **(1)** it is stimulated by its neighbors through the action (8.4.1), or **(2)** it randomly generates its own spontaneous excitation. The **probability of spontaneous excitation** is denoted by **PSE**, and the **magnitude** of this excitation is denoted by **MSE**.

All of these factors now define the dynamics of this ant colony. The colony is taken to move on a square lattice with **NS** sites on its sides and consists of **NA ants.** The controlling factor in this self-organization is the magnitude of the **density**

$$D = \frac{NA}{NS^2} \quad (D < 1) \quad (8.4.4)$$

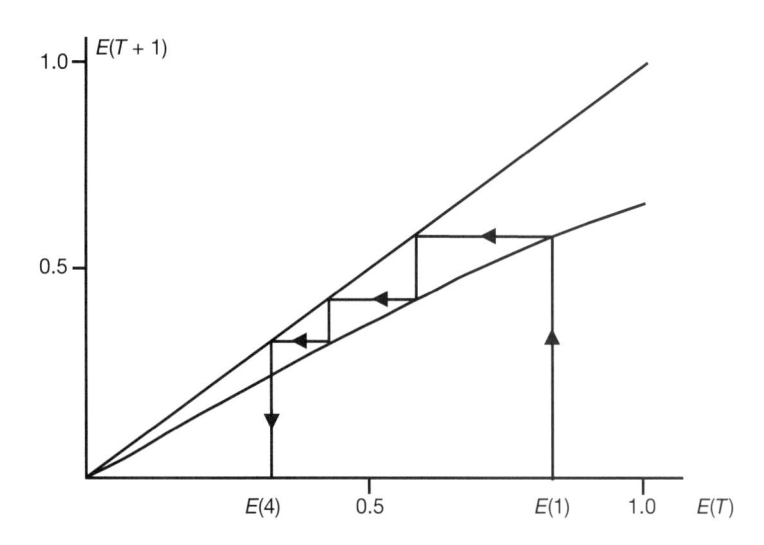

Fig. 8.14.
Several iterations of the map (8.4.3) are shown for the large value of $G = 0.8$, in order to clearly illustrate several steps.

which is less than one because there is no more than one ant per site.

Program 8-4A illustrates this type of dynamics for a **particular system** defined by the values

$$NS = 10, ATH = 10^{-15}, PSE = 0.01, MSE = 0.01, G = 0.05 \quad (8.4.5)$$

which are generally similar to **[9]**. The size of the lattice (**NS·NS** = 100 sites) has been limited in order to keep the computation time short, but **can easily be adjusted** at the beginning of the program. **Figure 8.15** shows an example of the percentage of excited ants $[E(I) > ATH]$ as a function of time (starting with none excited), when the density is **D = 0.1**. It can be seen that the percentage excitation is irregular, and generally low in value. By contrast, when **D = 0.7**, as shown in **Fig. 8.16**, the percentage of excitation is much more regular (periodic?), and generally much higher in value.

Explore 8.4A

Using **Program 8-4A,** first see how the examples in Figs. 8.15 and 8.16 vary with time, and **think about the questions** in the captions. Specifically, determine how **some features** of the **dynamics vary** as the **density is varied (using <I>/<D>).** Is there a value **D = D*** at which you feel that there is some **dynamic bifurcation?** If so, specify what is the feature, and for what "value" of **D***.

The graphics show the actual dynamics of the ants on the lattice, as well as the **percentage of excited ants,** as a function of time, along the bottom of the graphics. There is **also a "%" number displayed** on the graph, which is interesting information that you can figure out by observation, or by looking into the program.

Fig. 8.15.
One example of the percentage of excited ants as a function of time, when the density is **D = 0.1.** Does this pattern remain irregular for all times?

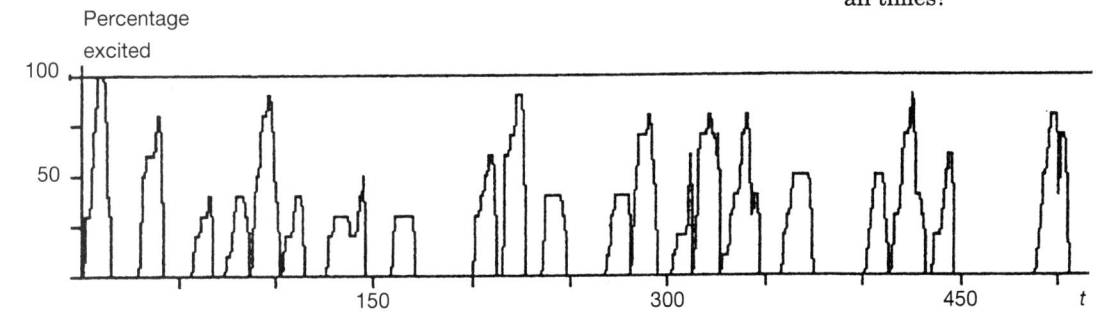

A Variety of Dynamics in Space and/or Time

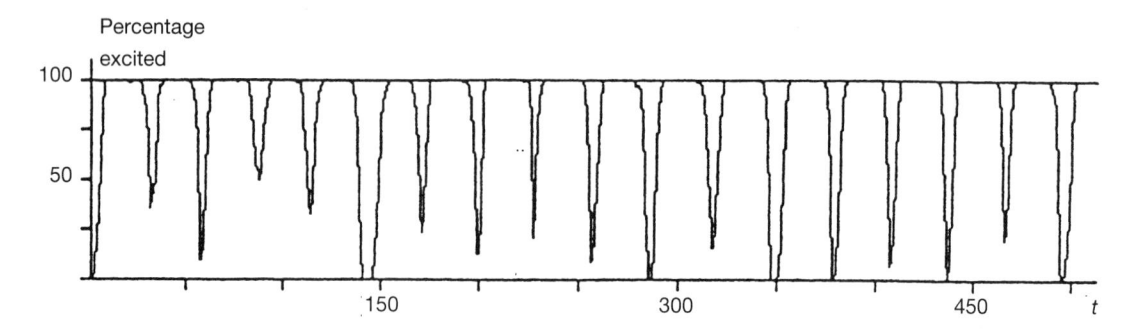

Fig. 8.16.
The percentage of excited ants when the density is $D = 0.7$. The pattern now has (periodic?) regularities, with much larger percentages of ants activated at the same time—a **"fairly self-organized"** behavior. How does this arise as D varies? **What** is "organized"?

Explore 8.4B

Observing a phenomenon, such as this bifurcation, and **understanding** it are two different things. To **understand** something (e.g., this bifurcation) requires the **discovery of some relationship** between the **fixed factors, (8.4.5),** which define the system in which the bifurcation occurs, and the **bifurcation value** of the density for that system. In other words, to **understand this phenomenon** requires some identification of a relationship of the form

$$D^* = F(NA, NS, ATH, PSE, MSE, G) \qquad (8.4.6)$$

See what you can **understand** by first selecting a smaller set of the "most interesting" parameters within the group in (8.4.6). What does "most interesting" mean? It means you have to think! Note that the parameters **NS** and **NA** specify the **environment** of each ant (they are **"exogenous"** parameters), whereas all the other parameters in (8.4.6) relate to the **internal dynamics** of the ants (and are referred to as **"endogenous"** parameters). These distinctions may be useful in the above explorations. Obviously, it would be nice to find **some** understanding that involves a **small set** of parameters.

Think 8.4B

For example, are **both** of the exogenous parameters important, or only the density? How is G related to the **memory duration** of an ant? How do you think this might influence the organizational properties? Check out your ideas using Program 8-4B. Likewise, consider the influence of one of the endogenous parameters that you find intuitive.

Other Social Insects

There are a variety of insects (ants, bees, wasps, termites, etc. **[10]**) that have been studied in order to understand how their various social (organizational) dynamics arise from individual interactions. Attempts are even being made to learn how to use their methods in a variety of technologies. A few references, from which more can be found, are **[3–8, 10]**.

References

1. G. E. Robinson. From Society to Genes with the Honeybee. *American Scientist, 86*(5), 456–462. Sept.–Oct., 1998.
2. Z-Y Huang and G. E. Robinson. Social Control of Division of Labor in Honey Bee Colonies. In C. Detrain et al. (Eds.), *Information Processing in Social Insects,* pp. 165–186. Birkhäuser, Basel, Switzerland, 1999.
3. T. Collett. Measuring Beeline to Food. *Science, 287,* 817, 4 Feb. 2000.
4. M. V. Srinivasan, S. Zhang, M. Altwein, and J. O. Tautz. Honeybee Navigation: Nature and Calibration of the "Odometer." *Science, 287,* 851–853, 4 Feb. 2000.
5. E. Bonabeau and G. Théraulaz. Smart Swarms. *Scientific American,* 73–79, March, 2000.
6. M. W. Moffett, Ants and Plants: Tree Fortresses, *National Geographic,* May, 2000; Friends and Foes, *National Geographic,* May 1999; A Profitable Partnership, Ibid. Feb. 1999.
7. B. Hölldobler and E. O. Wilson. *Journey to the Ants: A Story of Scientific Exploration.* Harvard University Press, Cambridge, MA, 1994.
8. B. Goodwin. *How the Leopard Changed Its Spots: The Evolution of Complexity.* pp. 68 ff. Charles Scribner, New York, 1994.
9. O. Miramontes, R. V. Solé, and B. C. Goodwin. Collective Behavior of Random-Activated Mobile Cellular Automata. *Physica, D 63,* 145–160, 1993.
10. C. Detrain, et al. (Eds.). *Information Processing in Social Insects*, Birkhäuser, Basel, Switzerland, 1999.

Computer Appendixes

These appendixes begin with **three basic elements:**

C0: How to **run the programs** on the disk

C1: How to make your own working space (your "computer palette") for **exploring questions** and **saving ideas** concerning these dynamics

C2: Reviewing the **Basic program commands** on the disk, which you can use to accomplish these explorations.

Following these are specialized topics on graphics, the Runge–Kutta method, printing and editing graphics, and the use of tonal representations of dynamics.

C0: INITIAL ACTIONS: ACCESSING DISK; RUNNING, TRANSFERRING BETWEEN PROGRAMS; EXITING DOS WHILE PROGRAMS RUN

The introductory programs FIRST, SECOND, THIRD, FOURTH, DETAILS, and DYNAMICS are discussed in some detail in **Section 1.4.** This appendix discusses how to **access the disk,** and then **any program on the disk; running that program;** how to **exit** from any program **at any time**; and, finally, how to **transfer to other programs.** Although this may be very familiar to many, **you should carefully note the instructions contained between the lines indicated by** *************.

****** **Fast Back and Forth** ******

These instructions begin on a Microsoft Windows page, but all programs are operated **within MS-DOS.** You can return to the Microsoft page **at any time** to carry out other activities, and return later to the programs **(even one that has remained running)** by using the **instructions at the end of this section.** Working in MS-DOS will give you more **flexibility to explore and create your own ideas, programs, and directories;** in other words, gain some freedom from our **"canned" world!**

****

To Open the Disk

The first order of business is to get onto the disk that is included with this book (**or** download from Wiley's ftp site), which contains all of the computer programs used in the **Explores** and **Appendixes.** To open the disk:

1. Insert the disk into the computer.

2. Click the cursor on "Start" (bottom left of the screen).

3. Run the cursor up to "Programs," then to the right, and down to "MS-DOS Prompt," and click on this. The screen will come up black with a white `C:\ Windows>`.

4. For later purposes it is useful to temporarily exit Windows. To do this, `<cd ..>`, where this notation means that you **type whatever is between < and >, and then strike the Enter key.** Notice that there is a **space following cd** (which stands for "change directory), before the two periods. You will **now see `C:\>`,** and you are outside of Windows (temporarily). To get to where the disk is in the computer, simply `<A:>`, and you will see `A:\>` (A being the computer drive where the disk is).

5. To see all the programs, `<dir/w>`, which gives you the **"directory"** for what is on this disk. There are around 63 Basic programs (designated by `.BAS`), and others labeled by `.EXE` and `.INI`, which is the software that is needed for the computer to run the written programs. The first number of a program is the **chapter,** and the next number and letter

designates the **section,** and its **order** within that section. The programs labeled C are used in the later numbered sections of this **Appendix A1.**

You will note that at the **bottom** of this directory are the **above-listed six programs,** even though they are the first to be used. This is the ordering imposed by the computer, which we will deal with shortly. The programs **FIRST through FOURTH** are the programs that you should **study** and **practice** with at first, as explained in the **Explores.** The programs, DETAILS and DYNAMICS are discussed in Section 1.4.

To Begin with a Particular Program

****** Caps Lock Key *****
In the **future interactions** with the programs you will need to use **capital letters,** so you might as well **turn on the "caps lock"** key (so that the **light** comes on), even if this is not needed all the time.
****

After doing that, to activate the **program FIRST** simply **<QB FIRST>** (notice that their is a space after QB). When you do this, a blue screen comes up with a lot of white type on it. This is an **"explanatory" program** for both you and the computer, for it contains a lot of explanatory comments that have nothing to do with the computer's running of the program. **Explanatory comments** always have an **apostrophe (') at their left,** which tells the computer to **ignore what follows on that line.** On the contrary, **you** should be sure to **understand** these **comments.** This process of exploring and understanding is discussed further in **Appendixes C1 and C2.**

To Run that Program

For the time being, to **run this program,** click the cursor at the top on **"Run,"** then click on **"Start."** From this point on, just follow the instructions that show up on the screen (they are written as **PRINT** statements in the program). If you ever want to **exit a program** (to go back to the written program on the blue screen), you can **always** do this by **[Ctrl + Pause],** where this notation means that you **press both** the keys labeled **"Ctrl"** and **"Pause"** at the **same time** (not the Enter key). Many times, you will be able to **exit a program** by simply **<X>** (i.e., strike the X key), but if this does not work at any time, use the above escape method.

****** [Ctrl + Pause] ******

The use of [Ctrl + Pause] is **very useful** if you want to go back to the program to see how some **dynamics, interaction from the keyboard,** or **screen presentation** is accomplished in the program (such as the above PRINT command). This gives you much more control for **exploring your ideas** than is ever given in any program. **Make use of it!**

****

To Change to Another Program

There are **two ways** that you **may be able to exit** a **running program,** either to change the written program or to go on to another program. The **first method** was **just noted above.** The **second way** that you will find in most programs is an instruction for you to **strike the key X** if you want to **exit from the program.** This will be denoted in the program by the **instruction** <X>.

****** "Running" Interactions from the Keyboard ******
In all **interactions from the keyboard, while a program is running,** the notation <Z> will mean to **only strike the key Z** (do not use Enter). This is the same as the notation [Z] used above. Note that these are **capital letters.**

****

Once you have returned to the program on the **blue screen,** to go to another program click on **File** and then on **Open Program.** You will see a list of programs, that only extend into Chapter 5. The other programs, including the present programs of interest, are further down in the directory. To get to these, click on **any** program, so that it is "dark-lighted," then go to the **"down arrow" key,** ↓, to the right of the Ctrl key, and **press on it.** You will see that it scrolls down through all of the programs in the directory, until it comes to the one you want (you can also use the ↑ key, if needed). Once there, either **press Enter,** or **click on OK.**

So now **begin with FIRST,** by <QB FIRST>, and then use the above procedure for any other programs in the future.

To Return Temporarily to the Windows Page

At **any time** you can return to the Windows page by using [**Ctrl + Esc**] on the **left side** of the keyboard. You can do this at **C>,** or **A>,** or at any program page **(blue screen),** or even **when a program is running.** When you do this, you will find **at the bottom** of the Windows page, a strange symbol, followed by, for example, **A: -QB** (or whatever, depending **where you used [Ctrl + Esc]**). If you click on this it will take you **directly back** to the place you used [Ctrl + Esc], **even your running program** (which was **not stopped** in this process). This is very handy if, for example, you want to send or read some e-mail, get on the Web, record your thoughts, or whatever.

Permanently Exiting MS-DOS

If you want to leave this programming world for an extended time, then at A:\> or C:\>, simply <Exit>. You will return to the Windows page, but will no longer find the **A:-QB** at the bottom of the page. To reestablish this MS-DOS connection, you will need to **return to the beginning** (opening the disk).

C1: A PLACE TO DEVELOP YOUR IDEAS: MAKING A DIRECTORY FOR SAVING YOUR PROGRAM COLLECTION

To **develop your own ideas,** you need a **computer palette** on which you can work, and save any number of **exploratory ideas.** To do this, you need to first set up **your "work space" on the computer** where you can copy any programs on the disk, in order to try some modifications or make permanent changes. This palette will be your **personal directory,** which you name. In this directory, you will be able to store any number of programs, and transfer your favorites back to the book's disk if you like (there is a lot of room). Finally, you will be able to store them on **your own disks,** and make **disk backups,** to guard against their future failures, which can, unfortunately, occur at times.

(1) Make a New Directory for Yourself on the Computer

If you are at any program (blue background), you can go back the drive A:\> by going to **File** and down to **Exit.** Once you are at A:\>, you get back to the computer by <C:>, in which case you find C:\> once again.

The first order of business is make your own directory, which you need to name (here we will use name "Mine"). To accomplish this, **<MKDIR MINE>** (note that there is a space following **mkdir,** which stands for **"make directory"**). If now you **<CD MINE>, you will find that C:\> MINE** will come up, and you will be in **your directory!** (The command **CD** stands for **"change directory."**) Now you have established some **independence!** If you **<DIR>** you will find no programs or directories (only some "computerese"). You can now **copy** any number of the programs from the **book disk,** and you can **modify them** in any way **without worry,** since the originals are still on the disk. More importantly, you can **explore making your own computer programs,** saving them in this directory.

(2) Transferring Program Copies to/from the Book's Disk

To **copy (not remove)** a program from the book's disk, **start in your directory** and go to the disk with **<A:>.** Once at **A:\>,** bring up the list of programs with **<dir/w>,** so that you get the name correctly. To copy a program to your directory use, for example, **<copy first.bas C: >,** and you will see **"1 file(s) copied"** come up on the screen, if you did everything correctly (note again the **two spaces** in the last command). Now if you **<C:>,** followed by **<dir>,** you will see **<FIRST.BAS>** in your directory. If you **<qb first>,** and you get that program on the blue screen, then you are all set to run it, as discussed in **Appendix C0.** (If you do not get this program, then you will need to **copy** the **two EXC programs,** and the **INI program** from the disk into your directory, exactly as above.)

(3) Modifying and Storing New Programs

Write down your ideas and questions as you run and look over the programs on the disk, even before you know how to program. Come back to these notes as you become more proficient in making some program changes.

When you bring up a program, you can **scroll through it rapidly** by either using the **keys Page Up, Page Dn, or ↑, ↓.** Practice this.

If you want to organize your programs at a later time, according to some topics or ideas, you may want to put these in separate directories within your directory. So if you have some new ideas, you might want to put them in a directory named "newideas" or "Eureka," or whatever. This can be done by again using **<MKDIR**

NEWIDEAS>, or <MKDIR EUREKA>, within your personal directory. If you <CD NEWIDEAS>, or <CD EUREKA>, then C:\> NEWIDEA or C:\> EUREKA will show on the screen. **To go back to** C:\> MINE, use <CD..>. The two periods always take you back to the directory in which the present directory is contained.

(4) Making a Copy of the Book's Disk

It is a smart idea at some time to make a copy of the book's disk, in case it fails at some time. To make this copy **onto the computer,** first exit your directory MINE by <cd ..>, which will take you back to A:\>. Next one needs to make a directory in which to store all of these programs, so <mkdir END>, and then <cd END> to get into this directory, C:\END>. Next go to A:\>, using <A:>. Now, rather than copying each of these 66 Basic and operating programs one at a time, it is only necessary to **<copy *.* C:\> (two spaces here).** The symbol *.* means any designation found before and after the period (and <copy 2*.* C:> would copy all the programs in chapter two). Following this, you will see each program sequentially come up on the screen, all ending with **"68 file(s) copied."** Using C: to return to END, and <dir/w> will exhibit all of these programs. To exit back to C:\>, use **<cd ..>,** and then <cd mine> to return home (this can be combined into <cd ..\mine>, with only one space). If you want to make a separate **disk copy of the book's disk,** return to the directory END, put a new formatted disk into the computer, and <copy *.* A:>. Boxes containing ten of these formatted high density disks can be bought cheaply at any computer supply store.

C2: BASIC COMMANDS, INTERACTIONS, AND SAVING PROGRAMS

The intention of this book is not to make you a great computer programmer! If I can program all the examples presented on the disk, and many more ideas that I have explored, let me assure you that you can do it also! I take some comfort in the purported fact that Richard Feynman, a singularly imaginative genius in physics, also only learned the programming language Basic used here. You will see that this is not only adequate, but indeed better than many fancier languages for the present purpose of exploring ideas. (The references to qbasic, quick basic, or qb in the programs has nothing to do with another language, but only with a compiling method used by the computer to speed things up, which I do not understand, or need to.)

Having made these points, let me introduce you to the few basic commands that will give you great flexibility to carry out computer explorations. Many more details can be found in note [1]. If you feel compelled to write these programs in other languages, you may obviously do so. However, you can do **more very useful interactions** between the **keyboard** and **running programs** with this simple language and some options offered by Microsoft than is possible with advanced languages. These include (**while the dynamic program continues to run**):

1. Changing the values of parameters
2. Rotating three-dimensional graphical coordinate systems
3. Clearing the screen from graphical "messes" and proceeding
4. Turning on and off dynamic sounds
5. Returning for new initial conditions new dynamic options
6. "Freezing" graphical figures and making printouts of them, with optional added notations
7. Returning **temporarily** to the Windows page for other activities while the program continues to run.

The ABC's of Programming Commands

Here the notation {***} indicates that *** appears in the present programs, or you can write this into future programs, **at the beginning of the program (all these appear as capital letters in the program):**

{**screen 9**}: or some number like that (try others) will put a certain **size font** on your screen.

{**dim**}: the "dimension" of each variable, may be needed to be specified. Thus, if $x(i)$ is a set of variables, $i = 1, 2, \ldots, N$, then N is called its **dimension.** This only iieeds to be included if $N > 10,$ but is often included anyway, as a reminder.

{**defdbl r, u-z**}: then all calculations are **double precision,** for the **variables r, u, v, w, x, y, z.** This means that the calculations use 17 digits of accuracy (prints to 16 digits of accuracy), as compared with the standard 7 digits of accuracy. Most of the time this type of accuracy is **not needed.**

{int x}: will take all variables $x(i)$ **integers** throughout the program, not adding spurious decimals from the computer.

{randomize timer}: this seeds the random number generator (used in the **{rnd z}** command below) with a new number each time the program is run, based on the present time. If you want to repeat the same sequence of random numbers in different cases, do not include this command.

In the recursive part of the program that generates the dynamics (the **recursive part** is frequently programmed in a **loop action,** enclosed between **From .... to ....** and **Return** commands; see Program **FIRST**):

{rnd z}: If you want a **random number** assigned to some parameter, say **z,** on **each iteration** of a recursive relation, then on a line in the program **{rnd z}**. In this case z will be assigned a value between **0 to 1,** with **uniform probability.** This sequence of "random" numbers will be **repeated** every time the program is run, unless the last command is introduced in the beginning.

{x = a + (b - a) * md}: This will give the **variable** x a random number **between the numbers (a, b),** since $x = b$ if $rnd = 1$, and $x = a$ if $rnd = 0$.

{A:}: at the **beginning of a line** gives a location in the program to which you may want the program to go or return to at some time. You can use any group of letters, such as **{ABC:}**. Note that you need the **colon** to keep it separated from what follows.

{goto A}: is the command for the program to go to the line labeled by **A:**

{cls}: will clear the screen when the program reaches that point. This removes old graphical lines or text that are no longer of interest, preparing the screen for new information. However, it will also clear graphical coordinates, which may need to be refreshed before new dynamics are shown. This can be accomplished with the following {gosub Z:}.

{gosub Z:}: This is a handy command when some information needs to be used in different locations. This will send the program to the line labeled **Z:,** where a program is locat-

273

ed, such as **drawing a coordinate system (see C3).** At the end of this program, the final line contains **{return},** and the computer returns to **wherever** the `gosub z:` is located.

{int x(i)}: within the program will give only the integer part of $x(i)$ (e.g., int (5.3) = 5, int (0.99) = 0).

{abs(x)}: gives the absolute value of x (so abs (−32.5) = 32.5).

{print x}, {print x,}, {print x;}, {locate 10,15: print x}: these are all different ways of printing the number x on the screen. If you want to see a **sequence of numbers** all together on the screen, use one of the **first three commands.** The first puts each number on a separate line, whereas the second and third spaces the numbers on a line differently, with **the third being the most compact.** The last command locates one value of x (overwriting the last value) at the **"10,15" location** on the screen at a line. To try these out, go to File and click on New Program, then type in and run this simple program. (Recall that anything on a line following an apostrophe is not read by the computer):

```
SCREEN 9
CLS
FOR K = 1 TO 10
X = K * 13.25 'try some other numbers, or K^2, and see what that does
PRINT X 'on this line use each of the above possibilities
'if you activate the next line, each iteration is slowed down for
'100000 counts
'FOR J = 1 TO 100000: NEXT 'you need this for the last example
NEXT K
```

This is a good time to **practice saving a program,** so let's save this one!

Saving a New Program

When you have a new program that you want to save, go to **File** and click on **Exit.** You will see a **question** "One or more loaded files are not saved. Save them now?" To save the program **click on Yes** (otherwise No). You will be asked for a File Name (that is, a program name). Make up **some name,** for example TRIALS, or whatever you like **(no more than eight letters),** and type this in the box. The ending .BAS will be added automatically. Finally click on <OK> and you will have a new program on the disk. To see it, either go to File and New Program, then scroll down to the end of the directory; or go to File and Exit, and then <dir/w> to see it.

If you have made your directory in C:\>, **and if** <C:> takes you back to that directory, you can **move this program** into your directory (and off the book's disk), by **starting at A:\>** and using <MOVE TRIALS.BAS C:>

Interacting from the Keyboard with a Running Program

See the end of the **program FIRST for some examples.**

{A$ = inkey$}: is the way you set up the keyboard so that you can **inject your wishes** while the dynamic **program is running** (e.g., **examples 1 to 5** above).

Following this last command, one can then use, for example:

{if A$ = "C" then cls}, to clear the screen of **everything,** every time you strike the key **C** on the keyboard (see {cls}).

{if A$ = "C" then goto A} will send the program to the line labeled **A:**. Notice the quotation marks, and that **all letters are capital.**

{if A$ = "" then goto B} sends the program to B: if **no key** is struck. This can be used to skip some lines if there is no input from the keyboard (such as a new command).

{if x<5 then goto B} will send the program to the line labeled **B:** if the value of x is less than 5 at that time during the dynamics; use x<= 5 if x can also equal 5.

References and Notes

1. Although I know of no resources for **studying dynamics** using the Basic language, one resource for an **introduction to Basic programming** is the collection of books by David A. Lien (CompuSoft, San Diego, CA, 1987):

   1. *MS-DOS: The Basic Course.* (Contains extensive details about the mechanics of dealing with Basic programs on MS-DOS.)

   2. *Learning Basic for Tandy Computers.* (**A very useful general** programming reference, regardless of the DOS computer!)

   3. ***Learning Microsoft Basic for the MacIntosh:*** *New 2.0 Version,*

275

and 2.1 Version. (This might well be a **very valuable resource for new types of dynamic innovations and explorations,** which is the **raison d'êtra of this book.** Unfortunately, I never learned about it in time for dynamic explorations.)

2. A resource for more advanced applications of Basic to **mathematical topics,** requiring Numerical Recipes Software, can be found in J. C. Sprott, *Numerical Recipes: Routines and Examples in Basic,* companion manual to *Numerical Recipes: The Art of Scientific Computing,* Cambridge University Press, New York, 1998.

C3: SOME GRAPHICAL INFORMATION

An important part of this simple programming is that you are free to make graphical **dynamic** pictures of **any type of activity** as it happens. Some of the variety of dynamics and graphics (**representing the history** over some extended period of time) used in the various programs are:

1. One variable and the time $[x(t), t]$ for discrete or continuous t, **some accompanied by tones** (Section 3.6)

2. Two variables $[x(t), y(t)]$, in a two-dimensional "phase space" (Chapters 4 and 7)

3. The two-dimensional projection of three variables $[x(t), y(t), z(t)]$ in a three-dimensional "phase space," which can be **arbitrarily rotated** (Section 8.3)

4. The collection of discrete intersection points of the dynamics in (3) when it passes through a plane in this space (a Poincaré map, which is uniquely informative), see Section 8.3

5. One variable, $x(t, c)$, for discrete times and a continuous range of values of a dynamic **parameter, c,** showing very different types of dynamics (periodic, chaotic, etc.; see **Chapter 3**)

6. Many randomly moving excitable "ants" interacting with each other on a plane (Section 8.4)

7. **Spatial patterns** of the **color-coded** "excitable, refractive, quiescent" states of interactive elements on a plane, **and others** that are more difficult to briefly describe (Section 8.1)

So let us begin with the graphics of (1) and (2). **Examples of both** of these can be found in the program **DYNAMICS.** Here we have one variable and the time, or two variables that we want to display graphically as the time changes. To do this, we need to tell the computer where to put points on the screen for either the values of [$x(t)$, t] or [$x(t)$, $y(t)$] over some range of t values. This involves two pieces of information, for which we need a two-dimensional "coordinate system." A typical computer screen has a range of roughly 0–636 spaces left to right and 0–330 lines top to bottom (this varies with computers, so experiment), and this information must fit within these limits. In this case, the center of the screen is at **(318, 165)**, but one can put the origin of a coordinate system any place that is useful. Run **Program C3** to explore these issues. **{pset(x, y), c}** will put a small dot at the location (x, y), with a **color c.**

In this simple programming, only a **few colors** are available. The **numerical code** for some **bright colors is** given at the beginning of **FIRST,** and in many other programs. All of the colors can be found in the program DETAILS.

- **{circle(x,y),s,c}** will put a circle of **size s** and color c centered at (x, y). Circles are easier to see than the dots. If you want to **remove a circle** at a later time without clearing the entire screen, use **{circle(x,y),s,0}.** (The "color" 0 is a blank space!) This is useful when **moving circles are used [see (6) above].**

- **{line(u,v)-(x,y),c}** is a command that draws a **straight line** between (u, v) and (x, y), using the color c (it will be white if you omit {,c}).

Continuous-Curve Representations of Dynamic Processes

If you are running a dynamic program in which $x(t)$ and $y(t)$ [or similarly, $x(t)$ and t] are changing with time, and you want to display this as a **connected curve** that joins values of [$x(t)$, $y(t)$] for sequential values of t, use the above straight line method, but with sequential values of [$x(t)$, $y(t)$]. Generally, for **any fixed value of dt,** this can be accomplished with **{line(x(t + dt), y(t+dt))-(x(t),y(t))}.** If dt is very small, the result will appear to be a **"continuous" curve (smooth).** In this case, in an iteration process, in order to obtain the values at ($t + dt$), you need to keep a temporary record of $x(t)$ (that is, for one iteration). To do this, you can set $ox = x(t)$, and $oy = y(t)$ at each time t, then when $x(t + dt)$ and $y(t + dt)$ **are determined;** use ox, oy for $x(t)$, $y(t)$ in the second part of the above command. **Illustrations** of these ideas can be found in the program **DYNAMICS.**

To add some **tone characterizations to these dynamics, see C7. To save a graphical picture, see C5.**

C4: THE FOURTH-ORDER RUNGE–KUTTA ITERATION METHOD

It was seen in Section 4.1 that the simple Euler approximation of ordinary differential equations may lead to unstable numerical results if there is more than one dynamic equation. (There are times, however, particularly if there is some damping in the dynamics, when this is not a problem, and the Euler approximation is useful for a "quick approximate" indication of the dynamic, as done in **Program 8-2a**). However, for an accurate evaluation of the dynamics of N ordinary differential equations, the most useful method is the following fourth-order Runge–Kutta method (involving four evaluations at each time step). If dt is the time step, this method is **accurate to order** $(dt)^5$, which is 10^{-10} if $dt = 0.01$. The fourth-order Runge–Kutta qbasic programs used in this book are variations of the one printed below. Note that this contains the graphical **subroutine** discussed in **C2, at {gosub Z:}.**

A **simpler** program, without the interactive commands and subroutine, is used in **Program C4.** You can use this program as a **skeletal outline** for adding any features. To make your **working copy of C4,** go to **File,** then **Save As,** give it a **new name** (e.g., NEWRK), and click on **OK.** If you have made your directory in C:\>, and if <C:> takes you back to that directory, you can move this program into your directory **from A:\>** by using <MOVE NEWRK.BAS C:>. Then you can begin experimenting with **new ideas!**

```
CLS
'RUNGKUT4.BAS
SCREEN 9
RANDOMIZE TIMER
PRINT" FOURTH-ORDER RUNGE-KUTTA INTEGRATION FOR N
PRINT"              FIRST-ORDER ODE"

N = 2:' THE NUMBER OF FIRST-ORDER EQUATIONS
DIM X(N),OX(N),Y(N),K(4, N)
DBL = 9:GRN = 10: LBL = 11: RED = 12
  MAG = 13: YEL = 14: "***** COLOR SELECTIONS

PRINT""
INPUT" GRAPHICS MAGNIFICATION =";M
MX = M: MY = .731 * M: 'Y-MAGNIFICATION ADJUSTED FOR GRAPHICS
PRINT""
```

C4: The Fourth-Order Runge–Kutta Iteration Method

```
    INPUT"        THE SIZE OF THE TIME STEP"; DT
    PRINT""

A:  CLS
    CT = 0
    FOR I = 1 TO N
    'PRINT""
    'PRINT " X("; I;") =",
    'INPUT X(I)
    X(I)=3
    NEXT I

    GOSUB B: 'GRAPHICS FOR THE COORDINATE SYSTEM
    LOCATE 1,15: PRINT "THIS DEMONSTRATION IS A REVERSING SPIRAL FLOW"
    PRINT" <C> TO CLEAR SCREEN; <N> FOR NEW INITIAL CONDITIONS"
C:  ******** FOURTH ORDER RUNGE-KUTTA ROUTINE ************
    A$ = INKEY$
    IF A$ = "C" THEN GOSUB B
    IF A$ = "N" THEN GOTO A
    CT = CT +1: 'ITERATION COUNT
    '*** Y(I) IS A DUMMY VARIABLE WHICH IS CHANGED AT EACH
    '*** OF THE FOUR ORDERS
    FOR I = 1 TO N
    Y(I) = X(I)
    NEXT I

    J = 0:'INDEX FOR THE FOUR ORDERS

D:  J = J + 1
      '**** THE N DYNAMIC FUNCTIONS ****
    K(J, 1) = DT*((Y(2)*Y(2) + Y(1)*Y(1)-1)*Y(2)-.2*Y(1))
    K(J, 2) = DT*(-(Y(2)*Y(2) + Y(1)*Y(1)-1)*Y(1)-.2*Y(2))
      ********************************

ON J GOTO E, F, G, H: '** THE FOUR ITERATIONS

E:  T = T + .5 *DT
    FOR I = 1 TO N
    Y(I) = X(I) + K(1,I)/2
    NEXT I
    GOTO D

F:  FOR I = 1 TO N
    Y(I) = X(I) + K(2,I)/2
```

279

Computer Appendixes

```
        NEXT I
        GOTO D

G:  T = T + .5 * DT
    FOR I = 1 TO N
    Y(I) = X(I) + K(3,I)
    NEXT I
    GOTO D

H:  FOR I = 1 TO N
    X(I) = X(I) + K(1,I)/6 + K(2,I)/3 + K(3,I)/3 + K(4,I)/6
    NEXT I

    IF CT I THEN GOTO I:' GRAPHIC LINE STARTS AFTER SECOND ITERATION
    LINE (320 + MX * X(1), 200 - MY * X(2))-(320 + MX * OX(1), 200 - MY * OX(2)), LBL
I:  FOR 1 = 1 TO N: OX(I) = X(I): NEXT I:' SAVES THE LAST X(I) FOR GRAPHIC LINE
    GOTO C

B:  '********** COORDINATE GRAPHICS ****************
    CLS
    NX = 300/MX: NY = 190/MY: 'NUMBER OF UNIT INTERVALS
    LINE (20, 200)-(620, 200), DBL
    LINE (320, 20)-(320, 380), DBL
    FOR I = 1 TO NX
    FOR J = 1 TO 4
    PSET (320 + MX * I, 200 - J), DBL
    PSET (320 - MX * I, 200 - J), DBL
    NEXT J
    NEXT I
    FOR J = 1 TO 4
    PSET (320, 400 - J), DBL
    NEXT J
    FOR I = 1 TO NY
    FOR J = 1 TO 3
    PSET (320 + J, 200 - MY * I), DBL
    PSET (320 + J, 200 + MY * I), DBL
    NEXT J
    NEXT I
    RETURN
```

C5: GRAPHICS PRINTOUT AND EDITING

If you have access to the Microsoft Paint program [check step (4) below], then it is reasonably easy to get a **printout and edit** any graphics that you have on the screen. The procedure used here (Windows 95) is as follows:

1. Run your usual qbasic program, producing the screen graphics, and when you see what you would like to print out, first freeze the dynamics by pressing the **"pause" key.**

2. Now if you **[Alt + Print Screen],** you should see the screen "jump," at which point you have recorded that picture.

3. Use **[Ctrl + Esc]** to temporarily return to the Windows page. You will later be able to return to the program and the picture by using the **A: -QB** button at the bottom of the screen.

4. From **Start** at the bottom left of the screen, go up to **Programs,** then way up to **Accessories,** and way down to **click on Paint**

5. Now go to **Edit** and click on **Paste.** Your figure should show up in the paint window.

6. To obtain a white background, rather than the present black background (saving ink!), go to **Image** and click on **Invert Colors.**

7. Return to **Image,** then **Attributes,** and select the **Black and White,** at least until you want to get fancy! (When you do, consult Help.) If you have used **bright colors (9–14)** for your screen graphics, they will all be turned into black. Some other colors may be lost in this process, and you will need to return to the program and change your graphic colors in order to get this printout. When you select Black and White, a message will warn you of this possible loss. If you have used only bright colors, then **enter Yes,** and you will have a black and white graphics picture. If you lose any of the text in your screen graphics, this can easily be replaced or modified. See (9) below.

8. If you want your figure to appear lengthwise on a page, go

to File, and Page Setup, and select Landscape. You can try other page setups in the future (see Help), but at first just get a simple printout.

9. **To edit the picture,** such as adding text, etc., see Help for more details. For example, **to add text** see **Help, Help Topics, Putting Text in Pictures.** Paint has a very **useful eraser (top left),** which you can use to remove all **unwanted screen text** (note that you can **change its size** at the bottom left). In particular, **click on A,** and drag the cursor around with the click engaged, generating a rectangle when you release the clicker. You **can now type text** inside. Or click on either of the **line buttons** and note the selection of thickness at the bottom. Experiment with these. If you don't like something that you did, go to **Edit** and **Undo** the **backstep** up to **three steps.** For more **disastrous** problems, go to **View** and **Clear Figure** (going back to a clear screen), and **return to (5).**

10. Finally go to **File,** and look under **Print Preview** (this does not always give an accurate reproduction, but try it anyway). If it looks good, then enter **Print.** You may want to select the single page command (1 to 1), to possibly avoid outputs of blank pages. You should get the printout on your printer

11. You can **save this picture** by going to **Edit** and **Save As.** Click the cursor at the left of the blue window, and type in a name, delete "untitled," but **leave on the ending .bmp** (for bit map), then **Save.**

C6: LET THERE BE DYNAMIC TONES!

Graphics use only our visual senses, and there are things to be learned and understood that are sometimes best done with the help of our auditory senses. Both of these can also introduce their own contributions to our aesthetic impressions, which may give us new ideas. The most extensive use of these tones is in **Program 3-6,** but you might want to explore its effects in other contexts. So let us look at some possibilities.

{sound F,D}. If you want to associate your dynamics with some tones, then try this. F denotes the **frequency** of the note, and D denotes the **duration** of the tone. The frequencies of notes from middle C on up are: C = 523, C# = 555, D = 587, D# = 623, E = 659, F = 698, F# = 741, G = 784, G# = 832, A = 880, A# = 934, B = 988, then to high C = 2·523, etc. This is particularly useful for the **tonal characterizations of dynamics.** For example, if $x(t)$ is a **dynamic variable,** and you would like to "hear" where it is, as well as see it in the graphics, you might put in the command `{F = 100 x(t)}` and `{D = 3}`, or other numbers. This will give a **"sliding" tone** if the dynamics is "continuous," or a **"jumping" tone** if the dynamics is discrete (as in Program 3-6). Note that the inclusion of all such commands **slows down the** processing of the dynamic **program.** With fast computers, this is often not an important problem, unless you want to do a lot of calculating. However, these tones can be **turned on and off** from the keyboard by using a command such as `{if A$ = "F" then S = 0}`, where D means F (no sound), then followed by `{if S = 0 then goto C}`, where the location `C:` is **below** the line with the above sound command, so no sound is produced. Then, by using something like `{if A$ = "N" then S = 1}` (meaning 0**N**), the sound command will not be skipped over, and the tones will return (slowing down the program, as you can clearly see in Program 3-6). Make sure you use the **"Interacting from the keyboard" instructions in C2.**

The **PLAY** command is a simple way to make **single-note music on the standard scale.** You need to specify the octave, which lies between 1 and 7 (middle C is in octave 3), if you just want the notes on a piano. From Program 3-6, we have this **Bach example:**

PLAY
"L8O4CDEGFFAGGO5CO4BO5CO4GECDEFGAGFEDECO3BO4CDO3GB
O4DFEDE"
PLAY
"L8O4CDEGFFAGGO5CO4BO5CO4GECDEO3AO4GL6FL5EL5DL4CL4O3
GL3O4CL2O3BL1O4C"

The notes are designated by the usual letters, with + or − for sharps and flats; e.g., C, C+ (or D−) etc. The **octaves** are designated by **O3, O4,** etc., and the **length** of the note by **Ln,** where the n indicates that the length of the note is **1/n the standard length.** For example, PLAY "L4O3CO2BAGL6O3DEL2F" plays the notes at the same length of 1/4, beginning with middle C, then drops down to the next octave for B, A, and G. It then returns to the middle octave, but now plays D and E with the length 1/6 and F with length 1/2. This is **just an illustration,** not a suggestion. I have no idea what this sounds like!

Mathematical Appendixes

These appendixes begin with an introduction to the concept and properties of derivatives, and the initial application to the basic functions $a \cdot t^n$. Next, the derivative is introduced in use to characterize the **linear approximation in D** of any function $F(x + D)$, which is approximately valid for small values of D. This is applied to the study of the **stability of fixed points** of one-dimensional maps, used in Chapter 3. The study of derivatives is then extended to **exponential** and **trigonometric functions** of x, which are important in many types of dynamics. Appendix M4 shows how we can get an approximation for how the **frequency** of a **nonlinear oscillator depends on the amplitude** of its oscillation, in contrast to the amplitude-independent frequency of the harmonic oscillator.

M1: PROPERTIES OF DERIVATIVES AND THE DERIVATIVE OF $a \cdot t^n$

For any function $F(t)$, the **derivative of $F(t)$** is defined as

$$\frac{dF(t)}{dt} = \lim \text{ as } dt \to 0 \; \frac{F(t + dt) - F(t)}{dt}$$

or, in simpler notation

$$\frac{dF(t)}{dt} = \lim_{dt \to 0} \frac{F(t + dt) - F(t)}{dt} \qquad \text{(M1.1)}$$

Mathematical Appendixes

Before considering specific function $F(t)$, several **basic properties of derivatives** should be noted.

P1: The first property is that, if **A is any constant** (independent of t) then

$$\frac{d[A \cdot F(t)]}{dt} = \frac{A \cdot dF(t)}{dt} \tag{M1.2}$$

This can be seen by inserting A into both terms in M1.1 and noting that it can be factored out of the limiting process, yielding the right side of (M1.2)

P2: The second property is that if $F(t)$ and $G(t)$ are any two functions, then

$$\frac{d[F(t) + G(t)]}{dt} = \frac{d[F(t)]}{dt} + \frac{d[G(t)]}{dt} \tag{M1.3}$$

or, in words, the derivative of the sum of two functions equals the sum of their derivatives. This follows from (M1.1) by noting that the numerator corresponding to the left side of (M1.3) can be separated into the sum of two parts, one involving only the function F and the other only the function G. This can then be broken up into the sum of two limits, which yields (M1.3)

These basic properties are frequently used, as we will see in Appendix M3. Now we turn to the derivative of $F(t) = a \cdot t^n$ (n = 1, 2, 3, ...). The example of $F(t) = a \cdot t^2$ is treated in detail in **Section 4.1,** showing that

$$\frac{d(a \cdot t^2)}{dt} = 2a \cdot t \tag{M1.4}$$

This can be readily generalized to find the derivative of any integer power of t,

$$F(t) = a \cdot t^n$$

In order to evaluate (M1.1) in this case, it is useful to use the so-called famous **binomial theorem,** which states that, for any n,

$$(x + dt)^n = t^n + dt \cdot n \cdot t^{(n-1)} + \text{(terms that are of order } dt^2) \tag{M1.5}$$

The last expression means that all remaining terms are proportional to dt^2 or dt^3, or dt^4, and so on, up to dt^n. Therefore, using (M1.5) with (M1.1) yields

$$\frac{d(a \cdot t^n)}{dt} = \lim_{dt \to 0} \frac{a(t + dt)^n - a \cdot t^n}{dt}$$

$$= \lim_{dt \to 0} = a(n-1)t^{(n-1)} + \frac{\text{(terms that are of order } dt^2)}{dt}$$

But the last terms go to zero, because all of the factors in the numerator go to zero faster than the denominator. So this leaves the general result

$$\frac{d(a \cdot t^n)}{dt} = a \cdot t^{(n-1)} \qquad \text{(M1.6)}$$

The result (M1.4) is the special case of $n = 2$, and it can be used also for $n = 1$. (M1.6) is very useful for evaluating the derivatives of other common functions. **Important examples** are given in **Appendix M3**.

M2: THE STABILITY OF FIXED POINTS OF A ONE-DIMENSIONAL MAP

In this Appendix we will look at the issue of whether a fixed point of a map is "stable." The **idea of "stability"** of a fixed point concerns what happens to the system if it is moved from the fixed-point state to a **nearby state** (a so-called **"perturbation"** of the original state). Does its **dynamics** tend to **return it** to the fixed point, or does it tend to carry it **further away** from the fixed point? If the **former** happens, the fixed point state is said to be **stable,** otherwise the fixed point is called an **unstable** state of the system.

So let us consider what happens to the system when it starts near a fixed point. Does it tend to move closer or further away from the fixed point in the next time step? **If it moves closer,** then the fixed point is **stable** (to small perturbations), because by reapplying this result for succeeding time steps they will all get successively closer to the fixed point.

If the dynamics is given by

$$x(t + 1) = F[x(t), c] \qquad \text{(M2.1)}$$

then $x(t) = x^*$ is a fixed point if $x(t + 1) = x(t)$, or, in other words, if x^* satisfies the equation

$$x^* = F(x^*, c) \qquad \text{(M2.2)}$$

Next we consider a point that starts out near x^* and see what happens to it in the next time step. Does it remain close to x^*, or

Mathematical Appendixes

does it move away from x^*? So we take an initial point (at $t = 0$) near x^* and write this as

$$x(0) = x^* + D \quad (D \ll 1) \quad \text{(M2.3)}$$

Here D is taken to be some very small number (**a "perturbation"** away from x^*). That means that we can consider only those terms in $F[x(0), c]$ that either do not depend on D, or only depend on D to the first power (ignoring D^2, D^3, and so on, since they are all very small compared to D). This is known as the **linear approximation** with respect to D.

In the case of the **logistic map**,

$$F[x(t), c] = c \cdot x(t)[1 - x(t)] \quad \text{(3.2.2)},$$

this gives for x at the next time step, $x(1)$, the approximate result

$$x(1) = c(x^* + D)[1 - (x^* + D)] \approx c \cdot x^*(1 - x^*) + cD[-x^* + (1 - x^*)]$$

or

$$x(1) \cong F(x^*, c) + c(1 - 2x^*)D \quad \text{(M2.4)}$$

However, we can now use the fact that for any fixed point $F(x^*, c) = x^*$, then (M2.4) can be put into the form

$$x(1) - x^* \cong c(1 - 2x^*)D = c(1 - 2x^*)[x(0) - x^*]$$

where the last equality follows from (M2.3).

Now we introduce the notation $|x - y|$ to represent the (positive) distance between the two points x and y [so, if $(x - y) = -3$, then $|-3| = +3$, which is called the **absolute value** of $(x - y)$]. Using this notation, $|x(0) - x^*|$ is the initial distance from the fixed point, and $|x(1) - x^*|$ is the distance after one time step. Therefore, we conclude that this fixed point is stable only if the latter distance is smaller than the initial distance. Therefore, the **fixed point x^* is stable** only if

$$1 > \frac{|x(1) - x^*|}{|x(0) - x^*|} = |c(1 - 2x^*)| \quad \text{(M2.5)}.$$

Notice that we are only concerned with the magnitude of the distance. We will **consider the relative signs** of the two middle terms a little **later.**

In Section (3.2) we found that there are two the fixed points for the logistic map that are given by (3.2.5), which we repeat here:

$$x^{**} = 0 \quad (\text{any } c > 0) \quad \text{(3.2.5a)}$$

$$x^* = 1 - (1/c). \quad (\text{only if } c > 1) \quad \text{(3.2.5b)}$$

In the case of the trivial fixed point, $x^{**} = 0$, (M2.5) shows that it is stable only if

$$1 > c \quad (x^{**} \text{ stable}) \tag{M2.6}$$

On the other hand, if we substitute the fixed point (3.2.5b) into (M2.5), we find that this fixed point is stable only if

$$1 > |c\{1 - 2[1 - (1/c)]\}| = |c - 2|$$

However, since c must be in the range $4 > c > 0$, we conclude that this is satisfied only if

$$3 > c > 1 \quad (x^* \text{ exists and is stable}) \tag{M2.7}$$

This is all that is required for **Section 3.A.**

Of course, the results (M2.6) and (M2.7) apply only to the **logistic map,** but the same method can be used for any map. **In general,** if we substitute (M2.4) into M(2.3), we obtain (noting again that $D \ll 1$)

$$x(1) = F(x^* + D) \approx F(x^*) + D\left(\frac{dF}{dx}\right)_{x^*} \tag{M2.8}$$

where higher powers of D are neglected (this is the **linear approximation** of what is known as the **Taylor Series expansion in D**). To illustrate this in the case of the logistic map, where $F = c \cdot x(1 - x)$, using the results in **Appendix M1,** we find that

$$\frac{dF}{dx} = c(1 - 2x) \tag{M2.9}$$

and using this in (M2.8), yields

$$x(1) = F(x^*, c) + D \cdot c(1 - 2x^*)$$

which is the same as **(M2.4).**

Now, **for any map,** using (M2.3) again, we can write (M2.8) in the form

$$[x(1) - x^*] = [x(0) - x^*]\left(\frac{dF}{dx}\right)_{x^*} \tag{M2.10}$$

and conclude that x^* is a stable fixed point provided that

$$1 > \frac{|x(1) - x^*|}{|x(0) - x^*|} = \left|\left(\frac{dF}{dx}\right)_{x^*}\right|.$$

Thus we have the general result that if x^* **is a fixed point of $F(x)$,** then it is a **stable fixed point (an attractor)** only if

$$1 > \left|\left(\frac{dF}{dx}\right)_{x^*}\right| \quad \text{(stable fixed point, } x^*\text{)} \quad \text{(M2.11a)}$$

$$\left|\left(\frac{dF}{dx}\right)_{x^*}\right| > 1 \quad \text{(unstable fixed point, } x^*\text{)} \quad \text{(M2.11b)}$$

This general result will be used in **Section 3.4.**

▬ Think M2A

Draw a line and mark some points x^* and $x(0)$. Using **(M2.10)**, **qualitatively** determine how $x(1)$, $x(2)$, ... approach x^* in the stable case, for the **two possible signs of** $(dF/dx)_{x^*}$

▬ Think M2B

If you want a challenge, determine what is special about the case when $|(dF/dx)_{x^*}| = 0$. Note that in this case (M2.8) must be modified to contain the first nonzero term, which is the next term in the Taylor expansion, $\frac{1}{2}D^2(d^2F/dx^2)_{x^*}$. Now what about this dependence on D of this **modified** (M2.8), compared to (M2.8)? What conditions replace (M2.11)? **How rapidly** would the stable dynamics approach x^* in this case, as given by (M2.10)? (This is known as a superstable situation; see p. 156 of my book, *Perspectives of Nonlinear Dynamics,* Vol. 1.)

M3: THE DERIVATIVES OF EXPONENTIAL AND TRIGONOMETRIC FUNCTIONS

An exponential function is defined by the infinite series of terms

$$\exp(x) \equiv e^x = 1 + x + \frac{x^2}{2!} + \frac{x^3}{3!} + \ldots . \quad \text{(M3.1)}$$

where

$$n! = n(n-1)(n-2)\ldots(2)(1) \quad \text{(M3.2)}$$

is called **"n factorial"** (so, for example, $5! = 120$). Note that $n!$ has the property that

$$\frac{n}{n!} = \frac{1}{(n-1)!} \quad \text{(M3.3)}$$

which you can easily show by substituting (M3.2) into the left side,

M3: The Derivatives of Exponential and Trigonometric Functions

and canceling n in the numerator and denominator. By using these properties, and those in Appendix M1, it is easy to show the rather amazing fact that

$$\frac{d(e^x)}{dx} = e^x \quad \text{(M3.4)}$$

This says that taking the derivative of exp(x), defined by M3.1, **does not change the function!**

Let us see how this can happen. When we differentiate the right side of (M3.1), we are differentiating a sum of different functions. But we know from the **property P2** in **Appendix Ml** that this equals the sum of the derivatives of each function. That is, the derivative of the right side of (M3.1) equals

$$\frac{d(1)}{dx} + \frac{d(x)}{dx} + \frac{d}{dx}\left(\frac{x^2}{2!}\right) + \frac{d}{dx}\left(\frac{x^3}{3!}\right) + \frac{d}{dx}\left(\frac{x^4}{4!}\right) + \ldots$$

Now we use the result found in Appendix Ml

$$\frac{d(a \cdot t^n)}{dt} = a \cdot n \cdot t^{n-1} \quad \text{(M1.6)}$$

and see that the previous sum of terms equals

$$0 + 1 + \left(\frac{2}{2!}\right)x + \left(\frac{3}{3!}\right)x^2 + \left(\frac{4}{4!}\right)x^3 + \ldots$$

But if we use the property of the factorial, (M3.2), this expression is precisely the right side of (M1.1). In other words, the derivative of exp(x) is again exp(x), **proving (M3.4).** You can see that this is possible because there are an infinite number of factors in exp(x), hence losing the first term in the last expression does not decrease the number of terms in the sum! [1]

Now let us go a step further and break up the sum on the right side of (M3. 1) into **two infinite (!) sums**—one that contains only **even powers of x,** and only those of **odd powers of x**—and define two new **hyperbolic functions**

$$\exp(x) = \left[1 + \frac{x^2}{2!} + \frac{x^4}{4!} + \ldots\right] + \left[x + \frac{x^3}{3!} + \frac{x^5}{5!} \ldots\right]$$

$$= \text{(definition)} \quad \cosh(x) \quad + \quad \sinh(x)$$

(pronounced "cosch" and "sinch"). It is easy to verify that these functions are related to the exponential function by

Mathematical Appendixes

$$\sinh(x) = \frac{(e^x - e^{-x})}{2}; \quad \cosh(x) = \frac{(e^x + e^{-x})}{2}$$

and using the result (M3.4) and the property P2 in Ml, one sees that

$$\frac{d[\sinh(x)]}{dx} = \cosh(x); \quad \frac{d[\cosh(x)]}{dx} = \sinh(x)$$

Now if we use the same device as in Appendix Ml, and **replace x by (at)** (where **a is any constant**), these yield the relationships

$$\frac{d[\sinh(at)]}{dt} = a \cdot \cosh(at); \quad \frac{d[\cosh(at)]}{dt} = a \cdot \sinh(at) \quad (M3.5)$$

In order to get **trigonometric functions** we introduce the "imaginary" number $i = \sqrt{-1}$ (i.e., the square root of -1, so $i^2 = -1$). The **trigonometric functions** are now **defined** by the relationships

$$\cos(x) \equiv \cosh(ix); \sin(x) \equiv -i \cdot \sinh(ix) \quad (M3.6)$$

It is easy to verify that these trigonometric functions contain only the "real" numbers.

▮ Think

After all these definitions, it would be useful for you to show that the infinite series of terms that defines the two very important trigonometric functions are:

$$\sin(x) = x - \frac{x^3}{3!} + \frac{x^5}{5!} - \frac{x^7}{7!} + \ldots$$
$$\cos(x) = 1 - \frac{x^2}{2!} + \frac{x^4}{4!} - \frac{x^6}{6!} + \ldots \quad (M3.7)$$

Note the alternating signs, which result from the $i = -1$ factors in (M3.6) and the previous definitions of the two hyperbolic functions.

Now using (M3.6) and (M3.5), along with the property P1 in Appendix Ml, we obtain

$$\frac{d[\cos(t)]}{dt} = i \cdot \sinh(it) = -\sin(t)$$

$$\frac{d[\sin(t)]}{dt} = -i[i \cdot \cosh(it)] = \cos(t)$$

Using again the substitution of ax for t, we obtain the **final result**

$$\frac{d[\cos(ax)]}{dx} = -a \cdot \sin(ax)$$
$$\frac{d[\sin(ax)]}{dx} = a \cdot \cos(ax) \tag{M3.8}$$

Note

1. Georg Cantor introduced the idea that **two infinite sets** have the **same number of terms** if each member of one set can be associated (related to) one member of the other set. This is referred to as a **one-to-one correspondence.** Clearly, you can do that for e^x and $d(e^x)/dx$ since their infinite sets are identical. But the number of terms in the two **different** infinite sets of $\sin(x)$ and $\cos(x)$ in (M3.7) is also the same, since they can clearly be put into a one-to-one correspondence. But strange things occur for infinite sets. For example, the infinite set of integers, which is only a portion (a subset) of the set of all rational numbers, nonetheless can be put into a one-to-one correspondence with all rational numbers. So part of an **infinite** set can have just as many members as the full set! However, neither of these sets have as many members as all of the real numbers. But calculus uses this set of numbers, so this represents much more information than we can ever **know in science,** or can **ever compute** in any finite number of iterations. For a very readable account of Cantor's constructions, and issues arising from it, along with limitations within mathematics, read M. Kline, *Mathematics: The Loss of Certainty*, Oxford University Press, Oxford, 1980, e.g., p. 280.

M4: THE NONLINEAR FREQUENCY–AMPLITUDE RELATIONSHIP

We begin with the **nonlinear Duffing oscillator,** (4.3.2), when there is **no damping,** $\mu = 0$, which is

$$\frac{dx}{dt} = v$$
$$m \cdot \frac{dv}{dt} = -\kappa x - \alpha x^3 \tag{4.3.2a}$$

or, which we temporarily modify by introducing $\beta = \alpha/m$ and $\omega^2 = \kappa/m$

Mathematical Appendixes

$$\frac{dv}{dt} = -\omega^2 x - \beta x^3 \tag{4.3.2b}$$

We know that **if β = 0,** a harmonic oscillator solution of these equations is given by (4.2.8):

$$v(t) = -(A\omega)\sin(\omega t) \qquad x = A\cdot\cos(\omega t) \tag{4.2.8}$$

where ω is given above by

$$\omega = \sqrt{\frac{\kappa}{m}} \tag{4.2.9}$$

Now, **if β > 0,** the solution of (4.3.2) is still periodic in time, but we do not know the frequency. Let us call the **unknown frequency** Ω, which for small values of β can be approximated by

$$\Omega \approx \omega + \beta\Omega^*(A) \qquad (\beta \ll 1) \tag{M4.1}$$

where $\Omega^*(A)$ is the **unknown** dependence on A. To this same approximation we take this unknown frequency Ω to be related to x and v by the modified form of (4.2.8)

$$v(t) \approx -(A\Omega)\sin(\Omega t); \qquad x = A\cdot\cos(\Omega t) \tag{M4.2}$$

which is an exact solution if $\beta = 0$. We see that this is also an exact solution of $dx/dt = v$ in (4.3.2), regardless of the value of Ω. So next we look at the remaining equation of motion, (4.3.2b), and substitute the expressions from (M4.2), then use the trigonometric identity $\cos^3(\Omega t) = \tfrac{3}{4}\cos(\Omega t) + \tfrac{1}{4}\cos(3\Omega t)$, to obtain

$$-(\Omega^2 A)\cos(\Omega t) = -\omega^2 A\cos(\Omega t) - \beta A^3[\tfrac{3}{4}\cos(\Omega t) + \tfrac{1}{4}\cos(3\Omega t)].$$

This equality is clearly not satisfied, because of the different time dependence in $\cos(3\Omega t)$ [showing that (M4.2) is not an exact solution]. This means that the approximation (M4.2) does not account for this faster time dependence. Moreover, the amplitude of this term is proportional to $(\beta \ll 1)$, so we simply ignore this small faster time dependence in this approximation. Next we cancel out the time dependence, $\cos(\Omega t)$, and divide by $-A$, to obtain the approximate relationship

$$\Omega^2 = \omega^2 + \beta\cdot\tfrac{3}{4}A^2$$

Substituting (M4.1) for Ω, we obtain

$$\omega^2 + 2\omega\beta\Omega^*(A) + [\beta\Omega^*(A)]^2 = \omega^2 + \beta\cdot\tfrac{3}{4}A^2$$

Since β is assumed to be small, we ignore the term containing β^2, which finally yields the approximate result for the **amplitude-dependent correction** to the harmonic frequency in (M4.1)

$$\Omega^*(A) = \frac{3A^2}{8\omega}$$

Inserting this back into (M4.1) and restoring the original parameters $\omega^2 = \kappa/m$, and $\beta = \alpha/m$ yields the **predicted dependence of the frequency on the amplitude and other parameters**

$$\Omega = \sqrt{\frac{\kappa}{m}} + \frac{3\alpha A^2}{8\sqrt{\kappa m}} \qquad \textbf{(M4.3)}$$

This theoretical prediction will be compared with the actual dynamics in **Explore 4.3D,** to establish the limitations of the present approximations.

M5: THE STABILITY OF FIXED POINTS OF TWO-DIMENSIONAL ORDINARY DIFFERENTIAL EQUATIONS

Consider a system of two ordinary differential equations

$$\frac{dx}{dt} = F(x, y); \qquad \frac{dy}{dt} = G(x, y) \qquad \textbf{(M.5.1)}$$

that have a fixed point, which we can take to be $x = 0, y = 0$, so

$$F(0, 0) = 0; \qquad G(0, 0) = 0$$

For values of x and y **near this fixed point,** the equations (M5. 1) can be approximated by the **linear equations**

$$\frac{dx}{dt} = ax + by; \qquad \frac{dy}{dt} = cx + dy \qquad \textbf{(M5.2)}$$

where (a, b, c, d) are constants, that depend on the functions in (M5.1), and $(ad - bc \neq 0)$, for otherwise some quadratic terms in (x, y) need to be included (a similar problem arises in Think M2B). The **fixed point is stable** if, for **any initial (x, y)** the solutions of (M5.2) **tend toward the origin.**

Now (M.5.2) has **special solutions**

$$x = x(0)\exp(st); \qquad y = y(0)\exp(st) \qquad \textbf{(M5.3)}$$

provided that the constant in the exponent, *s*, satisfies

$$x(0)s = ax(0) + by(0)$$

$$y(0)s = cx(0) + dy(0)$$

for any [*x*(0), *y*(0)]. This follows from substituting the functions (M5.3) into (M5.2), recalling from **M3** that $d \cdot \exp(st)/dt = s \cdot \exp(st)$, and canceling the common exponential function. If you write these in the form

$$x(0)(s - a) = by(0); \quad y(0)(s - d) = cx(0)$$

and eliminate *x*(0) and *y*(0), then a little algebra shows that *s* must satisfy

$$s^2 - ps + q = 0, \quad \text{where } p \equiv a + d, q \equiv (ad - bc)$$

Therefore, there are **two possible values of *s*,** given by

$$s = \frac{p}{2} \pm \tfrac{1}{2}\sqrt{p^2 - 4q} \qquad (M5.4)$$

A **linear combination** of these **two exponential functions,** with two arbitrary coefficients (to satisfy any initial conditions), is the **general solution** of (M5.2) (it is a solution due to the **property P2 in Ml**). Therefore, if **one *s*** has a **positive real part,** the fixed point is **unstable,** and stable if **all *s*** have a **negative real part.** This leads to the final results:

If $p > 0, q > 0$; unstable

$p^2 - 4q > 0$ (node); $\quad p^2 - 4q < 0$ (focus)

If $p < 0, q > 0$: stable

$p^2 - 4q > 0$ (node); $\quad p^2 - 4q < 0$ (focus)

If $q < 0$: unstable (saddle point)

If $p = 0$ and $q > 0$ the equations M5.2 yield a **harmonic oscillator.** This fixed point is **called elliptic** (sometimes a center). The **stability of (M5.1)** is then **governed by the nonlinear terms** in the functions ***F* and *G*.** Similar, nonlinear terms govern the stability if $q = 0$.

The square-root term in (M5.4) is imaginary if $p^2 - 4q < 0$, in which case ***s* is a complex number,** so there is the combination of oscillatory and exponential dynamic behavior, and the fixed point is called a **focus.**

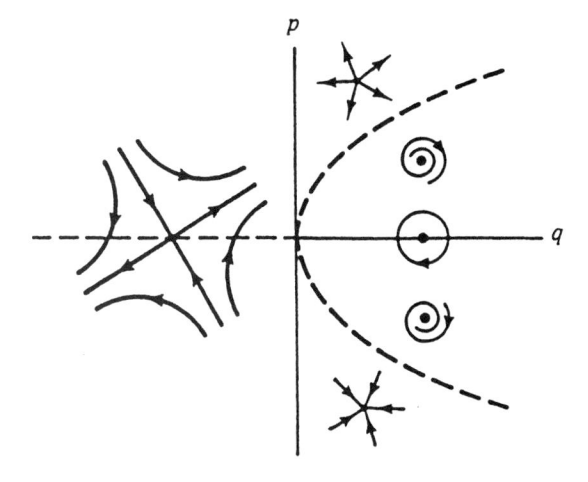

M5: The Stability of Fixed Points of Two-Dimensional Ordinary Differential Equations

Fig. M5.
This figure superimposes the phase space dynamics near the fixed point, on to the parameter space (p, q).

All of these types of dynamics (in the phase space) is illustrated in the parameter space (p, q) in **Fig. M5**.

Think

Label the types of fixed point in each region of Fig. M5, using the above information. Compare this with **Fig. 4.11 in Section 4.4.**

Index

The letter "r" following a page number refers the reader to a bibliographic reference.

Action potential (neurons), 170
Adaptations:
   biology, bodily functions, culture, physiology, 207–211
   marine organisms, 232r
   references, 232–233
Anderson, Philip
   science advancement, 67r
Ant colony dynamics:
   self-organization, 257–263r
Approximations:
   linear, 212
   see, Euler; Runge–Kutta
Archimedes, 7–10, 15, 18, 19r
Assumptions, models:
   human character, 35
   implicit, 157, 160
   simplicity, 26–27, 65–67
   Think 1.5F, 41
Asymptotic sets, 87, 89–92, 95
Attractors:
   Fixed points, 80

Period 2^n, 85, 88
   See also, Basins; Lorenz; Chaotic (strange); Limit cycles.
Autocatalytic oscillators, 168–183
   environmental flux, 161–166
   pacemakers, 229, 176–181

Backward dynamics, 136–137
Bank dynamics, 20
   "Safe," 21, 27
   "Swift," 24
Basins of attraction, 120
   Explore 7.3B, 225
   interwoven, 137, 142
   multiple, 142
Bee colony dynamics, 263, 10r
Bernoulli, Daniel:
   fluid dynamics, 12, 18, 19r
Bifurcations, 81–85,
   linked limit cycles, 252–256
   points, 82–89, 130

   symmetry breaking, 221–223
   in three dimensions, 243–256
Biology, 54, 55r, 242r
   diversity, 42, 43
   see, Ecology; Mayr.
Biological clocks, *see,* Pacemakers.
Bistability:
   cognitive phenomena, 219
   gene-regulatory networks, 219
   oscillators, 215–219
Blood circulation, 201
   See also, Harvey.
Bohm, David, 56r
   quantum formulation of dynamic orbits, 53
Boltzmann, Ludwig, 51, 54r
   probabililty (distribution) functions, 52–53
Bonhoeffer–van der Pol (BvP) neuron model, 171
   heart, 176–181
Brain (mind), 168, 176r

Index

Cantor, Georg:
  "comb," 193–194
  sets, 191–194, 293
Cardiac dynamics, see Heart.
Catastrophe set:
  cusp, 218, 219
  K in Figures 7.3, 7.4, 218, 219
"Catastrophe" theory, 219r
Cells:
  internal dynamics, 5r
  general, 176r
Chaos, 102, 219
  bounded sensitivity, 98, 99
  deterministic, 53–54, 97–98
  early examples, 58
  ecology and evolution, 236r, 242r
  health, 181r
  logistic map, 89ff
  solar system, 55r, 188,
  See also, Spatial, Poincaré
Chaotic (strange) attractor, 225–226,
  mathematics, 257r
  See, Lorenz.
Chemical dynamics, 242r
  oscillations, 236
Circadian rhythms, 228, 226–233, 232r
Cognitive aspects, 3
  multistable cognitive phenomena, 219
  See also, Bank, "Swift"; Culture; Pythagorus; Understand.
Compound interest, 23–24; see Exponential growth.
Computer:
  disk, 28–35, 265–284
  early dynamic discoveries:
    Fermi–Pasta–Ulam, 57
    Hénon–Heiles, 58
    Edward Lorenz, 58
    Robert May, 58
  programs, 28–35, 265–284
    Macintosh, 35
  See, Runge–Kutta.
Computational, mathematical empirical information, 56–63, 157
Constants of motion, 128, 134, 135, 139
Copernicus, 11, 19r, 27
Cultural influences:

on population dynamics, 41–42
on science, 11, 18, 48–54r
Darwin, Charles, 15, 50, 55r
  law of Higgledy-Piggledy, 67
  views of, 54–55r
  modern influence, see Mayr
Deterministic:
  equations, 52, 53
  views of nature, 49–54r
Differential equations, 18
  ordinary (ODE) 47, 47, 105–115
  partial (PDE), 47
  second order, 115–116
Diffusion, see Reaction-diffusion.
Dimensions:
  fractal capacity, 190
  topological, 189
Diversity of nature, 1–5r, 43
Diseases, infectious, 43r, 161
  dynamics of, 242r
Double-well potential, 133–138,
Duffing nonlinear oscillator, 212
Duhem, Pierre:
  empirical and mathematical determinism, 52–53, 56
  his or Duhem's lesson, 52
Dynamics:
  discrete time (maps) 72–75
  fast/slow, 167
  insights, 43–44
  intermittency, 96
  nature's diversity, 1–3: see Cells.
  semiperiodic, 95
  sensitivity, 243
  tonal representation, 93–100, 181, 282–283
  variables:
    dummy, 123
    dynamic, 21
    independent, 21
  See also, Models; Nondeterministic; Time Scales; Tonal.

Ecology:
  conservation, 152r
  fragility, 42, 43
  general dynamics, 152–153r
  population dynamics, 71–103
  universal classes, 100–103
Entrainment, 228–231
Environmental influences, 161–168,

176–181, 207–211r, 226–233;
  See Cultural systems
Euler approximation of differential equations, 109, 133, 114, 122, 151–152
Evolution:
  life, 5r, see Darwin.
  science of dynamics, 6–20. 45–63
  of the universe, 5r
Excitable systems:
  ant colonies, 257–263
    excitation-diffusion, 235–242
  social insects, 257, 263r
  neurons, 168–181
Exercise types:
  Explore and Think, 4
Existence and uniqueness of solutions, 48, 121, 122
Exponential:
  function, 113, 209–293
  growth, 22, 23, 40, 41, see Sensitivity; Resource limits.
Extended phase space, 224–226

Fermi–Pasta–Ulam phenomenon
  discovery, 57, 62r, 63r
Feynman, Richard, 5r, 114r
  on continuum dyanmics, 109
  on science, 59
Finite resources, 25, 27, 40–42, 73
  nonlinear equations, 25
  Think 1.5F, 41:
Fixed points:
  Figure 4.9, 135
  stable (attractor/unstable (repellor), 83, 85
  types, 129
  See also, Stability.
Frequency, angular, 118; see Oscillators.
Fluid dynamic understandings. See Archimedes; Bernoulli; Harvey; Lorenz.
Forced damped oscillators:
  chaos, fractals, 225–226
  hysteresis, 216–219
  moderate (nonlinear), 211–215
  strong (bistable), 215–219
  strong (very chaotic), 225–226
  weak (linear), 208–211
Fractals, 189–205, 251r

Index

capacity dimension, 190
forced dynamics, 225–226
Lorenz attractor, 247, 250
music, 196
physical?, 195–196
Sierpinski sieve, 195
Functions, 21; *see* Boltzman;
  Probability distributions;
  Schrödinger, quantum wave.

Galileo, Galilei, 15–16, 19–20r, 139, 143
Genetic dynamics, 219
  See Mendel.
Glial cells, brain, 170
Graphics, 32, 33, 78–80, 85–92,276–278, 281; *see* Three dimensional.
Gravitation, "universal law," 17

Harmonic oscillator, 115–124
Harvey, William, blood circulation, 12
Heart, 176–181
  arrhythmias, 180, 181r, 242r
  excitation network, 177
Higgledy-Piggledy dynamics, 67–70, 194–195
  in space, 257–263
History of scientific understanding of dynamics, 6–20, 45–66
Holistic dynamic concepts, 14–15, 121, 129, 168
Holistic understanding, 15, 189–197, 211–226
  Kepler's relationships, 14
  tonal sequences, 93–100
  See also, Archimedes, Harvey;
    Bernoulli; constants of motion; Higgledy-Piggledy;
    Poincaré.
Homoclinic/heteroclinic limit points, 138r,
  Figure 4.9, 128
  Figure 4.12, 135
Host–parasite interactions, 153–161
Human population:
  Cities, countries, limitations, 39–42, 42–43

Explore 1–5B, 37
See, Verhulst, Malthus
Hysteresis effect, 216–218, 220

Ice ages, complex dynamic causes, 211r
Independent variables, 21
Infinitesimal time steps:
  Feynman, 109
  Newton, 105–115
Information:
  three distinct sources, 56–63, 155–156
  scientific understanding, 63–67
Intermittency dynamics, 96–97

Kepler, Johannes, 10, 13–15, 17, 19, 52, 121
  first heliocentric theory, 13
  see Holistic understanding
Koch fractal, 192–193

Leonardo da Vinci, 11–12, 18
Limit sets, *see,* Attractors, homoclinic/heteroclinic.
Limit cycles, 161–168
  Liénard, 167
  linked in phase space, 252–257
  linked in physical space, 257r
  van der Pol, 166
Linear:
  equations, 21
  growth, 22–23
  stability, 155, 287–290, 295–297
Living systems,
  human, 35–43
  insect, 72–75, 100–104, 226–234, 235–242, 257–264
  variety, 145–182
  vis-à-vis inanimate, 199–200, 219
Logically distinct information, 56–62, 156
Lorenz, Edward: *The Essence of Chaos,* 251r
  butterfly misunderstanding, 244–245
  Think 8.2A, 244–245
  dynamic sensitivity, 24,
  meteorological model, 243–251

strange, chaotic attractor, Figure 8.6, 247
Lotka–Volterra, *see* Predator–prey.
  dynamics, 151
  equations, 146
Lyapunov exponent, 164–188
  computational, 186–187
  planetary system, 188
  sensitivity to initial conditions, 185
Malthus, Robert:
  and Darwin, 51
  and policies, 112
  social dynamics, 50
Mathematics, 287–290
  difference/differential equations, 105–115
  dynamic discoveries, 51–53, 55r
  See, Poincaré; Topology; Newton
Maps:
  discrete dynamics, 72–76
  logistic, 76–78
  universal classes, 73, 100–103r
  See, Poincaré, maps.
Mayr, Ernst:
  biology, 54r
  Darwin's influence, 55r
  evolution, 55r
Maxwell, James Clerk:
  electromagnetic theory of light, 47
Mendel, Gregor:
  genetic dynamics, 50, 67
Meteorology, *see* Lorenz.
Models:
  bases, 26–27, 36
  deterministic, 53–54
  general biological, 161r
  types, 35–43, 63–67, 72–76, 100–104, 145–161
  See, Assumptions.
Moore, J. A., *Science as a Way of Knowing,* 54r

Neurons, 168–176
  action potential, 170
  data source, 175
  excitations, 172
  Hodgkin–Huxley/Bonhoeffer–van der Pol models, 171
  neurotransmitters, 170
  noise inputs, 173, 175
  quiescent state, 173

301

Index

Neurons *(continued)*
 structure, 169
Newton, Issac, 10, 16–18
 differential equations, 45–48,
  105–106
 planetary dynamics, 20r, 114r
Nicholson–Bailey model, 153–154
 mathematical analysis, 155–156
 physical understanding, 158, 160
Nondeterministic dyanamics, 25, 36,
  39, 44
 probabilisitic, 40
 See, Genetic; Higgledy-Piggledy;
  Intermittency.
Nonlinear:
 dynamics, 37, 56–57
 equations, 25, 36, 45–49
 limited resources, 25–27, 40–42, 73
 See, Ulam.
Nonlinear oscillators, 125–133
 frequency, 131, 287–290
 hard/soft, 126–127
 relaxation, 167
 See, Autocatalytic.
Nullclines, 167, 173–174

Occam, William of:
 Occam's razor, 26–27, 44
Orbits in phase space, 33, 119–121,
  127, 135
 extended phase space, 225
 Figure 4.9, 128
 Figure 4.12, 135
 homoclinic/heteroclinic, 138
Oscillators:
 frequency comparison, 131–132
 See, Bonhoffer–van der Pol; Double-
  well; Duffing; Forced; van der
  Pol; Nonlinear; Harmonic

Pacemakers, 168–182, 227–233
Patterns, in various spaces:
 colliding with spiral patterns, 241
 Figure 8.2 (a) and (b), 238–239
 Figure 8.12, 258
 target, 239
 on toroidal surface, 240
 references and notes, 242, 263
 See, Dynamics.
Period of dynamics, 14, 118

theory, computed, empirical,
  131–132
 pendulum, 139
 See, Kepler's holism.
Phase portraits, 119, 121, 128, 135,
  142
Phase space, 33, 34
 extended (toroidal), 224, 225
  Figures 7.7, 7.8, 223–224
 Liénard, 167
  Explore 8.2 A, 237
  Explore 8.2B, 239
  Figures 8.6, 8.9–8.11, 247,
   252–255
 three dimensional, 243–257
  Explore 8.3A–C, 255–256
 wrap-around, 141–143
 See, Orbits.
Poincaré, Henri, 51–53, 226–232,
  243–252
 deterministic chaos, 53
 holistic aspects, 121
 maps, 222–226, 249, 250
 Poincaré–Birkhoff theorem, 52, 53,
  55r
 surface of section, 223–226
 See, Topological concepts;
  Understanding.
Population dynamics, *see,* Ecology,
  human
Predator–prey, 145–153
 lynx and hare, 148–149
Probabilistic dynamics, 37, 38, 40,
  44
 See, Boltzmann; Darwin; Mendel;
  Schrödinger.
Psychological aspects, 7, 36–37
Ptolemy, 10, 11, 27
Pythagorus of Samos, 6–7, 15, 18, 47

Quantum mechanics:
 dynamics, 53, 56r
 See, Bohm; Schrödinger.

Reaction–diffusion
  (excitation–diffusion), 235–243
 excited, refractory, quiescent states,
  235–236
 three dimensions, 242r
Recursive relations:

additive, 29
 linear, 2, 43
 Newton, 45
 nonlinear, 25
Relations, *see,* Holistic; Poincaré;
  Recursive; Understanding.
Resonance response, 209
 in education, 210
Resources, 72–76, 100–104
 environmental, 158
 in finance, *see,* Banks.
 limited in nature, Fig. 1.6, 26; *see*
  Populations, nonlinear
Rotations of phase space, *see* Phase
  space, three dimensional.
Runge–Kutta approximation method,
  123–124, 278–280

Scaling variables, 149, 154
 topological equivalence, 150
Schrödinger, Erwin, 47, 53
 Copenhagen interpretation and
  dynamics, 56r
 quantum mechanical wave
  function, 53
 see, Bohm.
Science:
 changing foundations, 55r, 59–62,
  63r, 66–67r
 Feynman characterization, 5r, 59
 Magritte on reality, 65
 uncertainty, 56r
 See, Evolution, Moore.
Self-organizational dynamics,
  257–265
Semiperiodic dynamics, 95
Sensitivity of dynamics:
 empirical accuracy, predictive
  accuracy and time scales, 98,
  Figure 3.13, 99
 Lorenz, 243–244
 Think 8.2A, 244–246
 See, Dynamics; Lyaponov.
Slime mold dynamics, 236, 242r
Social insects, 257–265,
 ants, 163r
 bees, 236r
 variety, 236, 263
Solutions of equations, 22
 existence and uniqueness, 48, 49,
  121, 122, 221–223

Index

Solutions, examples:
    damped harmonic, 120
    harmonic, 117
    nonlinear, 113
Spatial dynamics, 159–160r, 235–252, 257–265
    chaos, Figure 8.3B, 241
    references and notes, 242
    See, Lorenz; Patterns.
Species, 145–163
    interactions, 152
Stability, 128, 135, 248
    of fixed points, 287–290, 295–297
    of limit cycles, 252–257
Strange attractor, *see,* Chaotic attractor.
Symmetry-breaking bifurcations, 221, 252–257
Systems and environments, 197–206, 211r, 226–234

Thales of Miletus (Western science), 6, 19r
Time scales, of dynamics, 3–4, 5r, 41, 98, 198–200, 227
    infinite, 47, 211r

Tonal characterizations:
    of periodic, semiperiodic, chaotic, intermittent, sensitive dynamics, 93–99
    of physiological dynamics, 181
    programing, 282–283
Topological concepts:
    bifurcation points, 130
    equivalence, 82, 129, 130
    rubber sheet, 130
    See, Poincaré.
Topological understandings, *see,* Duhem; Poincaré; Catastrophe.
Turbulent excitable dynamics:
    in space, 241
    in time, 261
Tycho Brahe, 12–13, 19r
    parallax discoveries, 12

Ulam, Stanislaw: 62r
    nonlinear systems/nonelephant zoology, 57
Understanding, scientific: 3, 4, 18–19r, 59–62, 63r, 152, 160–161
    symbolic, 251r

See, Anderson; Cultural; Duhem; Feynman; Holistic; Mayr; Moore; Poincaré; Relations.
Universal classes, of population dynamics, 100–104
Uniqueness of solutions, *see,* Solutions of equations.

van der Pol oscillator:
    stable limit cycle, 165; *see* Autocatalytic oscillators
Verhulst, Pierre-Francois, human population theory, 112–115

Weather, *see,* Meteorology.
Western science, origins, 19r; *see,* Thales of Miletus
Wheeler, John Archibald, 70r; *see,* Higgledy-Piggledy.
Windows, 91–92
    Figure 3.9, 90
    Figure 3.10, 92
    Figure 3.11, 95
    intermittency, 96–97, Fig. 3.12, 96

Zeitgebers (time givers), 228–233

CUSTOMER NOTE: IF THIS BOOK IS ACCOMPANIED BY SOFTWARE, PLEASE READ THE FOLLOWING BEFORE OPENING THE PACKAGE.

This software contains files to help you utilize the models described in the accompanying book. By opening the package, you are agreeing to be bound by the following agreement:

This software product is protected by copyright and all rights are reserved by the author and John Wiley & Sons, Inc. You are licensed to use this software on a single computer. Copying the software to another medium or format for use on a single computer does not violate the U. S. Copyright Law. Copying the software for any other purpose is a violation of the U. S. Copyright Law.

This software product is sold as is without warranty of any kind, either express or implied, including but not limited to the implied warranty of merchantability and fitness for a particular purpose. Neither Wiley nor its dealers or distributors assumes any liability of any alleged or actual damages arising from the use of or the inability to use this software. (Some states do not allow the exclusion of implied warranties, so the exclusion may not apply to you.)

WILEY